UNATTENDED RADIATION SENSOR SYSTEMS FOR REMOTE APPLICATIONS

In 1995, the majestic spiral galaxy NGC 4414 was imaged by the Hubble Space Telescope as part of the HST Key Project on the Extragalactic Distance Scale. An international team of astronomers, led by Dr. Wendy Freedman of the Observatories of the Carnegie Institution of Washington, observed this galaxy on 13 different occasions over the course of two months.

UNATTENDED RADIATION SENSOR SYSTEMS FOR REMOTE APPLICATIONS

Washington, D.C.   15–17 April 2002

EDITORS

Jacob I. Trombka
NASA/Goddard Space Flight Center
Greenbelt, Maryland

David P. Spears
U.S. Department of Energy
Washington, D.C.

Pamela H. Solomon
The Catholic University of America
Washington, D.C.

SPONSORING ORGANIZATIONS
NASA Headquarters, NASA/Goddard Space Flight Center
U.S. Department of Energy
U.S. Department of Justice
U.S. Department of State, Nonproliferation and
    Arms Control Technology Working Group

Melville, New York, 2002
AIP CONFERENCE PROCEEDINGS ■ VOLUME 632

Editors:

Jacob I. Trombka
NASA/Goddard Space Flight Center
Code 691
Greenbelt, MD 20771
USA
E-mail: u1ijit@lepvx3.gsfc.nasa.gov

Pamela H. Solomon
The Catholic University of America
Washington, D.C. 20064
and
NASA/Goddard Space Flight Center
Mail Code 690.2
Greenbelt, MD 20771
USA

E-mail: Pamela.Solomon@gsfc.nasa.gov

David P. Spears
U.S. Department of Energy, NN20
Office of Research and Engineering
1000 Independence Avenue, GH068
Washington, D.C. 20585
USA

E-mail: david.spears@hq.doe.gov

L.C. Catalog Card No. 2002112403
ISBN 0-7354-0087-3
ISSN 0094-243X
Printed in the United States of America

CONTENTS

TERRESTRIAL APPLICATIONS

ENVIRONMENTAL APPLICATIONS

DETECTORS

DATA NETWORK SYSTEMS

COMMERCIALIZING NEW TECHNOLOGIES

ACADEMIC PERSPECTIVES

FORENSIC APPLICATIONS

APPENDICES

PREFACE

The past decade has provided remarkable advances in the field of radiation sensor systems, and born witness to new and innovative applications of these technologies. Remote sensor systems have been used for areas as diverse as the study of asteroids in space to playing a significant role in counter-terrorism efforts. As the field of sensor development continues to expand rapidly, we judged it timely and important to examine the current state of unattended sensor technologies and to explore potential requirements and applications for such systems for the future. To address these issues, a workshop entitled "Unattended Radiation Sensor Systems for Remote Applications" (URSSRA) was held in Washington D.C. April 15-17, 2002. The workshop was sponsored jointly by National Aeronautics and Space Administration Headquarters, (NASA) Office of Space Science, NASA Goddard Space Flight Center, U.S. Department of Energy, U.S. Department of Justice, National Institute of Justice, and U.S. Department of State, Nonproliferation and Arms Control Technology Working Group.

The workshop's primary objective was to foster communication among workers in government, private industry, national laboratories, and academia. We envisioned a setting that would stimulate cross-disciplinary discussions concerning the development and use of remote sensor systems. Over 150 individuals representing all of the target areas participated in very fruitful discussions, both in the open sessions and in the casual settings in and around the Carnegie Institution. The wide-ranging dialog conducted over the three days of the workshop explored concerns from the critical issues of technology development to the need to educate and train the scientists and engineers who are needed to implement these increasingly important governmental and private objectives in the future. Presentations at the workshop addressed the areas of astrophysics, nuclear non-proliferation, emergency response, airborne remote sensing, safeguards, waste management, and forensic science, encompassing both applications and system design.

A major source of concern among workshop participants centered on the lack of inter-governmental cooperation in developing and producing the systems that will be required for future applications. Current budgetary constraints dictate that federal agencies investigate alternatives to direct funding of their innovation efforts, and seek cooperative programs with those requiring similar technologies for differing objectives. Through cooperative programs, technologies for future operations can be developed and tested for wider ranging objectives. By demonstrating that these technologies have broader applications and wider markets, private industry would be stimulated to produce systems at more competitive prices, resulting in cost savings for federal and state governments, and for private institutions. Although there are a number of governmental programs whose objective is to commercialize sponsored technologies, discussions revealed that an efficient method for bringing these technologies to commercially viable production has not yet evolved. Means to make these endeavors less cumbersome to the commercial sector need to be defined and implemented.

The rapidly declining number of U.S. students enrolled in both undergraduate and graduate programs in science and engineering was identified and addressed as an increasing concern. As the number of students enrolled in such studies continues to fall, existing class populations reflect a heavy percentage of foreign students who do not plan to apply their expertise in the United States once their education is complete. A lack of technical personnel prepared for the future will negatively affect the development and implementation of remote sensor technologies vitally important to areas including homeland security, domestic and internal safeguards, nuclear nonproliferation and monitoring, and the exploration of space. The problem of a negative image held toward science and technology by students, particularly in the area of nuclear studies, and expanding technology's perceived harmful effect on the environment was viewed as a detriment to the goal of preparing innovators for the future. Positive, focused scientific and technical curricula need to be introduced and implemented in schools beginning in the early elementary and middle school years if we are to engage the minds of our brightest students in solving the increasingly complex technical and scientific problems that lie ahead.

Finally, an ongoing concern across all sectors was expressed repeatedly over the consistently dwindling funds dedicated to basic research. The major funding in present day research tends to be applied toward solving immediate goals. Panic-driven programs devoted solely to crises management usurps the thoughtful process that provides the time, energy and resources required for contemplating and anticipating future needs, and deriving the solutions which will secure that uncertain future.

Acknowledgments

We wish to acknowledge the outstanding support provided by Christine Shannon and Gorgiana Alonzo of the Lawrence Livermore National Laboratory, Elaine Mullen, Debbie Bush and Nancy Sprinkel of the MITRE Corporation, Rachel Smith of the Oak Ridge Institute for Science and Education, Mona Drexler of the NASA Goddard Space Flight Center, and Sharon Bassin of the Carnegie Institution of Washington, without which this meeting's success would not have been possible.

NUCLEAR NON-PROLIFERATION AND HOMELAND SECURITY

Low Cost, Low Power, High Sensitivity, Real Time Neutron Detection Microsystem

David W. Peterson and Benny H. Rose

Sandia National Laboratories[†]
1515 Eubank Blvd.
Albuquerque, NM 87185

Abstract. A Si array neutron detector is proposed based on commercial CCD and CMOS sensor technology coupled with a thin film neutron conversion coating. System sensitivity is estimated for a baseline device containing a single array and various schemes to increase detection probability by simple area scaling and stacking are discussed. Some possible use scenarios are discussed involving static and moving sources. Likely neutron source fluxes for weapons grade and commercial grade nuclear material are estimated along with expected intensities of cosmic background neutrons which would establish a noise floor to detection limits.

INTRODUCTION

There are essentially only two sources of terrestrial neutrons: neutrons originating from interactions between the atmosphere and galactic and solar cosmic rays and neutrons emitted from nuclear materials. The detection of excess neutrons above the normal cosmic ray induced background is, therefore, a cause for investigation. Over the last few years, Sandia has developed and demonstrated a neutron sensitive micro-detector based on commercial Si Charge Coupled Device (CCD) technology that has been coated with a ^{10}B neutron conversion film. The absorption path of a single α particle created by one thermal neutron interaction with ^{10}B will create at least 100,000 electron-hole pairs in a typical sensor array pixel, far more than enough to initiate the "on" state in a device sensitive to 10's or 100's of electrons per pixel. Neutron sensitivity is primarily determined by the conversion efficiency of the ^{10}B film and the probability that emitted ionizing particles have the correct trajectory to enter the sensor array. In a typical configuration, 3–4 neutrons out of every 100 entering the ^{10}B coating will result in an α particle entering the active region of the sensor array where the probability of detection is essentially unity. We will discuss ways in which neutron detection probability can be increased to as high as 25% through moderator back-scattering and chip stacking. Also discussed is cosmic ray induced background interference and ways it can be minimized.

We present several designs applying this detector to a number of potential applications, such as dosimetry, and portal or "pass-by" monitoring where either the detector or source are moving. One design contains a detector, control chip, battery,

[†] Sandia is a multiprogram laboratory operated by Sandia Corporation, a Lockheed Martin Company, for the United States Department of Energy under contract DE-AC04-94AL85000.

and wireless communication link which can be used for long-term remote sensing with "wake-up" interrogation by a distant data gathering station. The pass-by scenario is particularly difficult given the $1/r^2$ drop-off of intensity from an isotropic point source of neutrons and places a premium on area scalability of the detector array at reasonable cost.

Our research prototype is based on a Kodak Charge Coupled Device (CCD) sensor array that would not be cost effective to scale in area. A simple 3-transistor Complementary Metal Oxide Semiconductor (CMOS) active pixel array is proposed that could be tiled or fabricated monolithically at the wafer level at significantly reduced cost. Wafer fabrication has the additional benefit of facilitating wafer-level deposition of the ^{10}B conversion film for a chip stacking option.

NEUTRON ENVIRONMENT

Since this detection system is sensitive to ionizing particles from a conversion coating traversing a pixel's active region, it will be important to establish some estimate of the neutron count rate expected in the absence of other sources. This count rate is expected to originate from cosmic ray generated neutrons that interact with nuclei of atmospheric atoms. These neutrons lose energy in passing through the atmosphere, eventually coming to thermal energy as they reach equilibrium with their surroundings.

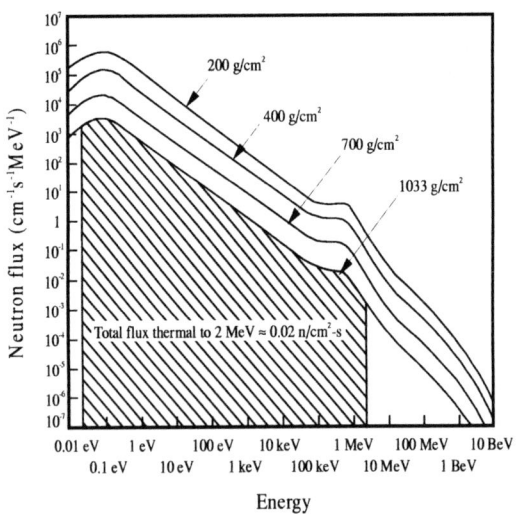

FIGURE 1. Neutron flux vs. altitude taken from Ziegler (ref. 1). Altitude is measured in air pressure, i.e., 1033 g/cm^2 is sea-level, 700 g/cm^2 is about 3.2 km (~2 miles). The hatched area under the curve represents the total flux at sea-level of neutrons from 2 MeV to thermal (~0.0259 eV) and is estimated to be approximately 0.02 n/cm^2-s.

The cosmic ray induced neutron flux as a function of altitude has been comprehensively investigated by Ziegler [1]. The energy distribution of secondary neutrons for different altitudes is shown in Fig. 1. The secondary neutron flux increases exponentially with altitude up to 15 km, the location of peak cosmic ray intensity (Pfotzer Maximum). For example, the neutron flux in Denver, Colorado (5280 ft above sea level) is 3.88 times that in New York City and Leadville, Colorado (10,430 ft above sea level) is 11.0 times the New York City flux. Graphical integration of Fig. 1 in the energy band shown (hatched area) gives an estimate of ~0.02 n/cm^2/s at sea-level, or ~0.08 n/cm^2/s at Denver. A 5% efficient 1 cm^2 detector should record a background neutron count about every 17 minutes at sea level and every 4 minutes in Albuquerque or Denver.

Special nuclear material is a group of materials defined in Department of Energy (DOE) orders[1], which includes all plutonium isotopes and uranium enriched in the isotopes ^{233}U and ^{235}U. The elements ^{239}Pu and ^{235}U are the primary fissile materials used in making nuclear weapons, but neither produces a significant number of neutrons while in a subcritical configuration. Detection of these materials in non-proliferation applications requires a neutron probe beam to create neutrons. Because of the high sensitivity and area scalability of ^{10}B coated Si array detectors, these isotopes might be detectable with reduced probe beam intensity.

There are other isotopes of plutonium created in the fission process, primarily ^{238}Pu and ^{240}Pu, that do produce a significant number of neutrons. Plutonium created during the fission process of a LWR (Light Water Reactor) can be chemically isolated and removed as part of the fuel reprocessing cycle. In this form, the plutonium is not considered weapons grade, but rather commercial grade. To create weapons grade plutonium, the spent reactor fuel needs to be enriched in the isotope ^{239}Pu. Using conventional methods the expected neutron emission rate from one kg of weapons grade Pu can be shown to be approximately 64,000 n/s [2,3].

Commercial grade plutonium, although not as enriched in ^{239}Pu as weapons grade material, is a nonproliferation concern. The amount of commercial grade metallic plutonium required to make a critical mass is approximately twice that of weapons grade plutonium. One kg of commercial grade plutonium could be expected to emit about 310,000 n/s. [2,3]. The unmoderated neutron emissions from both weapons and commercial grade material are in the 2 MeV range.

DETECTOR DESIGNS

The initial Sandia research prototype neutron detector used a CCD sensor array with an enriched boron film in close contact with its surface. The ^{10}B coating was applied to a thinned Si substrate by either sputtering or pulsed laser ablation. The substrate was then placed on the CCD with the boron coating next to the device. As described in more detail elsewhere [4], the device works by collecting charge generated by either the 1.47 MeV α particle or 0.840 MeV ^{7}Li nucleus created by the nuclear reaction between a thermal neutron and the ^{10}B nucleus. The α particle and ^{7}Li nucleus are emitted essentially isotropically in opposite directions. An α particle incident on a given pixel will generate far more charge in the pixel than necessary to

---

[1] DOE Order 461.1, "Packaging and Transfer or Transportation of Materials of National Security Interest," 09-29-00.

5

turn it "on". The range of a 1.47 MeV α particle in Si is about 5 μm and the depletion region depth of the CCD pixel is only about 1 μm. If the CCD collected only the charge generated by the α particle in the depletion region it would still be 8–10 times the noise floor of a typical cell. This effect is similar to single event upset in computer memories, where it has been determined that a cosmic ray penetrating a memory cell will set up a charge funnel wherein charge is collected from areas that are deeper than the thickness of the active layer of the memory [5]. Because of charge funneling phenomena, a pixel will collect electron-hole pairs from as deep as a few microns below the depletion layer. This is a specific advantage when compared to methods that use electrons as the ionization species. The stopping power of Si for α particles at 1.47 MeV is approximately 300 times that for electrons at energies of interest.

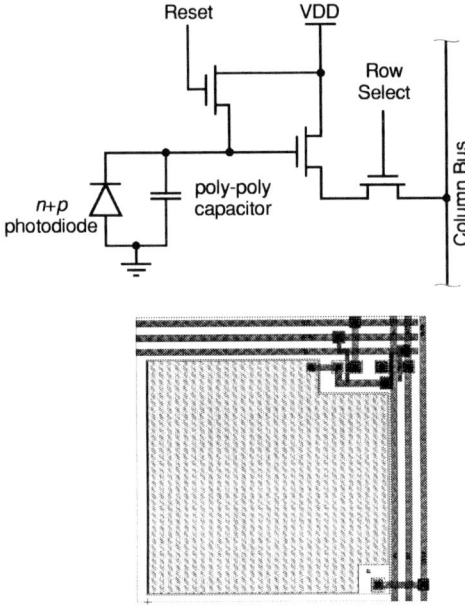

FIGURE 2. Simple three-transistor unit cell used in a CMOS sensor array.

Fig. 2 shows a simple three-transistor unit cell or pixel that is typical of conventional CMOS sensors. This particular design was developed for a satellite sensor project requiring a large area sensor array event detector and includes a poly-poly capacitor not normally used in commercial CMOS arrays. Each pixel contains selection and reset transistors and row-column interconnection traces that take up area that could be used for detection. The "fill factor" for a typical CMOS sensor array is around 30%, considerably less than a CCD, which can approach 100%. However, a custom CMOS design for neutron detection would combine large pixels with small state-of-the-art transistor and conductor features. A 100 μm square pixel, for example, would improve fill factor significantly, reduce system complexity with fewer total pixels, and result in a negligible decrease in sensitivity. A 10,000 pixel array (100 x 100) with 1 cm^2 of active region could be fabricated on a 12 to 14 mm square die.

Figure 3 contains a schematic drawing of a CMOS, three-level-metal sensor showing an active pixel with the ^{10}B neutron conversion coating. Also shown are the passivation and interlevel dielectric (ILD) layers, trajectories of an α particle and ^{7}Li nucleus from a typical neutron conversion event, and charge funnel created by the passage of an α particle through the depletion layer and into the substrate. It should be apparent that if passivation and ILD layers, with in this case a typical combined layer thickness of 6 μm, are present over the sensor array, the energy of both emitted particles will be dissipated before reaching the active region. These layers can be removed using a straight-forward etch back at the end of standard wafer processing, a process not feasible for commercial off-the-shelf CMOS sensor die.

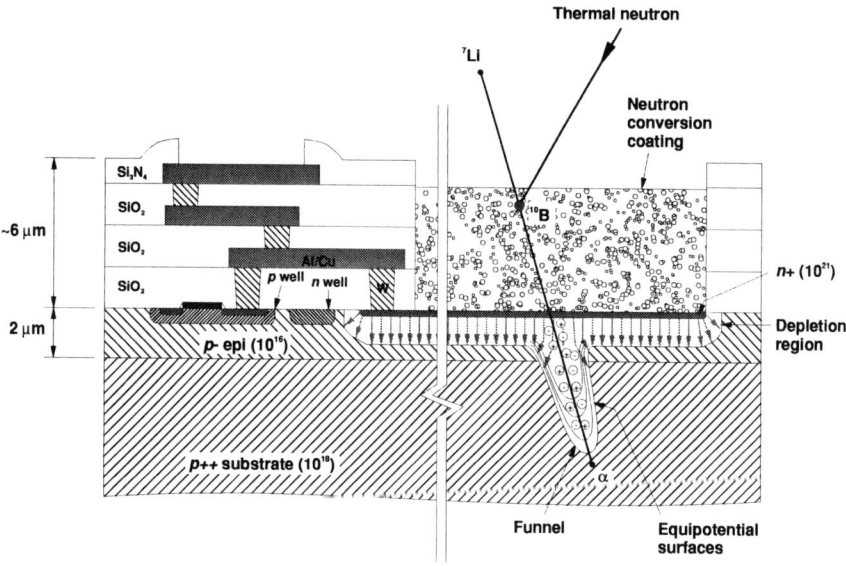

FIGURE 3. Cross-section of an *n+p* photodiode cell in typical sub-micron CMOS three level metal technology showing ILD thicknesses. Drawing is to scale vertically, but is compressed in the horizontal over the active region for clarity.

There are several ways a neutron detection microsystem can be realized using a Si CMOS-based sensor array. Fig. 4 contains a conceptual system design based on a standard micro controller such as the Motorola MC68. The MC68 contains all the functions needed to control the selection and clocking of raw data from the sensor array into an A/D converter within the micro controller. Event data are outputted to a display and optional wireless interface. In the case of a custom designed sensor array, it may be desirable to move some of the micro controller functions into the sensor array to reduce system size and complexity. A typical power subsystem would consist of four 1.5 V batteries or a single battery and boost converter, depending on volume, battery life constraints, and operating time requirements.

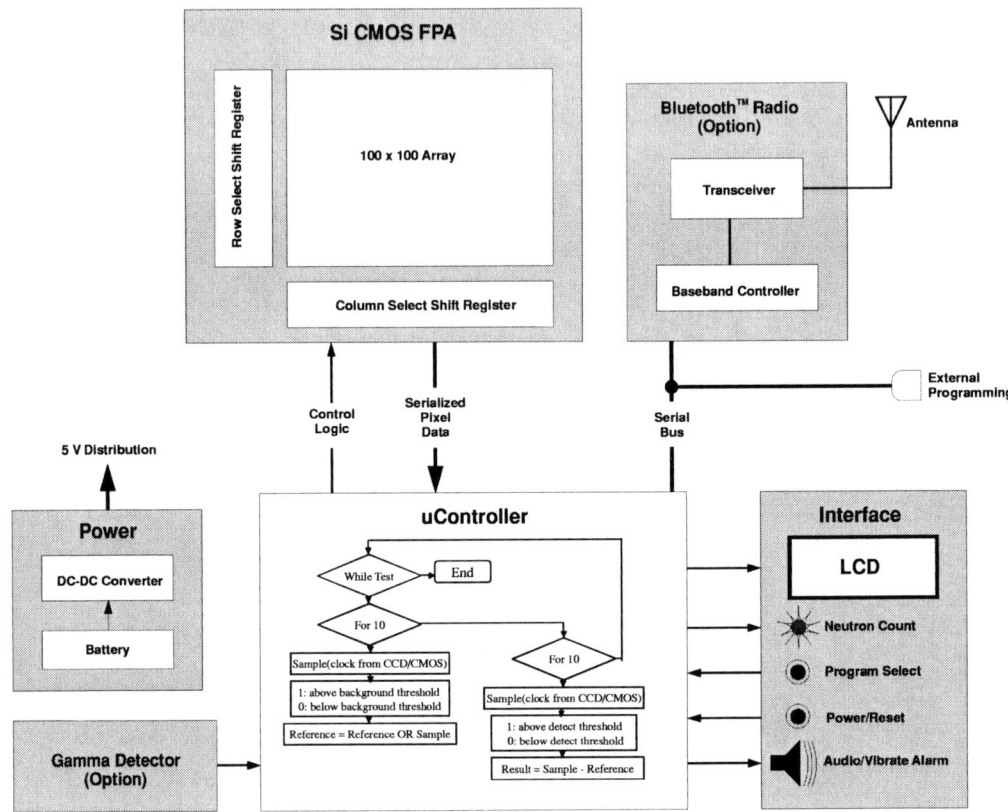

FIGURE 4. Block diagram for Si CMOS FPA microsystem.

PRINCIPLES OF DETECTION

The thermal neutron capture cross-section for ^{10}B is 3840 barns. The ^{10}B(n,α)^{7}Li nuclear reaction, after passing through a Li compound nucleus state, produces a 1.47 MeV α particle, an 840 KeV recoil Li nucleus and a 480 KeV gamma ray. This reaction occurs 94% of the time, and the remaining 6% produces a 1.77 MeV α particle and 1.015 Mev ^{7}Li and no gamma ray. The gamma ray Compton scattering interaction is so weak that large path lengths in the active region of the detector would be required to make the interaction probability appreciable. Stopping powers for electrons is small compared to α particles, and even protons with energy above 1 MeV should not upset the detector.

System sensitivity is defined as the number of ionizing species that form detectable charge in the CCD per incident thermal neutron, or the product of the probability of nuclear reaction and the probability of ionizing species particle collection. As shown in previous work [4], using the closed form solutions of McGregor [6] for the case of the α particle $(D < L)$, this sensitivity is given by,

$$S(D) = [(1 + 1/\Sigma L)(1 - \exp(-\Sigma D)) - D/L]/2, \qquad (1)$$

where D is the ^{10}B film thickness, L is the range of the 1.47 MeV α particle in boron, and Σ is the macroscopic absorption coefficient. The value of Σ is the product of the microscopic neutron capture cross-section per atom and the number density of boron nuclei in the film, or $\Sigma = \sigma\rho$. Since sensitivity is a trade-off between the probability of nuclear reaction, favoring thick films, and the probability of α particle collection, favoring thin films, the optimum thickness can be determined by differentiating Eq. (1), and solving for D in terms of L and Σ:

$$D = (1/\Sigma)\ln[\Sigma L + 1]. \qquad (2)$$

Using $L = 3.6$ μm, and $\Sigma = 0.05$ /μm, the optimum film thickness is 3.3 μm, giving a value for overall sensitivity of 4% for α particles alone. This will be somewhat greater when ^7Li nuclei are included. McGregor's closed form solution for the efficiency due to ^7Li nucleus collection is given by, $(D > L)$

$$S(D) = \exp(-\Sigma(D-L))[\{1 + (1/\Sigma L)\}\{1 - \exp(-\Sigma L)\} - 1]/2. \qquad (3)$$

Using Eq.(3) to calculate the efficiency contribution of the recoil Li nucleus, we obtain 0.019. This combined with the α particle contribution, giving a total ideal efficiency of $\approx 6\%$.

Our research prototype CCD-based system contains SiN and P-glass passivation layers which total about 3 μm thick. As a result, we estimate limited contribution from the ^7Li reaction product in our data. These layers also reduce collection of the α particles, adversely affecting our measured efficiency. To estimate the attenuating effect of a chip passivation layer between the boron conversion coating and the CCD, McGregor's expression for sensitivity is modified by adding an increment, d, to x, and replacing L by a sum $L_1 + L_2$. The range of the α particle in boron is now L_1, and in SiO_2 passivation is L_2. After performing the integrations, the result is an extra term in the first parenthesis of Eq. (1), where $(1 + 1/\Sigma L)$ is replaced by $(1 + 1/\Sigma L - d/L)$. and L is about 6 μm instead of 3.6 μm. With a passivation layer thickness $d = 3$ μm, the efficiency is reduced from 4% to 1.3%. Substantial increases in measured efficiency can be obtained by removing the unwanted layers of nitrides and oxides.

Monte Carlo calculations of radiation transport were done for detector efficiency with various thickness values of polyethylene both in front of and in back of the detector. It is estimated that a 58% increase in efficiency can be realized by adding a 3 cm moderator behind the detector to reflect neutrons back into the conversion coating. Assuming an unpassivated CCD active region and including both the contributions from ^7Li collection and additional efficiency from the back moderator reflections, the predicted ideal efficiency is about 7%.

MICROSYSTEM DESIGNS FOR SELECTED APPLICATION SCENARIOS

In this section we develop system design concepts that are relevant to some obvious application scenarios, such as personal dosimetry, pass through portals, and integrating units that can be temporarily attached to a container. In this way we can begin to understand the real system requirements and where the best cost/performance payoff is in terms of technical approaches.

DETECTOR WORN ON THE PERSON

Radiation workers could wear a neutron detector as an adjunct to gamma dosimetry in radiological zones to monitor total radiation exposure. This detector system could be integrated into a personal dosimeter that would provide total dose, neutron dose, and the rates associated with both. These dosimeters could prove useful to accident response groups for the nuclear power industry and groups responding to terrorist incidents that result in the spread of SNM.

The neutron detector sensitivity to individual neutron events and its natural discrimination against gamma and x-rays make this detector ideal for sensing the presence of a weak neutron source from possible SNM within shipping or transient storage facilities. Detectors worn by facility workers might indicate the presence of SNM as it transits through shipping and storage areas. Such a device might have some of the following characteristics:

- Maximum sensitivity to neutrons from ~2 MeV to thermal.
- Count, store, and read-back capability.
- Audible, visible, and perhaps a vibration alarm to an adjustable threshold.
- No bigger than a pager.
- Wireless option.
- Minimum 1-month battery life or a rechargeable battery.

For this application, the 3 cm or so of backside "reflector" could be eliminated, assuming that body mass will serve the same purpose. The resulting detector size is somewhere between a thick wristwatch and a thick pager depending primarily on battery size and wireless option.

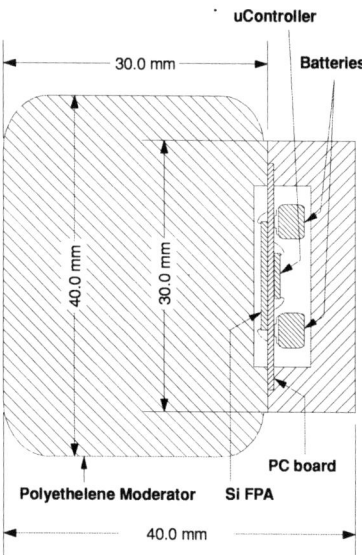

FIGURE. 5. Conceptual design for minimum (watch) sized neutron detector to be worn on the person. It contains a 1 cm² active region CMOS sensor array and optimum polyethylene moderator thickness of 3 cm.

The $1/r^2$ decrease in intensity from a neutron point source relative to cosmic background neutrons will determine the maximum effective distance of detection. One could expect one background count every 4–5 minutes from a 5% efficient, 1 cm² detector in an area of Denver devoid of natural or manmade shielding. For a 5:1 signal to noise ratio, a source that provides at least 1 count/min (cpm) at the detector could be assumed. Neglecting air absorption, scattering, and assuming thermalized neutrons, the flux from a neutron source S (n/s) at distance r (cm) can be expressed by the inverse square law relation,

$$\phi(r) = S/4\pi r^2. \tag{4}$$

Assuming a detector of efficiency ε and active area A placed in the flux at distance r, the measured counts will be,

$$C_{\text{fixed}}(r) = \phi(r)\varepsilon A = S\varepsilon A/4\pi r^2. \tag{5}$$

The detector efficiency and area are both adjustable design parameters. Assuming $A = 1$ cm², $\varepsilon = 0.05$, $S = 10^5$ n/s, and with an integration time of 60 s, Eq. (5) gives $r \approx 1.5$ m for distance between detector and source to obtain 5 counts. Increasing detector area, efficiency or integration time would increase both background counts and counts from the neutron point source (although some improvement in signal-to-noise may be realized with multiple detectors). Shielding from surrounding materials, particularly concrete walls and ceilings, should appreciably decrease cosmic background noise. Therefore the maximum detection distance for this scenario is limited primarily by cosmic ray background.

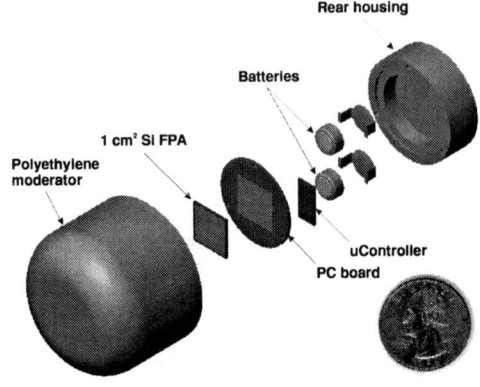

FIGURE 6. Exploded view of watch-sized neutron detector design concept.

Figure 5 contains a design concept of a watch-sized (30 mm diameter) unit using a ^{10}B coated, 1 cm^2 active region CCD or CMOS sensor array in which the bare die is directly attached to a 25 mm diameter PC board to minimize packaging volume. Except for direct chip attach of the sensor array, this design would use conventional PC board surface mount technology for assembly to minimize cost. Not shown is an audio/visual/vibratory alarm and the micro controller programming bus. The polyethylene moderator is 3 cm thick, based on Monte Carlo simulations, for optimum conversion of fission neutrons within the ^{10}B layer. This layout approaches the minimum volume possible for a detector, but doesn't have much space left for a wireless option or LCD readout, and is likely limited in operating time by the watch batteries. An exploded view of this design concept is shown in Fig. 6.

Integrating Detector (Detector moves with source)

In this scenario we consider a detector that is temporarily attached to a shipping container or trailer using a fixed magnet, suction cup, or other attachment device. The unit (or units) travels with the target for some defined interval, and is then removed for readout and subsequently reused. Characteristics of this unit include:

- Maximum sensitivity to neutrons from ~2 MeV to thermal.
- Count, store, and read-back capability.
- Wireless interface.
- Rechargeable battery with minimum 96 hour life.

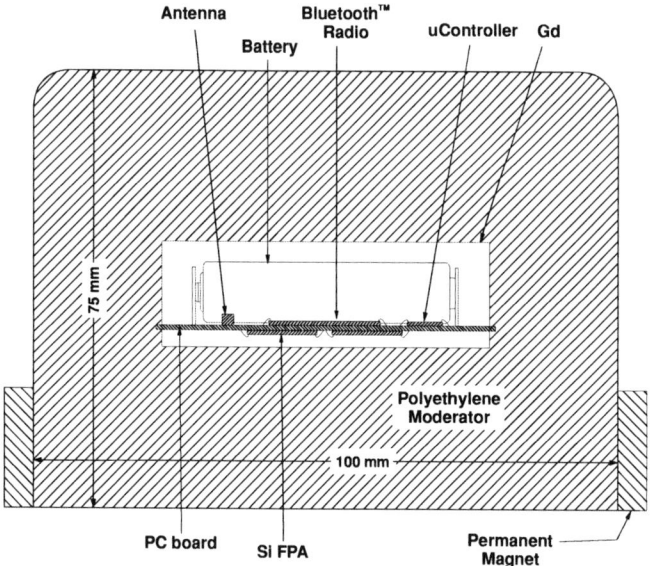

FIGURE 7. Conceptual design of an integrating detector for a pass-through portal, or temporary attachment to a container using a fixed magnet or suction cup. This unit could be optimized as shown for "look-down" sensing by incorporating an upper polyethylene moderator and Gd getter to absorb the vertical flux of neutrons due to cosmic rays.

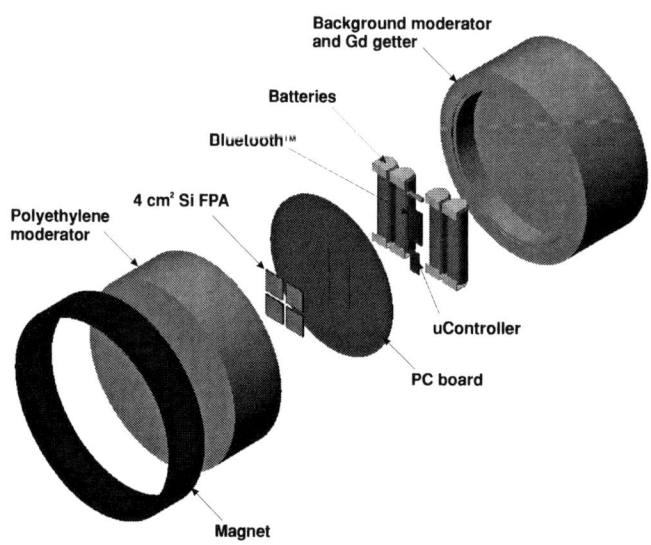

FIGURE 8. Exploded view of integrating detector.

13

Maximum range of detection will be limited by cosmic background noise in the same way as the pager or watch size unit. However, in this scenario the detector is placed intentionally close to a potential neutron point source and a polyethylene moderator and Gd neutron absorber surrounding the detector provide some shielding from the background neutron flux. In principle, the background count, estimated to occur every 4–5 minutes for the assumed detector design parameters and environment, could be eliminated or extended enough to be outside typical integration times by careful shielding. This would extend the effective range of detection for the watch and pager design beyond the 1.5 m estimated above, essentially limited by the integration time. For example, a 10 minute integration would produce one count from a source 4.9 m from the detector, the distance increasing by the square root of the time. A conceptual design of a wireless, portable, integrating detector, with reasonably achievable dimensions, is shown in Figs. 7 and 8.

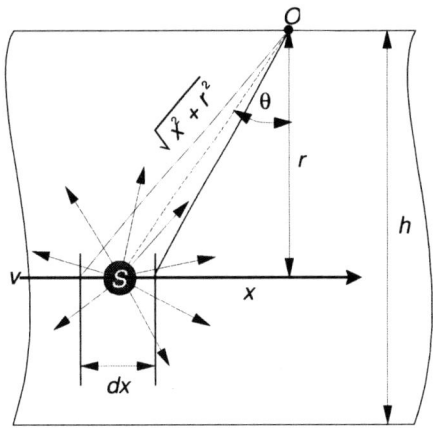

FIGURE 9. Isotropic point source S moving at speed v within confined boundary h.

Fixed Portal (Source passes by detector)

In this case, either the source or detector is moving in a straight line relative to the other with velocity v, and distance r at closest approach. Referring to Fig. 9, the flux density at the detector due to a point source at a distance x at its point of closest approach is given by $S/4\pi(x^2 + r^2)$. Since the source is in n/s, this expression gives n/cm^2/s. The count increment, ΔC, at the detector during the time interval taken by the source to travel the increment of path length dx is given by,

$$\Delta C = S\varepsilon A/[4\pi(x^2 + r^2)]dt \qquad (6)$$

Since $v = dx/dt$, a constant, $dt = (1/v)dx$, ΔC becomes,

$$\Delta C = (S\varepsilon A/4\pi v)[dx/(x^2 + r^2)] . \qquad (7)$$

The total number of counts in the detector from the point source passing at velocity v is given by the integral of Eq. (7) from $-\infty$ to $+\infty$ on x. The integral of the expression in square brackets is $(1/r)\arctan(x/r)$. Evaluating this expression yields π/r for the multiplier giving a count rate of,

$$C_{moving} = S\varepsilon A/4rv .\qquad(8)$$

Figure 10 is a parametric plot of the pass through velocity as a function of the distance of closest approach for different detector areas. The calculation assumes $C = 5$ counts, $S = 10^5$ n/s, and $\varepsilon = 5\%$ for the detector efficiency. For the case of a conveyer system with a 1 m^2 pass-through portal, a 4 cm^2 detector would pick up 5 counts from a point source passing through at about 4 ips (0.23 mph). A 40 cm^2 detector would see 5 counts at 3.3 fps or a little over 2.2 mph.

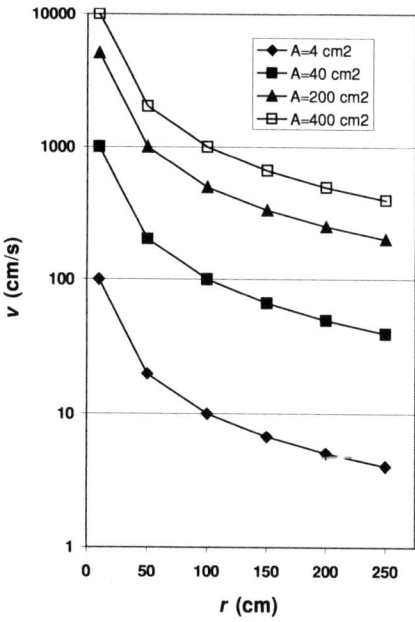

FIGURE 10. Plot velocity v as a function of distance to source r for various detector areas.

As a practical matter, Eq. (8) will estimate more counts than can be measured because the detector uses a look-down geometry as shown in Fig. 9. Neutrons coming from large distances are incident on the detector at large angles of incidence, making the effective detector area small relative to that assumed. The measured counts will probably more closely match that obtained from the angle of 45° on either side of the point of closest approach. Restricting the above calculation to $+\pi/4$ to $-\pi/4$ will give exactly half that given by Eq. (8).

DISCUSSIONS AND CONCLUSIONS

The Kodak CCD proof-of-principle experiments have shown that a commercial optical sensor array can be modified to detect neutrons converted into α particles and ^{7}Li nuclei by an overlying ^{10}B coating. This technology is not sensitive to *single* incident thermal neutrons, as over 85% will pass right through the conversion coating, but will likely detect ionizing species from a *single* neutron reaction. This is a significant improvement over large area PIN diode detectors. The scalability inherent in an array detector, and reduced anticipated system complexity and cost, make the CCD and CMOS sensor arrays preferable for portable neutron detection.

The maximum distance from which an SNM source neutron is detectable is the distance at which these neutrons are indistinguishable from background neutrons. This maximum distance must be less than the distance required for the $1/r^2$ dependence of the flux from a point source to be reduced to background levels. Since the detector must contain sufficient moderator to thermalize unmoderated 2 MeV neutrons, it will also count background neutrons from thermal to ~2MeV. Fortunately neutrons in this energy range can be absorbed or shielded using common construction materials.

Concrete is very effective in reducing background neutron flux, and proper placement of highly (thermal neutron) absorbent Gd can reduce or eliminate background neutrons incident from one complete hemisphere of a look-down detector. It will be necessary to verify experimentally current estimates of background neutron flux along with the efficacy of shielding these neutrons. This work is currently underway and is projected to be completed by the end of FY02.

REFERENCES

1. J.F. Ziegler, "Terrestrial cosmic rays," *IBM J. Res. Develop.*, **40** No. 1, 19-39 (1996).
2. R.A. Knief, *Nuclear Engineering – Theory and Technology of Commercial Nuclear Power*, 2nd Ed., Hemisphere Publishing Corp., Washington, 1992, pp. 167.
3. S. Glasstone and A. Sesonske, *Nuclear Reactor Engineering*, 3rd Ed., Van Norstrand Reinhold Inc., NY, 1981, pp. 544.
4. B.H. Rose, T.L. Hardin, D.W. Palmer, M.P. Siegal, T.L. Aselage, and D.L. Overmyer, "A New CCD-Based Thermal Neutron Detector," *Proc. Sensors Expo Spring 2001*, June 5-7, 2001, pp. 447.
5. G. C. Messenger and M. S. Ash, *The Effects of Radiation on Electronic Systems*, 2nd Edition, Van Nostrand Reinhold, N.Y., 1992, Chap.8.
6. D.S. McGregor, R.T. Klann, H.K. Gersch, Y.H. Yang, "Thin-film- coated bulk GaAs detectors for thermal and fast neutron measurements," *J. Nuc. Inst. And Meth.*, A **466** 126 (2001).

Monte Carlo Simulations of Prototype Radioxenon Beta-gamma Counting Systems

David Penn and Steven Biegalski

Veridian Systems Division, 1400 Key Boulevard, Arlington, Virginia 22209 USA

Abstract. One of the recently established methods for monitoring a Comprehensive nuclear Test Ban Treaty (CTBT) is the collection and analysis of radioxenon isotopes as collected and measured by International Monitoring System (IMS) stations. Radioxenon isotopes are produced from several sources including nuclear power generating stations, medical isotope production and nuclear test detonations. Each of the potential sources produces a characteristic energy spectrum that can interfere with that of another source. Therefore, to be able to accurately calculate isotopic concentrations, it is important to be able to distinguish between the different isotopes which share regions of interest in the beta-gamma energy domain. Veridian Systems has developed a method that utilizes the known resolution response from a particular detector system and Monte Carlo simulations to predict the interference from other isotopes into a region of interest. This allows the system to be setup with more well defined regions of interest and a better understanding of the interference parameters. This method can also lead to a more accurate determination of isotopic concentrations and a lower minimum detectable concentration. Along with determining the isotopic concentrations another Monte Carlo simulation has been developed that examines each particle track history in a simulation to determine the coincidence efficiency of the system. This efficiency is based on coincident interactions and deposited energy in both gamma and beta detectors from a single gamma source. The process and results for the Monte Carlo simulations are discussed and compared with experimental data.

BACKGROUND

The CTBT defines two methods of atmospheric monitoring for radiological analyses. The first method collects particulate matter and the second atmospheric xenon. The particulate method works well except in the case of subsurface detonations in which the particulate matter is not dispersed to the atmosphere. But, during a subsurface detonation some of the key elements that do escape to the atmosphere are the noble gases. Of these noble gases, the radioxenons are produced with an abundance and have a corresponding lifetime which makes them suitable for collection and radiological analyses(1).

Of the known xenon isomers and isotopes, the ones of primary interest are ^{135}Xe, ^{133}Xe, ^{133m}Xe, and ^{131m}Xe. These isotopes are typically produced by nuclear power reactors, nuclear fuel reprocessing, nuclear detonation, and medical isotope production. Each of these sources produces xenon in various quantities and the relative abundance of the different isotopes helps differentiate between the possible sources. The abundance of a xenon gas sample is determined by analyzing data

CP632, Unattended Radiation Sensor Systems for Remote Applications, edited by J. I. Trombka et al.
© 2002 American Institute of Physics 0-7354-0087-3/02/$19.00

produced with a nuclear counting system. The principle emissions from the xenons of interest are summarized in Table 1(1).

TABLE 1. Principle emissions for the radioxenons of interest.

	^{135}Xe	^{133}Xe	^{133m}Xe	^{131m}Xe
Half-life	9.10 h	5.24 d	2.19 d	11.84 d
Gamma-ray				
Energy (keV)	249.8	81.0	233.2	163.9
Abundance (%)	90	37.0	10.3	1.96
X-ray				
Energy (keV)	31	31	30	30
Abundance (%)	5.2	48.9	56.3	54
Beta spectrum				
Maximum energy (keV)	905	346		
Abundance (%)	96	99		
Conversion electron				
Energy (keV)	214	45	199	129
Abundance (%)	5.7	54	63.1	60.7

There are multiple methods for identifying radioxenons in a gas sample. These include high resolution gamma and x-ray spectroscopy and electron-photon coincidence spectroscopy. For this discussion the focus will be on Electron-Photon Coincidence spectroscopy Systems(EPCS) as used in the Automated Radioxenon Sampler Analyzer (ARSA). The general design of the atmospheric collection and xenon separation of the ARSA is beyond the scope of this work but an overview of the system can be found in prior work(1). The EPCS in the ARSA is composed of a NaI(Tl) scintillation detector surrounding a cylindrical gas sample cell made from a plastic scintillator. The scintillation detectors are optically isolated from each other and use a proprietary electronics assembly to perform the coincidence correlation. A conceptual drawing of the detector arrangement is shown in Figure 1.

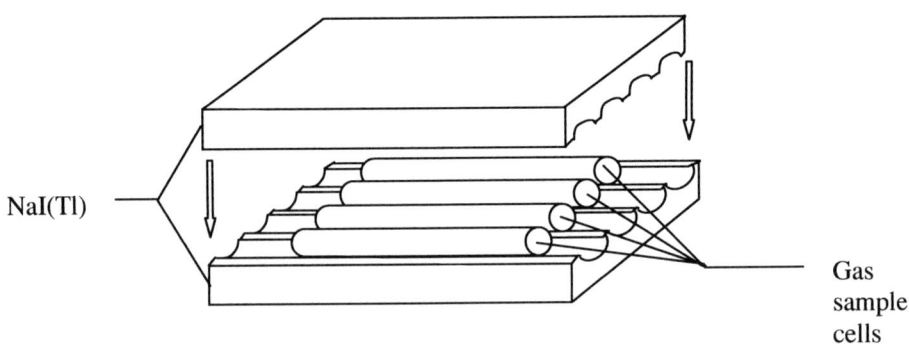

NaI(Tl)

Gas sample cells

FIGURE 1. Detector configuration.

The gas sample contained in the sample cells, primarily xenon, decays by emitting a photon in coincidence with at least one electron and the characteristic coincidence events for the xenon's are summarized in Table 2.

Table 2   Xenon characteristic coincident events.

	Photon	**Electron**
^{135}Xe	250 keV gamma	905 keV maximum energy beta
^{133}Xe	81 keV gamma	346 keV maximum energy beta
^{133m}Xe	30 keV x-ray	199 keV conversion electron
^{131m}Xe	30 keV x-ray	129 keV conversion electron

The number of detected coincident events are stored as a 2-d array by electron and photon energy.

For the data to be used to calculate isotopic concentrations the efficiency of the detector must be known as well as the amount of interference from one isotope to another i.e. x-rays produced from ^{133}Xe will also be seen in the ^{133m}Xe and ^{131m}Xe regions of interest. The amount of interference will depend on the concentrations of the interfering radionuclides, the regions of interest selected for each isotope, and consequently on detector resolution. Knowing all of these variables as accurately as possible is helpful in lowering the threshold of detectable activity. The Monte Carlo radiation transport code MCNP-4C is utilized to simulate the xenon sources in the detector system and calculate the efficiency of the detector and interference values.

The advantage of simulating the detector to determine the necessary constants needed is quite important from the viewpoint of cost, convenience, and time. Only the resolution needs to be experimentally measured for a simulation to be performed and the resolution can be determined with inexpensive, commercial, long-lived check sources. The system can be made so that the check sources are inserted automatically without human intervention. As the system ages, the resolution may degrade and by utilizing this method the necessary constants may be calculated remotely and updated. This method may also be used to reduce the cost and time required for the initial system setup by not requiring expensive and short lived xenon gas samples to determine the interference parameters. The following sections outline how a simulation was used to set up a detector and examine the ability of the system to identify the xenons when each are present with a moderate activity, 120 mBq/m^3 total (30 mBq/m^3 of air for each isomer).

SIMULATION AND POST PROCESSING FOR DETECTOR SETUP

There is only one system dependant parameter required to set up the simulation: the detector energy resolution. This is experimentally measured using a solid check source containing ^{154}Eu and ^{207}Bi. These two isotopes are convenient because they produce both electrons and photons, up to 1 MeV, in the energy range of interest (2).

The remaining work to be done to set up the simulation requires specifying the material properties, geometry, and source parameters as close as possible to the actual detector. For the 120 mBq/m^3 activity, each of the isotopes is simulated separately with a unique model for photons and electrons producing 8 simulations in all. Each

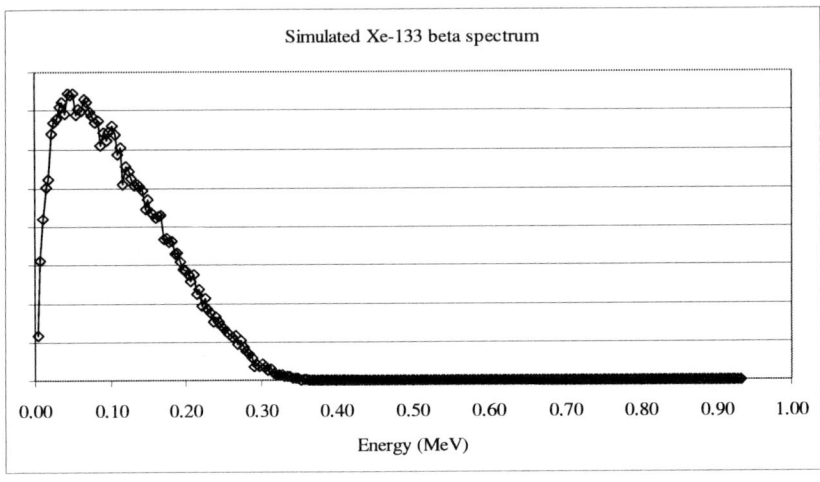

FIGURE 2 Simulated Xe-133 electron spectrum

source was modeled as a volume of pure xenon gas emitting isotropically inside the sample cell with decay abundances as reported in Table 1. The beta source distributions were modeled as histograms with bin probabilities determined from

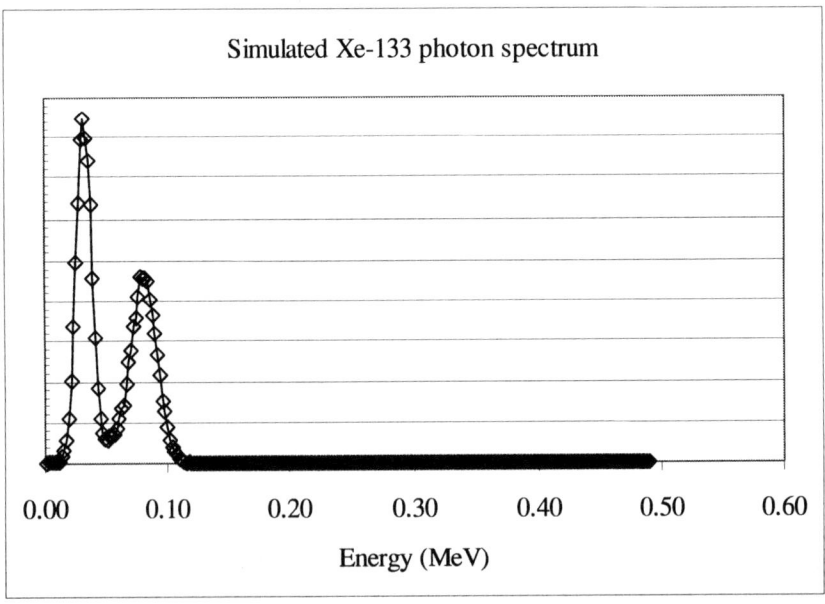

FIGURE 3 Simulated Xe-133 photon spectrum

point-wise data calculated with JEF-PC version 2.0 (3). Data was gathered for each simulation as pulse height arrays (histograms) for each detector volume. The experimental resolution data was used to determine the coefficients needed in MCNP-4C (5) to perform Gaussian energy broadening to the spectrum, thus providing a simulated spectrum with an energy resolution equivalent to the actual detector.

Once the simulations were completed, each took approximately 1 hour to run and collect all of the data, they could be folded together. This was done with a simple program written for Matlab (4). In this program the first step is to build the individual isotope arrays. This is done by determining whether there are more photon counts or electron counts. The smaller of the two will determine the total counts that will be in the coincidence spectrum. The spectrum of the larger of the two will be used to make a probability distribution for the smaller value. For instance, in the ^{133}Xe simulation there were approximately 36,000 counts for the beta spectrum, Figure 2, and 26,000 counts for the photon spectrum, Figure 3. Therefore, there will be 26,000 counts in the coincidence spectrum and the number of counts in each bin of the gamma spectrum will be distributed along the beta axis based on the probability distribution from the beta spectrum, Figure 4. For the test case the coincidence arrays were added together along with a typical background coincidence array to produce the final array, Figure 5. From the eight simulations there are now 5 coincidence arrays, four for the individual isotopes and one for the composite.

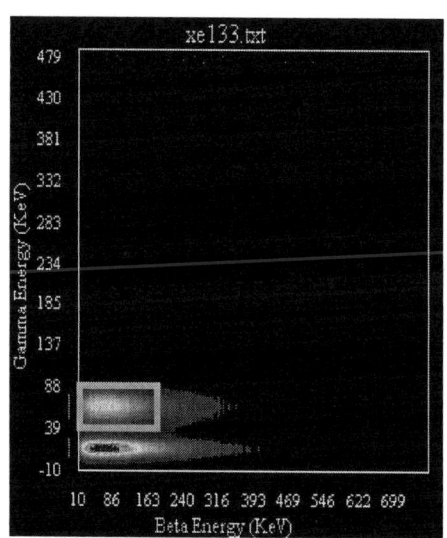

FIGURE 4 Simulated Xe-133 coincidence spectrum

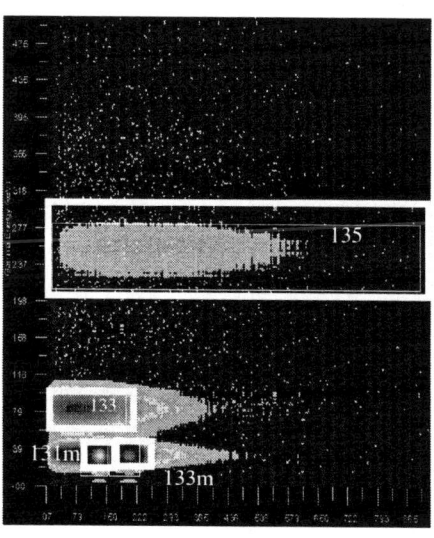

FIGURE 5 Simulated total coincidence spectrum

POST PROCESSING

With the arrays built the data can be analyzed to determine the relevant parameters needed for analyses during normal operation of the system. The first item to be determined are the regions of interest (ROI) for the individual isotopes. All other parameters will depend on the selected sizes of the ROI's. The individual isotope arrays are used to select a ROI to capture a significant amount of counts for each isotope, figure 4 shows an ROI selected for ^{133}Xe. Figure 5 displays the ROI's for each of the radioxenons. After all of the ROI's are defined the efficiency can be determined. This is done by comparing the number of coincidence events in an ROI to the number of events that were simulated.

The next set of parameters to be calculated are the isotopic interferences. This is done by comparing the number of coincidence events from one isotope in an ROI without interference to that in another isotope's ROI. An example would be the 30 keV x-ray in coincidence with a beta from ^{133}Xe interfering with the 30 keV x-ray in coincidence with a 190 keV conversion electron from ^{133m}Xe. Both of these events will be recorded in the same ROI.

A better understanding of the isotopic interference will allow for a larger region of interest, which results in better counting statistics and lower minimum detectable activities.

In the 120 mBq/m^3 example, all of the relevant parameters were calculated and formatted along with the composite array for input to the Center for Basic and Applied Research (CBAR).

The data pipeline at the CBAR automatically parses the information into an Oracle database and calculates the individual activity concentrations. The results are compared in Table 3 with the known values used in the simulation.

TABLE 3. Calculated Vs Known Xe Activity Concentration

Isotope	Known	Calculated
^{135}Xe	30 mBq/m^3	26.4 ± 0.8 mBq/m3
^{133}Xe	30 mBq/m^3	31.4 ± 2.6 mBq/m3
^{133m}Xe	30 mBq/m^3	36.2 ± 2.2 mBq/m3
^{131m}Xe	30 mBq/m^3	34.5 ± 1.9 mBq/m3

It should be emphasized that only the resolution of the detector need be measured to perform the calibration of the system and that it can be done with relatively long lived solid commercial sources. Simulations used in this manner add value to the overall operation of the system by allowing the operator to calibrate or re-calibrate by only performing a simple and quick experimental measurement. This may even be done remotely with an automated source insertion mechanism, thus eliminating the need for the operator to be present for calibration measurements. In the operation of a electron-photon coincidence system, the ability to optimize the system remotely may prove to be critical if there is significant fatigue with the detector system resulting in resolution degradation, especially if the system is installed at a location that is difficult to access.

SETUP AND SIMULATION FOR COINCIDENCE EFFICIENCY

The coincidence simulation is used to determine the coincidence efficiency of the overall detector system, detector and electronics. In this simulation, coincidence events are generated by 662 keV photons which interact in both the plastic and sodium iodide scintillators. The 662 keV photon source is replicating a commercially available ^{137}Cs solid check source. The position of the source is modeled as being external to the detector assembly but between the detector and a lead shield. The photons created by the decay of [137] Cs will create coincident events in the detector assembly through Compton scattering in the sodium iodide and Compton electrons produced in one of the plastic scintillators. As an example, Figure 9 shows an experimental spectrum taken with a [137] Cs source placed external to the detector in the same manner as is modeled here. For comparison, figure 6 shows the simulated spectrum.

A particular item of interest in the spectrum is a line that can be seen where the sum of the photon and electron energy is 662 keV, equal to the original photon energy. If this line is extended to intersect both axes it will cross at the 662 keV point, indicating that the full energy of the photon would have been deposited in a single detector volume. Another area of interest in the spectrum is at the photon energy 184 keV. This energy represents the photon backscatter detected in the sodium iodide block. Also there is a concentration of events with an electron energy at approximately 480 keV which corresponds to the Compton edge in the plastic scintillator spectra. The coincidence efficiency model is nearly identical to

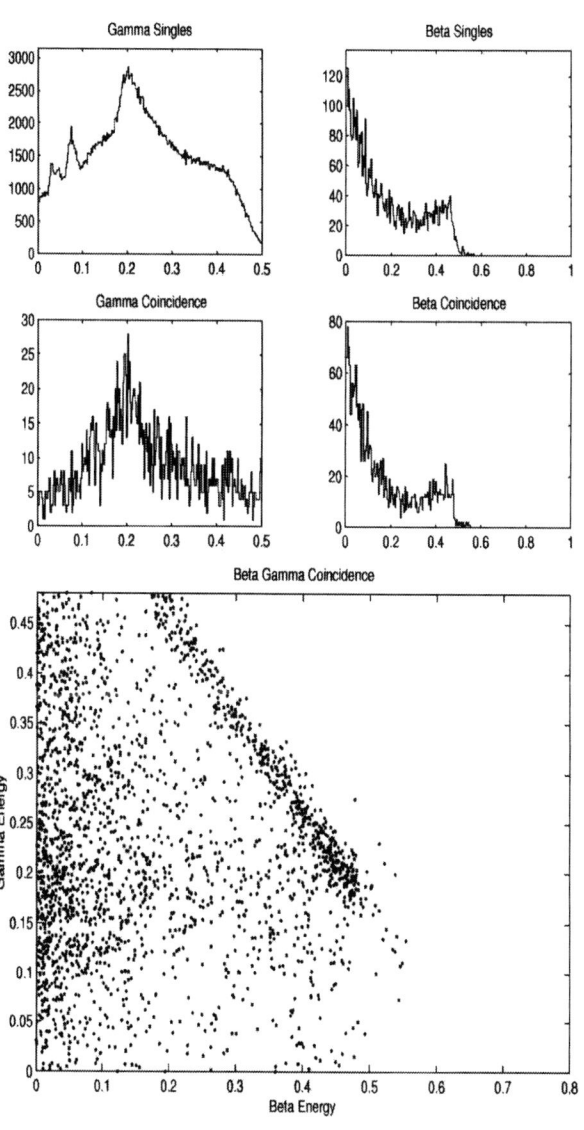

FIGURE 6 Simulated Cs-137 coincidence spectrum

the previous simulations except electrons are created and transported as well as a detailed treatment of Brehm-strahlung. Also, the individual particle track histories are recorded in a separate output file (PTRAC). The PTRAC file includes an extensive amount of information for each particle history and is used to determine the where, when, and how energy is deposited in the different detector volumes for each particle.

COINCIDENCE POST PROCESSING

For post processing, MATLAB is used to sort through and examine the PTRAC data files. Every interaction of each particle is examined to determine the type of interaction and the amount of energy deposited at that location. The energy deposited in each detector volume is summed and each particle history is examined to determine if there is a coincidence between any of the detectors. If so, the corresponding coincidence array is incremented. If not, the individual detector non-coincident spectrum is incremented. The post processing algorithm produces five arrays:

1. Plastic scintillator non-coincidence, (beta singles)
2. NaI(Tl) scintillator non-coincidence, (gamma singles)
3. Plastic gated NaI(Tl) coincidence, (gamma coincidence)
4. NaI(Tl) gated plastic coincidence, (beta coincidence)
5. Coincidence (beta gamma coincidence)

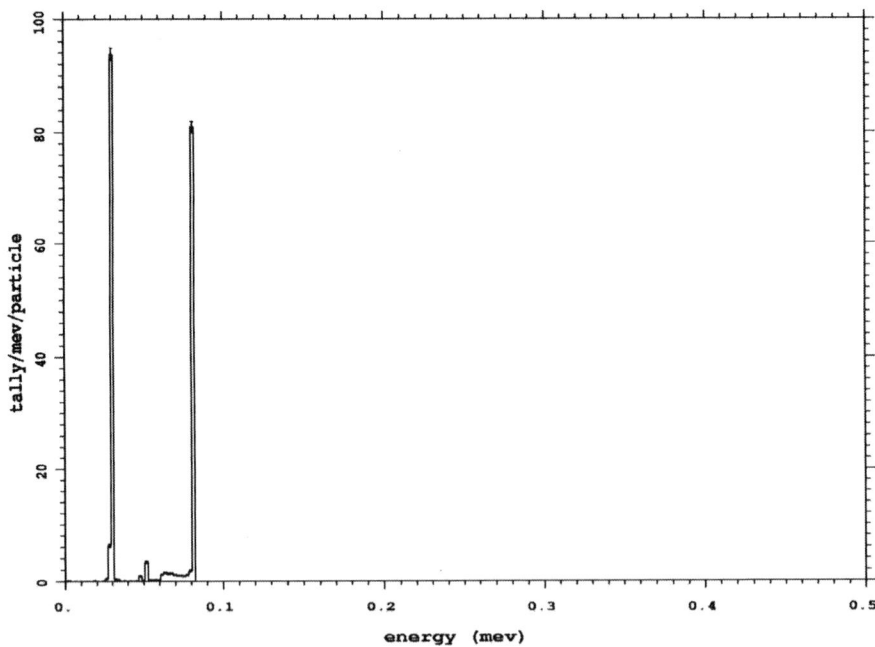

FIGURE 7 Simulation before Gaussian broadening

24

Each of the arrays is then modified, based on a Gaussian distribution, to be representative of the real detector resolution. This is done because the raw simulated spectra will have perfect energy resolution as can be seen in figure 7 for a ^{133}Xe simulation. The Gaussian energy broadening is done by examining each energy bin in a simulated array. The number of counts in each bin is then spread out over the spectrum by generating random numbers with a probability distribution similar to a Gaussian distribution centered at the energy bin and having a resolution corresponding to the experimentally measured value. This method conserves the total number of counts but reflects the detector energy resolution, see figure 8.

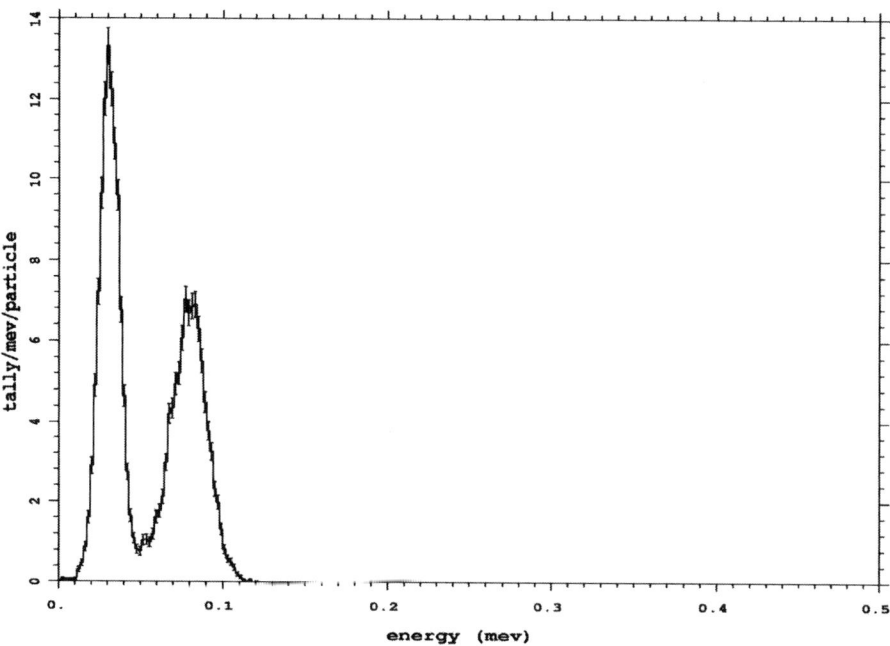

FIGURE 8 Simulation after Gaussian broadening

25

The reduced data for ^{137}Cs can be seen in Figure 6 and compared with experimental data, figure 9, taken with a real detector. When the number of coincident events is compared with the number of source events the model and experiment agree fairly

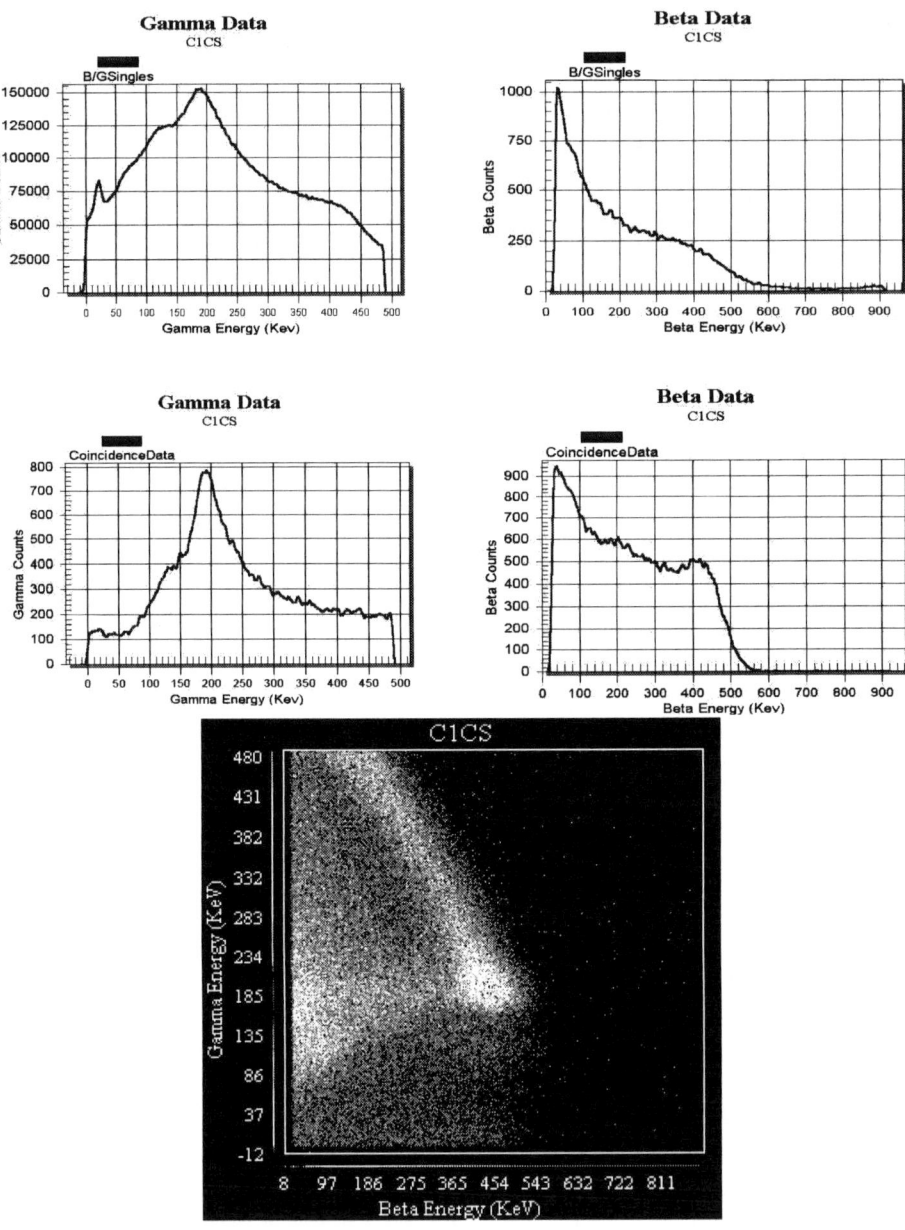

FIGURE 9 Experimental Cs-137 Spectrum

well, 94%, with the simulation giving a slightly higher number of counts. This is most likely due to slight differences between the real detector and the simulation geometry and materials.

The coincidence simulation was initiated to study the effect of different gamma sources external to the system and their possible use in calibration. From the preliminary simulations it appears that this technique may also find use in calculating the interference due to high photon background levels if needed. The largest obstacle with this method is dealing with the very large PTRAC file sizes. In the example shown here the file size was over 1 gigabyte which posed some problems after running several simulations

CONCLUSIONS

This is a work in progress and based on the initial results, it appears that a high fidelity model can be generated to simulate the response of a beta-gamma coincidence system in a short time frame. By utilizing the methodology explained here, many benefits can be realized in the set up and operation of beta-gamma coincidence systems including reduced time and cost. These methods can also be utilized for optimization and sensitivity analysis of the system with other isotopes.

REFERENCES

1. T. Bowyer et al, "*Xenon Radionuclides, Atmospheric Monitoring,*" in *Encyclopedia of Environmental Analysis and* Remediation, edited by R.A. Meyers, 1998, pp. 5299-5314
2. T. Bowyer, K. Abel, J. Hayes, J. McIntyre, P. Reeder, "*Calibration Procedure for Automated Radioxenon Sampler Analyzer*", PNNL-13070 Pacific Northwest National Laboratory, Richland, Washington, 1999
3. OECD Nuclear Energy Agency. (1997) *JEF-PC 2.0* France: Nuclear Energy Agency
4. Mathworks. (1999) *Matlab 5*
5. Los Alamos National Laboratory. (2001) MCNP-4C USA: RSICC

MGA Analysis on Elevated ^{238}Pu Samples

T.F. Wang, K. J. Moody, K. E. Raschke and W.D. Ruhter

Lawrence Livermore National Laboratory, Livermore, CA 94550, USA

Abstract. Plutonium gamma-ray data analysis, in the 100-keV region, using MGA has been improved to overcome the original maximum limit of 2% ^{238}Pu relative plutonium content in a sample in order to perform an analysis. MGA analysis results of elevated ^{238}Pu samples are compared to the results from mass spectrometry.

INTRODUCTION

The gamma-ray Multi-Group Analysis code[1,2,3] (MGA) developed at Lawrence Livermore National Laboratory is widely used in non-destructive isotopic evaluation of plutonium materials, especially for plutonium verification and accountability. The MGA code can analyze gamma-ray data collected using a planar Ge detector or data collected using a coaxial Ge detector, or both. The MGA code analyzes all three important regions of plutonium gamma-rays: the 100-keV, the 300-keV and the 600-keV, depending on the availability of the data in these regions. The 100-keV region contains all the plutonium isotopic information, however, the data in this region can be perturbed (i.e., homogeneity of the sample, shielding of the samples, etc.) due to low energy nature of these gamma-rays; the 300-keV and the 600-keV regions, though less perturbable, each only contains a subset of plutonium isotopic information.

Recently, because of the accelerated aging studies being done on plutonium, there is a need for non-destructive gamma-ray evaluation of elevated (> 2%) ^{238}Pu samples both in material accountability and material processing. Figure 1 shows the complicated processes[4] involved in producing elevated ^{238}Pu metals for the aging studies. These processes involve both chemical and mechanical treatments on the plutonium materials and associated chemical agents. During each process, constant monitoring of the plutonium material balance between the input materials and output products is required. Mass spectrometry, in general, can provide isotopic analysis of the products to high degree of accuracy. Unfortunately, how the sampling is done on the products is crucial, because mass spectrometry can only provide results that represent "microscopic" isotopic content of the products. On the contrary, the gamma-ray analysis results combined with calorimetry results, though not as accurate as mass spectrometry results, represent the "macroscopic" isotopic content of the products.

For non-destructive, gamma-ray isotopic analysis of the elevated ^{238}Pu products, the original MGA code could not be used on such high concentrations of ^{238}Pu. We have, therefore, modified the methodology used in the 100-keV region gamma-ray analysis

CP632, *Unattended Radiation Sensor Systems for Remote Applications,* edited by J. I. Trombka et al.
© 2002 American Institute of Physics 0-7354-0087-3/02/$19.00

of the original MGA analysis code to overcome this deficiency. In the paper, we will compare the results from this improved MGA code to the results from the mass spectrometry in a set of elevated (5 ~ 7 % ^{238}Pu) samples. We have also tested the ability of the code on the extremely high (> 40% ^{238}Pu) samples.

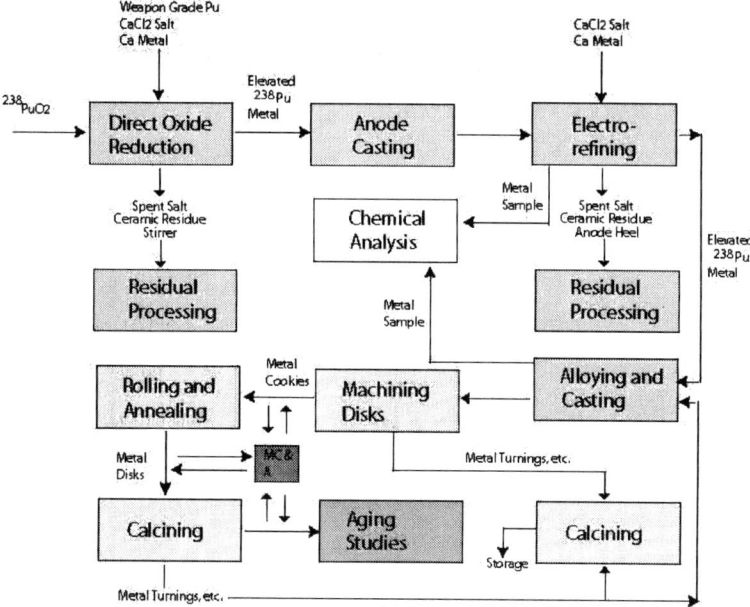

FIGURE 1. The chemical and mechanical processes involved in making elevated ^{238}Pu samples for aging studies

AN OVERVIEW OF THE MGA CODE

The MGA plutonium isotopic analysis code is unique for its ability to de-convolute the complicated, 100-keV x-ray and gamma-ray region to obtain the ratios of the Pu isotopes, as shown in Figure 2. Because the 100-keV region has most of the plutonium gamma-ray decay strength, MGA can determine the relative abundance of the plutonium isotopes with accuracy better than 1% using a high-resolution germanium detector in a few minutes of counting time. For confirmatory measurements that requiring few % isotopic accuracy, MGA can provide isotopic results within counting times of a minute.

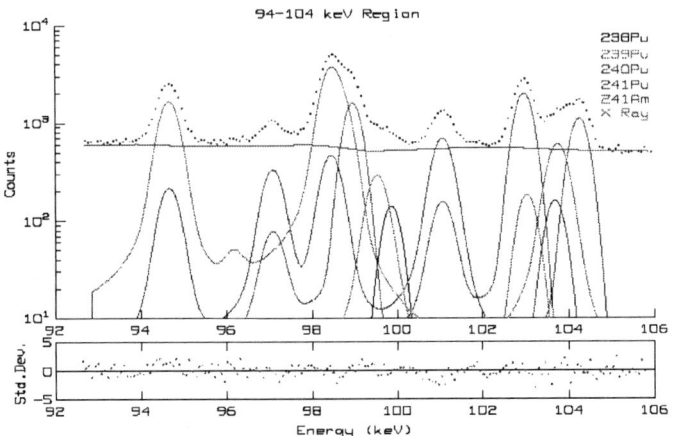

FIGURE 2. MGA analyzes the complicated 100-keV energy region of a weapon grade plutonium sample to obtain plutonium-^{241}Am isotopic information

The MGA was developed to analyze either a low-energy gamma-ray spectrum taken using a high-resolution planar HPGe detector for energies below 300 keV, or to analyze a low-energy spectrum combined with a high-energy spectrum (up to 1 MeV) taken with a HPGe coaxial detector in what we refer to as the two-detector analysis mode. The reason for using two detectors is simple: a high-energy spectrum taken using a coaxial HPGe detector will not provide sufficient energy resolution for 100-keV plutonium isotopic analysis, while the small planar HPGe used at low energies has inadequate high-energy gamma-ray detection efficiency. The "two-detector" mode MGA analysis was developed to improve the determination of the ^{241}Pu/^{239}Pu ratio in high burnup plutonium with the 300-keV regions from the high-energy spectrum. In high burnup plutonium, ^{241}Pu is dominant in the 100-keV region and the precision of the ^{241}Pu/^{239}Pu ratio from this region is reduced.

In general, MGA can provide accurate plutonium isotopic information for all burnups using the data collected with a high-resolution planar detector (below 300 keV). However, more refined isotopic results and additional isotopic information can be obtained from using higher-energy gamma rays (above 300 keV) collected with a high efficiency coaxial detector. For example, the ^{238}U abundance can be obtained from the 1001 keV peak, a more accurate analysis for ^{237}Np can be made, some fission products can be identified, and ^{241}Am inhomogeneities in the sample can also be determined. The two-detector mode of MGA is the only way to provide homogeneity information for a sample. It is worth noting that these two sets of data do not have to be collected at the same time under the same geometry.

Figure illustrates how the relative detection efficiency for the measurement is determined for the low-energy detector. To properly use the higher energy gamma-ray information, a separate intrinsic efficiency curve must be determined, as shown in Fig. 4. Like the low-energy curve in Fig. 3, the components of the efficiency are based on the physical processes involved in attenuating and detecting the gamma rays.

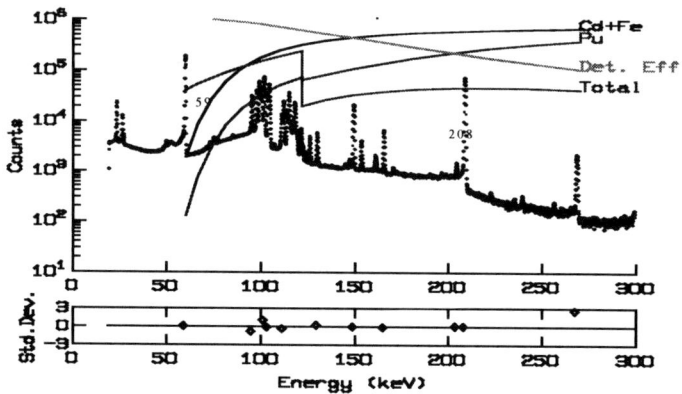

FIGURE 3. MGA uses physical attenuation corrections of both emission and absorption in the gamma ray interactions. This plot shows the three principal components (the detector efficiency, the absorption due to intermediate materials, and the self-absorption of plutonium) of that characterize the low-energy "intrinsic" efficiency curve.

The need for two detectors to measure high burnup plutonium is inconvenient for inspection organizations, especially with the recent introduction of the Safeguards type coaxial HPGe detectors that has adequate energy resolution for 100-keV gamma-rays and sufficient efficiency for 1-MeV gamma-rays. Therefore, MGA was upgraded to allow measurements of high burnup plutonium with a single high-resolution germanium detector[5].

The MGA can also operate in the MGAHI mode[6] to obtain isotopic information of moderate shielded plutonium sample when the 100-keV region gamma-ray information cannot be obtained.

31

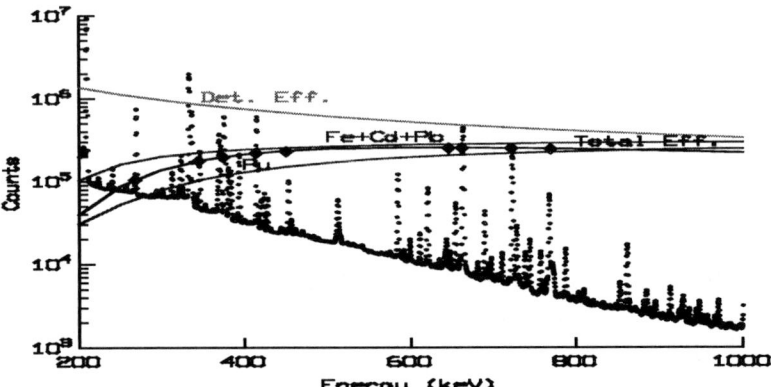

FIGURE 4. The same physical attenuation processes are used in analyzing the second detector data when running MGA in the two-detector mode. This plot shows the components affecting the efficiency curve. In general, there is a thin Pb absorber in front of the coaxial detector to reduce the count rate from the low-energy region of the spectrum (<200 keV).

MGA ANALYSIS AND MASS SPECTROMETRY ANALYSIS ON ELEVATED ^{238}PU SAMPLES

For elevated ^{238}Pu samples, the gamma rays in both 100-keV region and 150-keV region will be dominated by the decay of ^{238}Pu at 99.864 keV and 152.680 keV as shown in Figure 5. The ability of MGA analysis at the 100-keV region, which relies on the accurate energy calibration using the weighted peak heights of the 94.658-keV and 101.066-keV gamma rays, is perturbed by the "intrusion" of the 99.864-keV, 98.441-keV, and 94.658-keV ^{238}Pu gamma-rays. In ordinary plutonium samples, the intensity of the 98.441-keV and 94.658-keV gamma-rays is mostly from the ^{239}Pu decay.

32

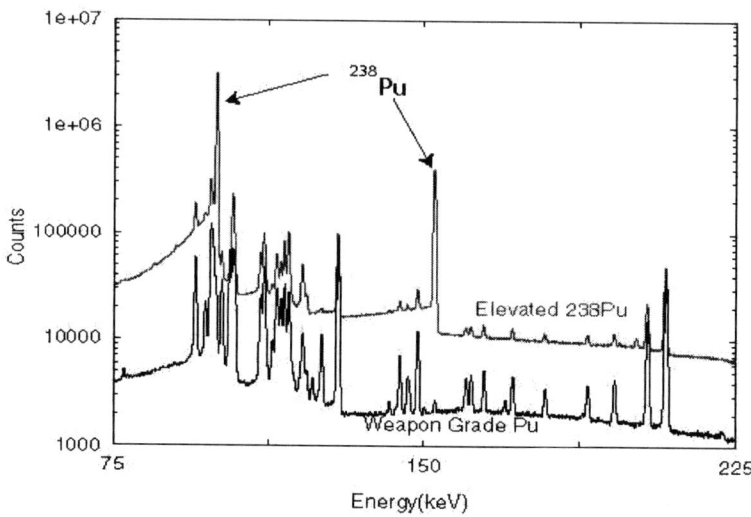

FIGURE 5. A comparison of the 100-keV region gamma-ray data between an elevated [238]Pu sample and a weapon grade Pu sample.

In order to do accurate energy calibration for the 100-keV region for elevated [238]Pu samples, we have included the weighted peak-height of the 99.864-keV into the energy calibration in the MGA code. However, this inclusion of the 99.864 peak height is meaningless for regular plutonium samples as can be seen from Figure 2 (i.e., it is impossible to obtain accurate peak-height information of this peak when the intensity is low). To be able to use this new energy calibration only for elevated [238]Pu samples, we have added a flag that uses the peak height ratio of 152.8keV ([238]Pu) and 129.29 keV ([239]Pu) to trigger the MGA code into the elevated [238]Pu mode. Figure 6 shows a screen dump of the modified MGA code fitting to one of the elevated [238]Pu samples.

33

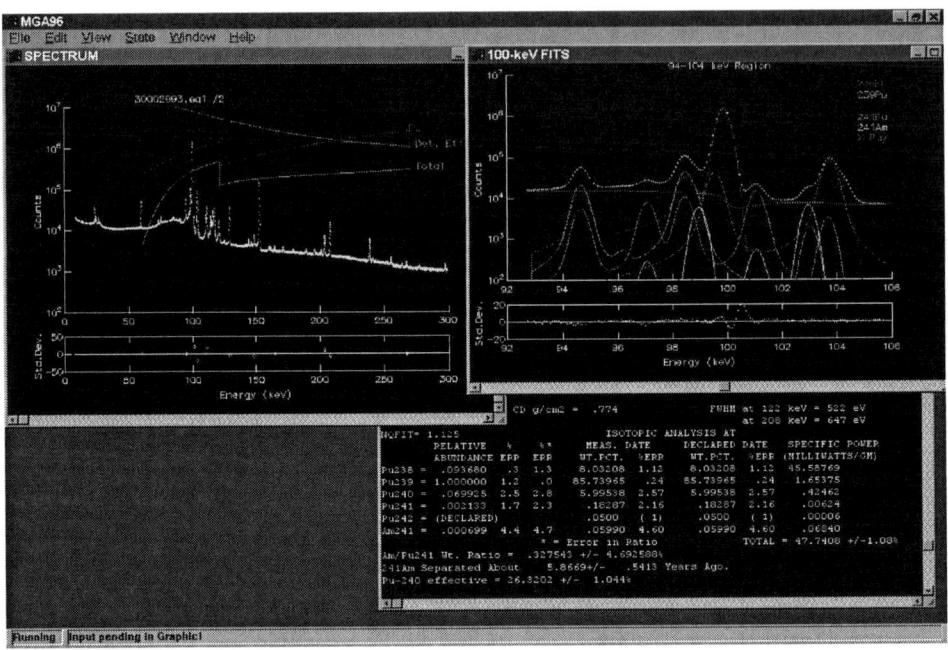

FIGURE 6. A screen dump of the modified MGA code while analyzing the "slug" component in the elevated ^{238}Pu process. On the upper left is the physical attenuation corrections displayed in graphical form; on the upper right is the spectral deconvolution of the 100-keV region. The output is at the bottom right.

Two different mass spectrometry instruments, the Inductivity Coupled Plasma Mass Spectrometry (ICPMS) and IsoProbe, were used in providing mass spectrometry results of the elevated samples. Careful chemical separations of ^{238}U from ^{238}Pu were performed on the samples as well as alpha particle detection using surface barrier detectors, prior to and after the separations, were also performed to reaffirm successful chemical separations, since the mass spectrometry cannot tell the difference between these two isotopes. Long duration gamma-ray counting on these samples after the chemical separations (due to small sample size used in the mass spectrometry measurements) was performed. Table 1 shows comparisons between the mass spectrometry results and modified MGA results for a set of elevated ^{238}Pu samples.

34

Table 1. Comparisons between the Mass Spectrometry and the MGA results. All numbers are in weight per cent, the ^{242}Pu weight per cent is assumed to be 0.05% for all samples. The averaged mass spectrometry results for ^{242}Pu is 0.05 +- 0.001 %.

Sample	Analysis mode	^{238}Pu
#1	MGA	7.33 (1.6%)
	Mass Spec.	7.1035(0.33%)
#2	MGA	7.15 (1.1%)
	Mass Spec.	7.1593 (0.33%)
#3	MGA	7.31 (1.15%)
	Mass Spec.	7.0865 (0.34%)
#4	MGA	7.17 (1.2%)
	Mass Spec.	7.1144 (0.34%)
#5	MGA	7.25 (1.2%)
	Mass Spec	7.0607 (0.03%)
#6	MGA	4.86 (1.2%)
	Mass Spec	4.83 (0.33%)

One important thing to note is that for the modified MGA to be able to provide good results, the knowledge of ^{242}Pu values for these "artificially made" elevated ^{238}Pu samples is required in the input. For regular samples of different burn-up times, MGA uses empirical parameterization of the other isotopes to obtain ^{242}Pu values.

For large mass (i.e., >100g) elevated ^{238}Pu samples, the modified MGA analysis has encountered some problems due to the packaging of the sample. The reason is understandable: the heat output of these samples is high, and large amounts of "heat-sink" materials were placed around these samples during the packaging. The gamma-rays in the 100-keV region, which were used in the modified MGA analysis, will not be attenuated isotropically because of the irregular shape of these heat dissipating materials. One possible solution to this problem is to use higher energy gamma-rays (i.e. lesser effect due to attenuation compare to the 100-keV region gamma-rays) in the 600- and 700- keV regions. At current, we are exploring this possibility.

CONCLUSION

The modified MGA algorithm in the 100-keV region has been tested in good agreement with the mass spectrometry results for small quantity of elevated ^{238}Pu samples. For larger ^{238}Pu samples, the packaging of the sample is crucial to ensure the success of the 100-keV region analysis. For some heterogeneous components produced in the process, we are using the active and passive tomography technique[7] to hopefully find a solution to the problem.

ACKNOWLEDGMENTS

We thank Dan Decman, Karen Dodson, Jackie Kenneally, Steve Kreek, Mark Mount, and Austin Prindle for their helps and discussions. This work was performed under the auspices of the U.S. Department of Energy by Lawrence Livermore National Laboratory under contract No. W-7405-Eng-48.

REFERENCES

1. W. E. Parker, T.F. Wang, D. Clark, W. M. Buckley, W. Romine and W. D. Ruhter, *Plutonium and Uranium Isotopic Analysis: Recent Developments of the MGA++ Code Suite,* Proceedings of the Sixth International Meeting on Facilities Operations - Safeguards Interface, pp. 192 - 197, American Nuclear Society, Jackson Hole, Wyoming, September 1999.
2. R. Keyser, T. Twomey, S. Haywood, W. E. Parker, T.F. Wang, D. Clark, K. Raschke, W. Romine, W. Buckley and W. Ruhter, *Recent Developments in the MGA++ Codes,* ESARDA Conference, Seville, Spain, May 1999.
3. R. Gunnink, W.D. Ruhter, MGA: *A Gamma-Ray Spectrum Analysis Code for Determining Plutonium Isotopic Abundances,* Volume I and II, Lawrence Livermore National Laboratory, Livermore, CA., UCRL-LR-103220, (1990).
4. Karen Dodson, private communication, (2001).
5. T.F. Wang, K. Raschke, and W.D. Ruhter, *Two-Detector Mode MGA Analysis of Plutonium Using a Single Ge Detector,* UCRL-JC-141234, presented at the 23rd Annual ESARDA Symposium on Safeguards and Nuclear Material Management, May 2001.
6. T.F. Wang, K.E. Raschke, W.D. Ruhter, S.A. Kreek, *MGAHI: A Plutonium Gamma-Ray Isotopic Analysis Code for Nondestructive Evaluations,* ANS Transactions, **81**, 234 (1999).
7. T.F. Wang, H.E. Martz, G.P. Roberson, E.A. Henry, W.D. Ruhter, L.O.Hester, *Three Dimensional imaging of a Molten-Salt-Extracted Plutonium Button using Both Active and Passive Gamma-ray Tomography,* Nuclear Instrument and Method in Physics Research, **A353** (1994).

Applications of Noble Gas Radiation Detectors to Counter-terrorism*

Peter E. Vanier

Brookhaven National Laboratory, Safeguards and Arms Control Division, Building 197C, Upton, NY 11973

Leon Forman

Ion Focus Technology, Inc., Chemistry Building, SUNY at Stony Brook, Stony Brook, NY 11794-34005

Abstract. Radiation detectors are essential tools in the detection, analysis and disposition of potential terrorist devices containing hazardous radioactive and/or fissionable materials. For applications where stand-off distance and source shielding are limiting factors, large detectors have advantages over small ones. The ability to distinguish between Special Nuclear Materials and false-positive signals from natural or man-made benign sources is also important. Ionization chambers containing compressed noble gases, notably xenon and helium-3, can be scaled up to very large sizes, improving the solid angle for acceptance of radiation from a distant source. Gamma spectrometers using Xe have a factor of three better energy resolution than NaI scintillators, allowing better discrimination between radioisotopes. Xenon detectors can be constructed so as to have extremely low leakage currents, enabling them to operate for long periods of time on batteries or solar cells. They are not sensitive to fluctuations in ambient temperature, and are therefore suitable for deployment in outdoor locations. Position-sensitive ^{3}He chambers have been built as large as 3000 cm^2, and with spatial resolution of less than 1 mm. Combined with coded apertures made of cadmium, they can be used to create images of thermal neutron sources. The natural background of spallation neutrons from cosmic rays generates a very low count rate, so this instrument could be quite effective at identifying a man-made source, such as a spontaneous fission source (Pu) in contact with a moderator (high explosive).

INTRODUCTION

The problem of detecting, localizing and characterizing an unknown, hidden radioactive source is a challenging one that has been worked on by many experts for over 50 years[1]. Standard commercial detectors have been optimized by years of development, and work very well under most circumstances for which they were designed. However, each type of detector has specific limitations imposed by physical properties of the components as well as design choices made for particular applications. In the light of recent heightened concern about nuclear terrorism, it may

---

* This work was performed under the auspices of the U.S. Department of Energy, Contract No. DE-AC02-98CH10886.

CP632, *Unattended Radiation Sensor Systems for Remote Applications*, edited by J. I. Trombka et al.
© 2002 American Institute of Physics 0-7354-0087-3/02/$19.00

be fruitful to take a fresh look at some alternative technologies that may not yet have been fully developed or exploited for the express purpose of intercepting illicit traffic in radioactive materials. In order to have a significant impact, new instruments would have to demonstrate clear advantages over traditional gamma detectors such as plastic scintillators, sodium iodide crystal scintillators and high purity germanium semiconductor crystals, and neutron detectors such as moderated He-3 or BF_3 tubes. For the most part, this paper will consider the case of an unattended detector in a remote location, designed to send data to a centralized monitor, but will not address the communications aspects of such a system.

REQUIRED DETECTOR CHARACTERISTICS

Area of Detector

The scenarios considered in countering terrorism may involve weak sources, possibly shielded, to be detected at maximum standoff distances, and therefore the radiation fluxes to be detected can be very low. A major consideration in the choice of a detector is the requirement to count a number of events that is statistically distinguishable from background. Assuming a localized source, the inverse square law argues for large-area detectors in order to intercept a significant number of gammas and neutrons. This fact can be quantified by considering a point source emitting 10^6 gammas per second. For example, a 1-cm^2 crystal placed 1 meter from the source will intercept about 8 photons per second. If the source were 100 meters away, the detector would need an area of 1 square meter to be exposed to the same number of photons, which would also be decreased by attenuation in the air. As the distance increases to the kilometer range, attenuation becomes the dominant factor, and increasing the area of the detector to 10 m x 10 m would not be sufficient to give the same count rate from the source. Also, the background count rate would increase with the detector area, and would dominate the signal. Semiconductor detectors are limited in size to a few centimeters because of the difficulties of growing large high-quality crystals. Both scintillators and compressed gas ionization chambers have some chance of being deployed in square meter dimensions.

Resolution

In the detection of gamma rays, the source spectrum consists of a number of sharp photopeaks corresponding to the characteristic gamma energies of the source isotopes. These signals are superposed on a broad continuum of Compton-scattering events where some of the original gamma ray energy is lost, either in the source, in the detector or in the intervening materials. The gross count rate of Compton events, as measured by a low-resolution detector such as a scintillator, can be used as an indicator of the presence of a radioactive source. However, to identify the isotope in question and distinguish it from other sources in the natural background requires a significant number of counts in the detected peaks. With high-resolution detectors such as Ge, the peaks (created by full deposition of the photon energy into photoelectrons) are narrow, with all the counts occurring in a few channels, so the peak-to-Compton ratio is favorable. With low-resolution detectors such as NaI, the peaks are broad, and the peak-to-Compton ratio is relatively poor. Plastic scintillators usually are dominated by the Compton events that contain little information about the source materials. Emerging technologies with intermediate resolution (about 2% FWHM at 662 keV) include CdZnTe crystals and compressed xenon.

In the case of semiconductors or gas ionization chambers, the photoelectrons must drift across the distance to the anode without significant losses by recombination. The variability in losses leads to a degradation of resolution. The electrode geometry should be designed so that the pulse duration at the anode is minimized, in order to minimize the noise integrated by the preamplifier. This is sometimes achieved by the construction of Frisch grids in gas detectors and coplanar grids on CZT.

In the case of scintillators, the detection is a two-step process. The ionized electrons excite optical transitions and the emitted light is detected by a photomultiplier, which generates a current pulse. The scintillator must be transparent to the emitted light, while the photocathode of the photomultiplier must absorb that light efficiently and excite a number of free electrons proportional to the energy deposited. For a given gamma energy, since the number of electrons generated at the photocathode is much smaller than the number of photoelectrons that would be produced in a semiconductor device, the statistical variations in the resulting current pulse are much greater. This results in poor resolution for scintillator detectors. Other factors affecting resolution include the variations in light collection efficiency with the position of the initial event.

Efficiency

The intrinsic detection efficiency is defined as the number of events counted divided by the number of incident gammas or neutrons. Several factors contribute to the efficiency, including the probability of interaction of the radiation by various mechanisms and the probability of collection of the resulting ionized electrons into an amplified pulse. For high efficiency, the thickness of the absorbing medium of the detector must be comparable to the mean free path for interaction at the particular energy of incident radiation. At the same time, the thickness should be small enough for the electrons to be detected without significant losses.

Neutron detectors based on high-pressure ^{3}He have efficiency greater than 50% for thermal neutrons, which have a high cross-section for the nuclear reaction that produces a proton and a triton, with the release of 764 keV. They have very low sensitivity to gammas and to fast neutrons. If there is material containing hydrogen close to a fission source (or close to the He detector) a fraction of the emitted neutrons will be thermalized and detected.

Imaging Capability

Some types of gamma and neutron detectors can be made into imaging systems that measure the direction of the incident radiation. The coded aperture approach relies on a mask in front of the detector to create a unique shadow on the detector face, such that the position of a distant point source can be deduced from the displacement of the pattern. Such coded-aperture instruments allow much of the random background to be ignored because it is widely distributed, while pinpointing a localized source with a single pixel. The counts in this pixel are proportional to T, the total number of counts detected over the whole area of the mask. In the construction of the real image from the mask distribution, all the data are used to calculate the intensity of each image pixel, so the random dark field counts are proportional to $\sqrt{T}$ by Poisson statistics. Therefore, if a large enough number of counts is accumulated, it is possible to tell whether the brightest pixel is statistically significant compared to the expected random noise image. With a single, non-imaging detector it may be much more difficult to determine whether a real source exists or whether there is an increased background. Spatial variations in neutron background may exist because of cosmic ray spallation in materials of differing density. Temporal variations can occur because of variations in solar activity and position.

COMPRESSED XENON GAMMA SPECTROMETERS

The internal electrode structure of a prototype compressed xenon gamma detector built at Brookhaven National Laboratory is shown in Figure 1. The construction and performance parameters of the detector are described by Mahler and co-authors.[2] This design contains planar electrodes with a planar Frisch grid, which facilitates the study of electron transport and recombination in Xe. Other designs have coaxial electrodes, which offer certain advantages for field-deployed detectors.

FIGURE 1. The internal electrode structure of a prototype compressed xenon detector built at BNL.

The BNL detector has remained remarkably stable for over 5 years, with no noticeable change in resolution or efficiency. This means that there has been no significant degradation of the purity of the xenon, either by leaks or by outgassing. There has not been any loss of xenon mass either. This type of detector is therefore quite suitable for unattended operation in remote locations for extended periods of time. The power requirement for the chamber is extremely low, because xenon has very high resistivity, and dark currents are negligible compared to the signals detected.

The useful spectral range for the xenon detector is about 60 keV to about 2 MeV for full-energy photopeaks. Near the lower limit, set by preamplifier noise, the titanium window and the dead layer of xenon at the end of the chamber cause some attenuation in the signals compared to NaI detectors. However, in the region of 186 keV, the dominant emission from unshielded U-235, the efficiency is as good as any detector of comparable area (see Figure 2). NaI detectors do not have sufficient resolution to identify this signature unambiguously.

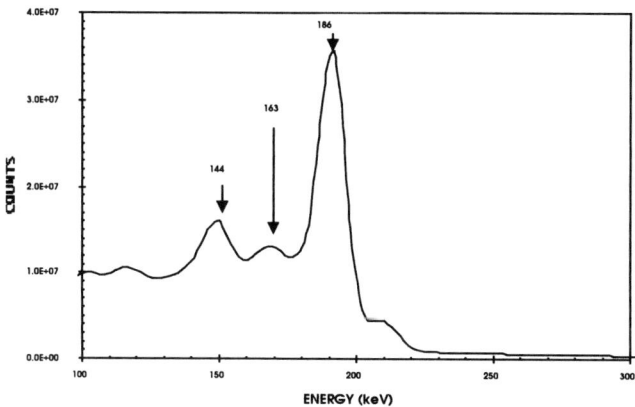

FIGURE 2. Low energy section of gamma spectrum of thin HEU foil measured with BNL xenon detector.

As with all gamma detectors, the intrinsic efficiency decreases at higher energies, as the photoelectric cross-section drops exponentially. The useful limit at high energy is approached when the ionization tracks produced by the primary photoelectric events are too long for efficient collection of the charge to take place. Then either (a) some of the energy is lost to the structural components or (b) the charge cannot be collected at the anode during the shaping time of the preamplifier, due to the low drift mobility of electrons in xenon. This effect (the ballistic deficit) results in a loss of resolution beyond the $E^{-1/2}$ dependence predicted by Poisson statistics. The BNL detector shows such line-broadening at 1.3 MeV, and is unable to display a peak at all at 2.6 MeV, which is the energy of a characteristic emission from U-232, a common impurity in highly enriched uranium.

Recent experiments have demonstrated a way to overcome this shortcoming by making use of the pair-production mechanism, the cross section of which increases with energy. When electron-positron pairs are produced, their total kinetic energy is less than the gamma energy by 1022 keV, and the remainder is shared between both charged particles, which will have shorter ionization tracks than a single photoelectron of the same energy. When the positron annihilates, the two 511 keV gammas that are emitted can escape from the detector without creating further ionization and the net charge can be collected without excessive ballistic deficit. The resulting "double-escape" peak from U-232 occurs at 1592 keV, and is quite prominent, even though the photopeak at 2614 keV cannot be measured (see Figure 3). This signature is important because all materials have a low absorption cross-section at an energy of 2.6 MeV. Therefore this gamma ray is not easily shielded, as illustrated by the second spectrum in Figure 3, where the source was shielded by 1 cm of Pb, and the double-escape peak is still quite measurable. There is another photopeak in the U-232 spectrum that is very close in energy to the 1592 keV peak and may contribute some counts, but is not separately resolved by the xenon detector. The ubiquitous background line at 1461 keV from K-40 is clearly resolved by xenon, but would not be separated by NaI.

Future development of large-volume xenon detectors will probably take the form of of many tubes combined into arrays (see Figure 4). A single large container would need very thick walls to withstand the internal pressure of ~60 bars. The total sensitivity to low energy gammas is proportional to the surface area of the array, and the sensitivity to high energy gammas is proportional to the mass of xenon. One other advantage of an array would be the capability to perform coincidence experiments (to enhance the escape peaks) or anticoincidences (to reject multiple Compton scattering).

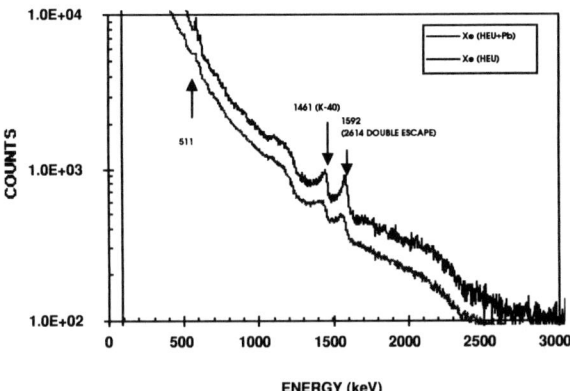

FIGURE 3. The high energy portion of a gamma spectrum of HEU foil showing the U-232 double-escape peak resolved from the K-40 background peak.

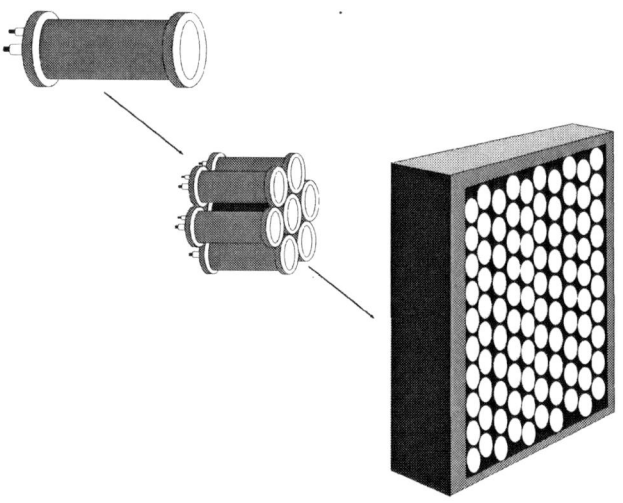

FIGURE 4. Possible future development of arrays of compressed xenon tubes with very high sensitivity.

HE-3 POSITION-SENSITIVE NEUTRON DETECTORS

One of the most sensitive thermal neutron detection isotopes is high pressure He-3. Moreover, because neutron interaction with He-3 produces energetic charged particles, it is possible to construct a position-sensitive device by collecting charge on crossed arrays of wires. BNL has produced some of the largest He-3 position-sensitive detectors, up to 3000 cm^2, with spatial resolution better than 1 mm (see Figure 5). Additional details about the design and performance of these detectors can be obtained from the BNL website at www.inst.bnl.gov.

The authors of this paper have constructed a coded-aperture imaging device consisting of a 350-cm^2 He-3 position-sensitive detector and a cadmium uniformly-redundant array mask[3]. With this device, it was possible to detect and locate a surrogate improvised nuclear device at a distance as great as 60 meters from the detector. A typical image of a distant point source is shown in Figure 6. This type of device would be a useful unattended sensor technique for monitoring harbors or other points of entry. One challenge is to increase the size of the detector to perhaps 4 m^2, which would reduce the data acquisition time by a factor of 100 compared to the existing imager. Such a device would not need to have sub-millimeter resolution, and therefore could operate at lower pressure and than the typical 10 bars used in the He-3 chambers.

FIGURE 5. Position-sensitive thermal neutron detectors developed at BNL. A coded aperture imaging system was built using the 20 cm x 20 cm chamber.

FIGURE 6. Image of a small thermal neutron source detected at a distance of 20 meters with a coded-aperture neutron imaging system.

CONCLUSIONS

In order to be useful at distances greater than arm's length, radiation detectors should be constructed with large areas for maximum sensitivity. Compressed noble gas ionization chambers offer a practical means to achieve such dimensions, and can be designed to detect either gamma rays or neutrons. Distinguishing between a terrorist nuclear or radiological weapon and natural background requires adequate energy resolution of gamma ray spectra, which can be accomplished with xenon detectors. It is also useful to create images so as to indicate localized sources that stand out against the background. This process has been demonstrated with thermal neutrons using a He-3 chamber. These inert gas detectors operate at room temperature for long periods of time with modest power requirements and a high degree of stability. They are therefore very appropriate for remote unattended monitoring systems.

ACKNOWLEDGMENTS

The authors are grateful for the expert assistance of the staff of the BNL Instrumentation Division, including Graham Smith, Bo Yu, Don Makowiecki, and Neil Schaknowski.

REFERENCES

1. Knoll, Glenn F., *"Radiation Detection and Measurement"*, John Wiley and Sons, New York, Third Edition, 1999.
2. Mahler, G.J., Yu, B., Smith, G.C., Kane, W.R., and Lemley, J.R., *IEEE Trans. Nuc. Science* **NS-45**, 1029-1033 (1998).
3. Vanier, P. E., and Forman, L., *Proceedings of the Institute of Nuclear Materials Management*, 37th Annual Meeting , Naples, FL (1996).

Detection of Materials of Interest to NonProliferation: A Novel Approach

Frederic Ze* Bernhard R. Tittmann[†], and P. M. Lenahan[†]

* Lawrence Livermore National Lab, Livermore, California.
[†] The Department of Engineering Science and Mechanics,
Pennsylvania State University, University Park, Pennsylvania

Abstract. We propose the development of a novel detector that can locate and identify materials of interest to Nuclear Arms Non Proliferation. The device will combine nuclear acoustic resonance (NAR) with superconducting quantum interference device (SQUID) widely used in nuclear magnetic resonance (NMR), geophysics, nondestructive evaluations, and biomagnetism, to name only few. NAR works like NMR. Thus resonant absorption (of applied ultrasonic energy) by a nuclear spin system occurs when the ultrasonic frequency is equal to the appropriate frequency separations between the magnetic nuclear energy levels. Ultrasonic energy couples to the nuclear spin system via spin-phonon interaction. The resulting nuclear acoustic resonance can be detected via the changes in (a) ultrasonic attenuation, (b) ultrasonic velocity, (c) material magnetization, (d) or nuclear magnetic susceptibility, all of which carries "intrinsic and unique signatures" of the material under investigation. The device's sensitivity and penetration depth (into metals) will be enhanced by incorporating SQUID technology into the design. We will present the details of interaction physics and outline a plan of action needed to successfully transform the concepts into a practical detector.

INTRODUCTION

We propose the development of a novel device to detect and identify smuggled materials that might be used in a nuclear weapons program or for terrorist activities. Its design will be based on Nuclear Acoustic Resonance (**NAR**) principles, and will incorporates **SQUID** (superconducting quantum interference device) to facilitate detection through metal barriers. The objective is to develop a detector which exhibit the following attributes:

- It will detect intentionally shielded fissile materials, as well as other materials of consequences to Non Proliferation. That means the device will be applicable to all materials of interest to Non Proliferation. The main objective will be to develop a device that can detect and identify fissile material even if steps had been taken by would-be smugglers to completely shield the material as a radiation source. Thus a successful development of the device could enormously strengthen our nation's ability to thwart the spread and smuggling of materials intended for the production of weapons of mass destruction.

- The device will be highly sensitive. The SQUID method has been shown to detect

CP632, *Unattended Radiation Sensor Systems for Remote Applications*, edited by J. I. Trombka et al.
© 2002 American Institute of Physics 0-7354-0087-3/02/$19.00

extremely minute changes induced in the material's properties by an applied field in a wide range of scientific disciplines such as acoustics, nuclear magnetic resonance, geophysics and biology. So the incorporation of SQUID technology into its design is expected to greatly enhance the device's sensitivity.

- The intended probing technique is expected to be highly penetrating into enclosures and into bulk materials of interest. It has the potential to probe things deeply buried into solid or liquid material structures, including metals and their alloys, a key attribute desired for the detection and identification of objects hidden inside thick metal boxes.

- The incorporation of the SQUID technique into the design will help detect those metallic spin systems that would be otherwise hard to access with room temperature NAR, like nuclei with negligibly small gyromagnetic ratios, metals, alloys, and metal hydrides in which spectral lines are broadened by quadrupolar or demagnetization effects; and spin-lattice systems with long relaxation times because the maximum signal to noise ratio with SQUID is independent of relaxation times.

The objectives of this article are: (1) to give a brief summary of how nuclear acoustic resonance works; (2) to briefly describe the two particular NAR interactions which we will attempt to explore and exploit during a feasibility study; (3) to show the theoretical derivation of measurable quantities from which unique signatures of materials can be inferred; (4) and to quickly summarize the NAR SQUID detection technique.

The actual design of a prototype device will be undertaken only after a feasibility study has been completed, including computer simulations and their validation in the laboratory. The latter step will require significant upgrading and adaptation of laboratory equipment now available a the Pennsylvania State University.

Nuclear Acoustic Resonance Explained

The material used in this summary was drawn from the listed references, and no claim to the originality of the concepts and principles is herein made by the authors. Our aim is not to propose a device based on a new theory. Our wish is to show that a well developed, researched and tested branch of physics already exist, and should be exploited for the development of a new detection technique to fight strategic material smuggling.

Although the first paper on NAR was published as far back as 1956 [1], the theory of nuclear acoustic resonance is still relatively unknown to most physicists and engineers. At some point in the history of material characterization research it became clear that RF sonic waves would be a good substitute for RF electromagnetic waves when working with conductors. Thus nuclear acoustic resonance works like nuclear magnetic resonance (NMR): (1) it induces shifts in the target material's nuclear spin energy levels in the presence of an externally applied constant magnetic field; (2) the energy levels affected are the same for NAR and NMR; (3) the respective energy shifts are not equal, but closely related.

More precisely when a RF frequency acoustic wave is launched into a material object, it couples to the material lattice where phonons of the same frequency as the acoustic wave are created. Phonon propagation in turn perturbs both the inter-nuclear distances and the overall charge distribution around nuclear sites. The induced asymmetry in charge distribution creates electric field gradients which nuclear spins interact with via magnetic dipole moments and electric multipole moments, leading to the splitting of spin energy levels in the presence of a static magnetic field similarly to the nuclear magnetic resonance.

Two NAR Coupling Mechanisms Makes It Appealing for a Material Detector Development

Acoustic energy can couple to a material's nuclear spin system via various electric and magnetic interactions, including (1) electric quadrupole interactions in which the acoustic wave couples to the electric quadrupole moments of the nuclear pins [2-5]; (2) magnetic dipole (or Alpher-Rubin) interactions [6,7]] in which the acoustic wave couples to the magnetic moments of the nuclear spins; (3) hexadecapole interactions which operate via acoustic wave coupling to the electric hexadecapole (16-pole) moments of the nuclear spins [8,9]; (4) or magnetic dipole-dipole interactions [10], in which acoustic wave couples to material via magnetic interactions among nuclear spin themselves.

These couplings have each its own characteristic values for the observable interaction intensities, angular dependence, line widths and line shapes, given the same experimental conditions. In terms of signal intensities, the most important nuclear spin-phonon interactions in NAR are the electric quadrupole interaction and the magnetic dipole interaction (or Alpher-Rubin coupling). These are the two mechanisms we want to explore during feasibility studies:

The electric quadrupole interaction is highly effective in terms of coupling acoustic waves to nuclear systems with spin quantum number $I > 1/2$. On the other hand, the Alpher-Rubin (magnetic dipole) interaction is quite efficient in metals. In fact it depends on the electromagnetic interaction of conduction electrons with the acoustic wave in the presence of an externally applied static magnetic field. Furthermore, because the coupling is to the magnetic dipole moments of the nuclear spins, dipolar coupling is for all values of the nuclear spin quantum number, including $I = 1/2$.

There are many other nuclear spin-phonon coupling mechanisms, but the above two are more appealing in terms of material characterization applications.

Theoretical Derivation for Measurable Quantities

We will now give a review of the derivations for the measurables[1]. In practice NAR coupling can be experimentally quantified two ways: (1) either by measuring induced changed in acoustic wave attenuation and dispersion, allowing one to infer corresponding changes in magnetic susceptibilities, (2) or by directly measuring NAR

---

[1] Again we repeat that the material shown come from the listed references, all of which are AIP publications.

induced changes in material magnetization. In either case all the variables involved can be theoretically derived and numerically predicted. In the next few sections we summarize the derivations from published studies done in this respect over the past few decades. Interested readers can find more details in the listed references. These derivations by no means covers the entire field of nuclear acoustic resonance. In our project we are concerned only with those aspects of the theory we would like to utilize during the detector design and development.

1 Acoustic Absorption Coefficient [9]

The power (per unit area) generated in a medium by a propagating wave can be written as

$$P_o = \varepsilon v \, \omega \tag{1}$$

where ε is the energy density of the wave in the material, and v is the propagation speed. In the case of acoustic wave, acoustic energy transfer to the material lattice results in the creation of n_p acoustic quanta (or phonons), each with the same frequency as the driving sonic

Thus the energy per unit volume in the material due to wave propagation is then

$$\varepsilon = \frac{n_p h v}{V} \tag{2}$$

where V is the sample's volume. The resulting power density per unit area can therefore be expressed as:

$$P_o = \frac{n_p h v}{V} v \tag{3}$$

Under resonant conditions, a fraction P_n of this power couples to the material's nuclear spin system, causing a transition between spin states m and m'. This corresponds to

$$P_n = \Delta n h W_{m,m'} \tag{4}$$

where $hv = E_m - E_{m'}$ is the energy level shift between the two spin states as the result of energy exchanged between the acoustic wave and the spin system via the material lattice, $\Delta n = n_m - n_{m'}$ is the equilibrium population difference per unit volume between m and m', and $W_{m,m'}$ is the probability per unit time that a spin makes a transition from an initial state m to a final state m'. Assuming no saturation, a Boltzmann distribution of spins can be assumed, and in the $hv \ll \kappa_B T$ limit, the population difference can be written as

$$\Delta n = n_m - n_{m'} = [\frac{N e^{-E_m/\kappa_B T}}{\sum_{m'=-I}^{I} e^{-E_{m'}/\kappa_B T}}][e^{hv/\kappa_B T} - 1] \approx \frac{N}{(2I+1)} \frac{hv}{\kappa_B T} \tag{5}$$

The coupled acoustic power density can then be written as

$$P_n = \frac{N}{(2I+1)} \frac{(h\nu)^2}{\kappa_B T} \sum_m W_{m,m'} \tag{6}$$

Defining the acoustic power absorption coefficient via $2\alpha_n = \frac{P_n}{P_o}$ [9], we then have:

$$\alpha_n = (\frac{N}{2I+1})(\frac{h\nu}{\kappa_B T})\frac{V}{2n_p \mathbf{v}} \sum_m W_{m,m'} \tag{7}$$

The probability per unit time is found via Fermi Golden Rule, that is,

$$W_{m,m'} = \frac{1}{4\hbar^2} \left| \langle p',m'|H_{s-p}|p,m \rangle \right|^2 g(\nu) \tag{8}$$

where p and p' = the initial and the final phonon states, respectively;
$\quad\quad\quad$ H_{s-p} = the spin-phonon interaction's Hamiltonian;
$\quad\quad\quad$ $g(\nu)$ = the spin absorption line shape function such that it

$\quad\quad\quad\quad\quad\quad$ peaks near the resonance frequency ν_{res}, and obeys the

$\quad\quad\quad\quad\quad\quad$ normalization requirement

$$\int_0^\infty g(\nu)d\nu = 1 \tag{9}$$

After substitution, the acoustic absorption coefficient is written as

$$\alpha_n = (\frac{N}{2I+1})(\frac{\pi\nu}{\kappa_B T})(\frac{V}{4\hbar n_p \mathbf{v}})g(\nu)\sum_m \left| \langle p',m|H_{s-p}|p,m \rangle \right|^2 \tag{10}$$

The calculation of the matrix elements $\langle p',m|H_{s-p}|p,m \rangle$ of the nuclear spin-phonon interaction, requires the knowledge of (1) the energy levels of the nuclear spin system of the material under investigation, (2) and the particular physical interaction responsible for the coupling of the acoustic wave to the nuclear spin of that material.

Energy levels can be determined from the knowledge of the nuclear properties of the material, namely the spin quantum number I, the nuclear magnetic dipole moment, the nuclear electric quadrupole moment, the experimentally determined static quadrupole interaction, and the magnitude and direction of the externally applied static magnetic field.

The second requirement involves explicit expressions for the dynamic interactions associated with NAR as will be illustrated later.

2 NAR Spin Susceptibility [2,7,9]

The scheme involves correlating the external ultrasonic perturbation Hamiltonian with experimental parameters. A generalized linear response of a density operator $\rho(t)$ to an external force $f(t)$ is obtained by applying the quantum mechanical version of Liouville's theorem on the density operator:

$$ i\hbar \frac{\partial \rho}{\partial t} = [H_o + H'(t), \rho(t)] \tag{11} $$

where H_o is the Hamiltonian of the unperturbed physical system. The Fourier transform of this equation is (keeping only the first order term):

$$ \rho(\omega) = \frac{1}{i\hbar} [\tilde{H}'(\omega), \rho_o] \tag{12} $$

where

$$ \tilde{H}'(\omega) = \int_0^\infty e^{-i\omega t} e^{-(i/\hbar)H_o t} H'(\omega) e^{(i/\hbar)H_o t} dt \tag{13} $$

and

$$ \rho_o = \text{the statistical operator at } t \to -\infty \tag{14} $$

It is then required that the inverse Fourier transform of the perturbation Hamiltonian (eq. 13) be compatible with the assumption of a linear response of $\rho(t)$ to the driving acoustic time-dependent force $f(t)$. That is, the inverse transform should be of the form:

$$ \tilde{H}'(\omega) = W(\omega) f(\omega) \tag{15} $$

where $W(\omega)$ is the Fourier transform of the response operator $W(\tau)$; and $W(\tau) = 0$ for $t < 0$ and is independent of $f(\tau)$. A generalized susceptibility is then defined as

$$ \chi_{H'}(\omega) = \frac{Tr\{\rho(\omega)H'(\omega)\}}{f(\omega)f(\omega)} \tag{16} $$

For NAR, $f(t)$ is in the form of a strain sensor ε. Further analysis then shows the denominator takes <u>two</u> forms, namely

$$ [f(\omega)]^2 = \rho_s v_a^2 \int_{(V_s)} d^3 r [\varepsilon'(r,\omega)]^2 \tag{17a} $$

and

$$ [f(\omega)]^2 = -\rho_s \int_{(V_s)} d^3 r [v_i(r,\omega)]^2 \tag{17b} $$

By substitution one can thus write the <u>two</u> cases for NAR induced material susceptibilities:

(a) $$\chi_{NAR}(\omega) = -\frac{1}{i\hbar} \frac{Tr\{[\tilde{H}'(\omega),\rho_0]H'(\omega)\}}{\rho_s v_a^2 \int\limits_{(V_s)} d^3r[\varepsilon'(r,\omega)]^2} \qquad (18a)$$

(b) and $$\chi_{NAR}(\omega) = \frac{1}{i\hbar} \frac{Tr\{[\tilde{H}'(\omega),\rho_0]H'(\omega)\}}{\rho_s \int\limits_{(V_s)} d^3r[v_i(r,\omega)]^2} \qquad (18b)$$

Case for an Acoustic Wave in a Metal

The appropriate spin-phonon Hamiltonian of an acoustic wave in a metal can be written as

$$H'(t) = H'_D(t) + H'_Q(t) \qquad (19)$$

where $$H'_D(t) = -\int d^3r\, m(r)b(r,t) \qquad (20)$$

and $$H_Q(t) = -\int d^3r\, Q(r):\nabla E(r,t) \qquad (21)$$

In these expressions, $H'_D(t)$ is the dynamic Alpher-Ruber term, and represents the coupling of nuclear spins to the acoustically induced magnetic field; $H_Q(t)$ is the dynamic electric quadrupole term which represents the coupling of nuclear spins to the acoustically induced electric field gradient; $m(r) =$ nuclear spin magnetization vector; $\nabla E(r,t)$ is electric field gradient vector; and $Q(r)$ is the tensor operator of the nuclear electric quadrupole moment density. So that one may now write

$$Tr\{[\tilde{H}'(\omega),\rho_o]H'(\omega)\} = \sum_{i,j=D\,or\,Q} Tr\{\tilde{H}'_i(\omega),\rho_o]H'_j(\omega)\} \qquad (22)$$

After lengthy algebra and substitution into equations 21 and 22, one obtains *a generalized expression for the NAR induced susceptibility in conductors*

$$\chi_{NAR}(\omega) = -\frac{1}{i\hbar}\left[\frac{(<[H'_D(\omega),\tilde{H}'_D(\omega)>_o + <[\tilde{H}'_Q(\omega),\tilde{H}'_Q(\omega)]>_o)}{\rho_s v_a^2 \int_{(V_s)} d^3r[\varepsilon'(r,\omega)]^2}\right.$$
$$\left.\frac{(<[H'_Q(\omega),\tilde{H}'_D(\omega)>_o + <[\tilde{H}'_Q(\omega),\tilde{H}'_D(\omega)]>_o)}{\rho_s v_a^2 \int_{(V_s)} d^3r[\varepsilon'(r,\omega)]^2}\right] \qquad (23)$$

Thus, there are <u>three</u> Contributions to the NAR-induced Susceptibility in a Conductor:
  (a) a contribution from NAR Dipole Coupling induced (or pure Alpher-Rubin) susceptibility given by

$$\chi_{NAR_{(D\,or\,A\text{-}R)}}(\omega) = -\frac{1}{i\hbar} \frac{<[H_D'(\omega), \tilde{H}_D'(\omega)]>_o}{\rho_s v_a^2 \int_{(V_s)} d^3r |\varepsilon'(r,\omega)|^2} \tag{24}$$

(b) a contribution from a "Pure dynamic Quadrupole" Coupling induced Susceptibility given by

$$\chi_{NAR_{(Q)}}(\omega) = -\frac{1}{i\hbar} \frac{<[H_Q'(\omega), \tilde{H}_Q'(\omega)]>_o}{\rho_s v_a^2 \int_{(V_s)} d^3r |\varepsilon'(r,\omega)|^2} \tag{25}$$

(c) and a contribution is due to an "Interference Susceptibility" between the dynamic Alpher-Rubin and Quadrupole, and is given by

$$\chi_{NAR_{(int)}}(\omega) = -\frac{1}{i\hbar} \frac{<[H_D'(\omega), \tilde{H}_Q'(\omega)] + [H_Q'(\omega), \tilde{H}_D'(\omega)]>_o}{\rho_s v_a^2 \int_{(V_s)} d^3r |\varepsilon'(r,\omega)|^2} \tag{26}$$

3 Explicit NAR Quadrupole Spin Susceptibility in a Metal [1,2,6,11]

The Hamiltonian for a NAR quadrupole interaction is given by

$$H_Q = \frac{e^2 qQ}{4I(2I-1)} \{3I_{z'}^2 - I(I+1) + \frac{1}{2}\eta(I_+^2 + I_-^2)\} \tag{27}$$

where $\quad I_\pm = (I_x \pm iI_y)$; and $\quad \eta = \dfrac{\partial^2 V/\partial x \partial x - \partial^2 V/\partial y \partial y}{\partial^2 V/\partial z \partial z}$ (28)

$$Q = \frac{I(2I-1)}{(2I+3)(I+1)}Q_o \quad \text{(with } Q_0 = \text{the intrinsic nuclear quadrupole moment} \tag{29a}$$

$$eq = \partial E_z/\partial z = \sum_{je_j}(3\cos^2\theta_j - 1)r_j^{-3} \tag{29b}$$

Equation 29a gives the magnitude of the axially symmetric field gradient due to all the charges e_j external to the nucleus, r_j and θ_j are the coordinates (with origin at nuclear center) of the element charge e_j. When the above quadrupole interaction Hamiltonian is used in the matrix element for NAR induced transition probability calculations, it is seen that both $\Delta m \pm 1$ and $\Delta m = \pm 2$ transitions are allowed in quadrupole interactions. And so in this case the NAR induced $\chi_{NAR_{(Q)}}$ is given by

$$\chi_{NAR_{(Q)}} = \chi_{NAR1} + \chi_{NAR2} \tag{30}$$

where $\quad \chi_{NAR1}(\omega) = D_1(\omega) \dfrac{\int_{(V_s)} d^3r V_{+1}(r,\omega)V_{-1}(r,\omega)}{\int_{(V_s)} d^3r |\varepsilon'(r,\omega)|^2}$ (31)

53

$$\chi_{NAR\,2}(\omega) = D_2(\omega) \frac{\int_{(V_s)} d^3 r V_{+2}(r,\omega) V_{-2}(r,\omega)}{\int_{(V_s)} d^3 r [\varepsilon'(r,\omega)]^2} \tag{32}$$

with
$$D_1(\omega) = \frac{\pi A^2}{16\hbar\rho_s v_a^2} \sum_{m=-I}^{I-1} \frac{N_m - N_{m+1}}{V_s} (2m+1)^2 f_I^2(m)[g(\omega - \omega_{m+1,m}) - g(\omega + \omega_{m+1,m})] \tag{33}$$

and
$$D_2(\omega) = \frac{\pi A^2}{16\hbar\rho_s v_a^2} \sum_{m=-I}^{I-2} \frac{N_m - N_{m+2}}{V_s} (2m+1)^2 f_I^2(m) f_I^2(I+m)[g(\omega - \omega_{m+2,m}) - g(\omega + \omega_{m+2,m})] \tag{34}$$

NAR Energy Levels Shifts in Quadrupole Interactions [9]

The NAR induced quadrupole-split energy levels are calculated using the above Hamiltonian for phonon-spin quadrupolar interactions:

$$E = E_m + \left(m \,|H_q|\, m\right) \tag{35a}$$

where the Hamiltonian is give by equation 27 above. After substitution and some algebra, we then have:

$$E = -m\hbar\omega_0 + [\hbar\omega_q \frac{e^2 qQ}{\hbar 2 I(2I-1)} \frac{3}{4}(3\cos^2\theta - 1)] \left[m^2 - \frac{1}{3} I(I+1) \right] \tag{35b}$$

where θ = the angle between the static field H_o and the axis of symmetry of q. In this derivation axial symmetry was assumed for simplicity. Thus NAR nuclear spin energy level shifts are relatively easy to estimate numerically, given the material properties, the transition rules as seen above, the experimental configuration, and the nature of the spin-phonon interaction of interest as shown in Figure one.

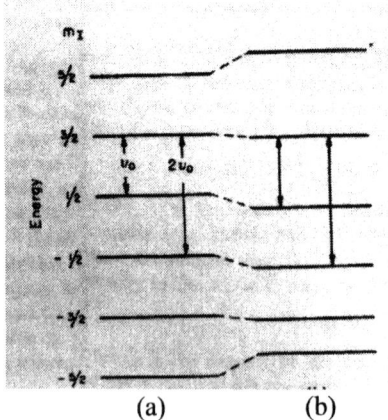

FIGURE 1. High field energy levels for I=5/2 (a) no s quadrupole interaction; (b) effect of quadrupole interaction. v_0 and $2 v$ corres pond to $\Delta m=\pm 1$ and ± 2 respectively. (Bolef and Menes, Phys. Rev. **114**, 6, p. 1442 (1959)).

5 Transition Probabilities for NAR Quadrupole Interactions [1,2,6,9].

Transition probabilities are similarly calculated using the same quadrupole interaction Hamiltonian given by equation (27):

$$W_{m,m'} = \frac{1}{4\hbar^2} \left| \left\langle p',m' \left| H_Q \right| p,m \right\rangle \right|^2 g(v) \tag{36}$$

Thus we have

$$W_Q = W_{m,m+1} = \frac{1}{4\hbar^2} \left| \left\langle p+1,m+1 \left| H_{Q_1} \right| p,m \right\rangle \right|^2 g(v)$$

$$= 6 \left(\frac{A}{4\hbar} \right)^2 \xi_\pm^2 g_{Q_1}(v) \, |V^{\pm 1}|^2 \tag{37}$$

$$W_{Q_2} = W_{m,m+2} = \frac{1}{4\hbar^2} \left| \left\langle p+2,m+2 \left| H_{Q_2} \right| p,m \right\rangle \right|^2 g(v)$$

$$= 6 \left(\frac{A}{4\hbar} \right)^2 \eta_\pm^2 g_{Q_2}(v) \, |V^{\pm 2}|^2 \tag{38}$$

where:

$$\xi_\pm = (2m \pm 1)[(I \pm m + 1)(I \mp m)]^{\frac{1}{2}} \tag{39a}$$

$$\eta_\pm = [(I \mp m)(I \mp m - 1)(I \pm m + 1)(I \pm m + 2)]^{\frac{1}{2}} \tag{39b}$$

$$A = \frac{eQ}{I(2I-1)} \tag{40}$$

$$V^{\pm 1} = \pm \frac{1}{4} \left(6^{\frac{1}{2}} \right) eq \sin\theta \cos\theta \, e^{\pm 2i\phi} \tag{41}$$

$$V^{\pm 2} = \frac{1}{8} \left(6^{\frac{1}{2}} \right) eq \sin^2\theta \, e^{\pm 2i\phi} \tag{42}$$

ϕ is the angle between the projection of symmetry axis on the xy-plane and the z-axis; $g_{Q_1(v)}$ and $g_{Q_2(v)}$ are the shape factors for the transitions $\Delta m = \pm 1$ and $\Delta m = \pm 2$, respectively. Other variables have been defined previously.

The relative signal levels are also predicted using the acoustic a formula developed earlier (equation 10), by substituting the quadrupole interaction Hamiltonian expression. The resulting relations are:

$$\alpha_{Q_1} = \frac{3\pi^2}{8} \frac{Ne^2 Q^2}{\rho v^3 \kappa_B T} v^2 g_{Q_1}(v) f_Q(I) \frac{|V^{\pm 1}|^2}{\varepsilon^2} \tag{43}$$

$$\alpha_{Q_2} = \frac{3\pi^2}{8} \frac{Ne^2Q^2}{\rho v^3 \kappa_B T} v^2 g_{Q_2}(v) f_Q(I) \frac{|V^{\pm 2}|^2}{\varepsilon^2} \tag{44}$$

$$\alpha_{NAR(A-R)} = \frac{\pi^2}{3} \frac{Nh^2}{\rho v^3 \kappa_B T} v^4 g_D(v) f_{NAR(N-A)}(I) \tag{45}$$

$$\text{with} \quad f_Q(I) = \frac{\sum_{m=-I}^{I} \xi^2(m)}{(2I)^2(2I-1)^2(2I+1)} \quad \text{and} \quad f_{NAR(N-A)}(I) = I(I+1) \tag{46}$$

The quantities $f_Q(I)$ and $f_{NAR(N-A)}$ predict the relative strengths of the Alpher-Rubin dipolar coupling and the quadrupolar coupling. In some materials the quadrupolar factor f_Q is much less than the NAR(A-R $f_{NAR(A-R)}$. The difference is quite significant because when used is made of the SQUID-NAR technique in metals, it is the Alpher-Rubin coupling which plays a major role. Another important factor in the Alpher-Rubin coupling is that the intensity is proportional to v^4 as shown in equation 43, which again add to the strength of signal intensity. On the other hand, signal intensities in quadrupolar interactions are proportional to the square of the nuclear electric quadrupole moments as shown in equations , which contribute to their relative strength when for materials with relatively large quadrupole moments.

Table 1 gives a summary of corresponding transition frequencies and their relative probabilities. Nuclear magnetic resonance transition frequencies and the relative probabilities are also shown in the table for comparison. In the table the frequency shifts are computed from the term $\quad \Delta = \frac{e^2 qQ}{h} \frac{(3\cos^2\theta - 1)}{8I(2I-1)} \tag{47}$

	NMR ($\Delta m = \pm 1$)		Acoustic ($\Delta m = \pm 1$)			Acoustic ($\Delta m = \pm 2$)	
Trans.	Trans. freq.	Trans. prob.	Trans. freq.	Trans. prob.	Trans.	Trans. freq.	Trans. prob.
$\frac{3}{2} \leftrightarrow \frac{5}{2}$	$v_0 + 12\Delta$	5	$v_0 + 12\Delta$	5	$\frac{1}{2} \leftrightarrow \frac{5}{2}$	$2v_0 + 18\Delta$	5
$\frac{1}{2} \leftrightarrow \frac{3}{2}$	$v_0 + 6\Delta$	8	$v_0 + 6\Delta$	2	$-\frac{1}{2} \leftrightarrow \frac{3}{2}$	$2v_0 + 6\Delta$	9
$-\frac{1}{2} \leftrightarrow \frac{1}{2}$	v_0	9					
$-\frac{3}{2} \leftrightarrow -\frac{1}{2}$	$v_0 - 6\Delta$	8	$v_0 - 6\Delta$	2	$-\frac{3}{2} \leftrightarrow \frac{1}{2}$	$2v_0 - 6\Delta$	9
$-\frac{5}{2} \leftrightarrow -\frac{3}{2}$	$v_0 - 12\Delta$	5	$v_0 - 12\Delta$	5	$-\frac{5}{2} \leftrightarrow -\frac{1}{2}$	$2v_0 - 18\Delta$	5

TABLE 1 Transition frequencies and probabilities in NAR for I = 5/2
(Bolef and Menes, Phys. Rev. 114, 6, p. 1442 (1959))

These previously published examples we have used have served to illustrate the **cause and effects relationships between the acoustic wave and the experimentally observable quantities** which **carry the unique signatures** of the material under investigation. The expose also serves to signify the need for us to make computer simulations an integral part of our expected material detector feasibility studies.

SQUID Detection of NAR induced magnetization change[2]

As stated in the introduction, a major feature in the proposed NAR-based detector will be the use of the SQUID as illustrated in Figure 2.

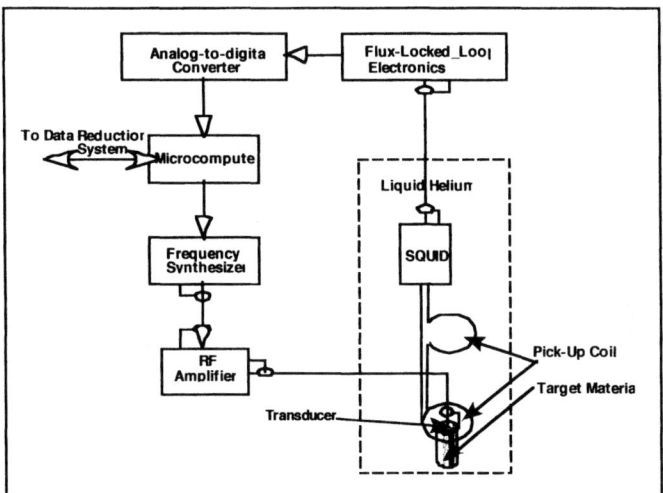

FIGURE 2. Schematic of a SDUID-acoustomagnetic spectrometer (see Pickens, Bolef, Holland, and Sundfors, Phys. Rev. B, **10**, 7, p.3644, (1984)).

The SQUID based NAR detection [2, 7-9,11-13] relies not on the observation acoustic radiation absorption or dispersion, but on the induced changes in the component of material magnetization which is parallel to an applied static magnetic field.

SQUID technique will confer to the device the ability to detect acoustically induced magnetization changes even if the latter were extremely weak. The induced transitions among the nuclear magnetic or nuclear quadrupole energy levels are the same whether they are observed via acoustic absorption/dispersion, or via changes in the z-component M_z of the induced magnetization M. Because the SQUID detected NAR technique relies on the detection of changes M_z, it is sensitive to the magnetic flux Φ itself rather than to the $d\Phi/dt$. The maximum magnetization available when M_z is varied from its equilibrium value M_o is given by

$$M_z = M_o = \chi_o H_o \qquad (48)$$

where χ_o the static nuclear susceptibility, and H_o is the applied magnetic field. This relation immediately links χ_o to the measurable quantity M_z. The maximum change in the magnetic flux Φ through a loop of area A is given by:

$$\Phi_{max} = (4\pi\chi_o H_o)(A\eta t_x) \qquad (49)$$

---

[2] (summary based on materials drawn from refs. 10-12)

where η is a filling factor ~ 1/2, and t_x is the flux transfer factor.

Thus rough estimates of the expected flux change Φ/Φ_o in several metals can be made at the radio frequency SQUID for acoustic coupling to the nuclear spins.

Figure 3 shows representative data using the SQUID system illustrated above.

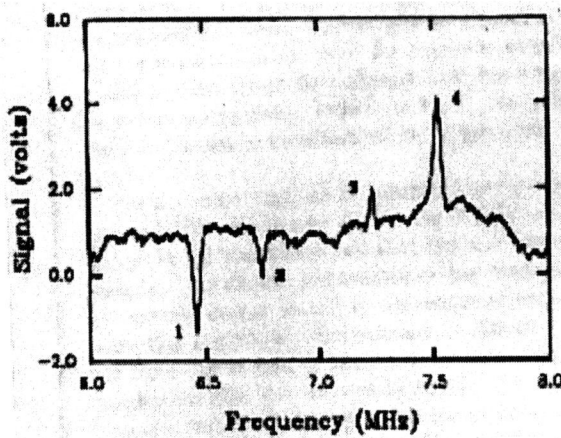

FIGURE 3. Illustration of a resonant response from a NAR SQUID characterization of element [123]Sb (see Pickens, Bolef, Holland, and Sundfors, Phys. Rev. B, **10**, 7, p.3646, (1984)).

Naturally the actual design of a prototype device will not be a carbon copy of the system illustrated here, and will be done only after a feasibility study has been completed. During the feasibility study, issues such as signal to noise ratios, spin-lattice relaxation times, the ambiance induced effects from electronic circuitry, etc., will be investigated in additions to signal levels issues, transducers design, transducers couplings, the automation of data acquisition and analysis, among others. More specifically, the feasibility studies will help us rationally address such questions as how NAR-induced signal intensities and changes in ultrasonic dispersion and attenuation, material magnetization, material susceptibility, signal to noise ratios, etcetera, scale with

- the ambient temperature
- the object size and geometry
- the specimen housing box material's Z-number
- the specimen housing box material's thickness
- the ordering and variation of layers in the specimen housing box's thickness
- the housing box fill materials around the specimen to mimic intentional material camouflage used to elude detection
- transducer sizes and models, and transducer/objects interfacing technique (including air gaps).

The exercise will require sustained computer simulations and their experimental validation. It is hoped that these activities give us a much needed insight that will enable us to rationally narrow down the choices not only for later lab tests, but also for the final detector specifications needed in phase 2 of the project. Additionally we would also like to do similar tests using **composite materials** and materials with **mixed isotopes** (as opposed to pure elements). For the fact is that most practical situations would require the technique to work for such materials. Most of the work for feasibility study will be performed at Penn State University in the Department or Engineering Science and Mechanics.

CONCLUSION

In summary, we want to do quality scientific work for a high risk scheme which, if successful, could meet a vital national security need. We have made explicit the physical theory upon which we want to base the new detector design. We have also spelled out the things we will do during the feasibility study to address all the practical and critical issues which only such a study can meaningfully and successfully address. We hope a case for the development of this new and unique detection technique has been made.

ACKNOWLEDGMENTS

Work performed under the auspices of the U.S. Department of Enerby by the University of California, Lawrence Livermore National Laboratory under Contract No. W-7405-Eng-48.

REFERENCES

Proctor, W. G., and Robinson, W. A., Physical Review, **104**, 5, 1344-1352 (1956).
Fedders, P. A., *Physical review* **B**, 1740-1743 (1973).
Fedders, P. A., *Physical Review*, **B12**, 2045-2048 (1975).
Mieher, R. L., Physical Review **125**, 1537-1551 (1962).
Fedders, P. A., *Physical Review*, **B10** 4510-4514 (1974).
Alpher, R. A., and Rubin, R. J., *J. Acoust. Soc. Am.* **26**, 452-453 (1954).
Müller, V., *Physical Letters* **60A**, 240-242 (1977).
Mahler, R. J., James, L. W., and Tantilla, W. H., *Phys. Rev. Letters* **16**, 259-261 (1966).
Bolef, D. I., and Menes, M., *Physical Review* **114**, 6, 1441-1451 (1959).
LuuKKala, M., *Ann. Acad. Sci. Fenn, AVI* **193**, 1-39 (1965).
Pickens, K. S., Mozurkewich, G., Bolef, D. I., and Sundfors, R. K., *Phys. Rev. Letters* **52**, 2, 156-159 (1984).
Pickens, K. S., Bolef, D. I., Holland, M. R., and Sundfors, R. K., *Physical Review* **B**, 30, 3644-3648 (1984).
Connor,C., Chang, J., Pines, A., *Rev. Sci. Instrum.* **61**, 1059-1063 (1990).

The Lost Source, Varying Backgrounds and Why Bigger May Not Be Better

K.P. Ziock and W.H. Goldstein

Lawrence Livermore National Laboratory
PO Box 808, Livermore, CA 94551

Abstract. The problem of finding a lost radioactive source is complicated by the variations in background as a function of position. The fact that the natural background varies on order of itself means that normal counting statistics do not apply. The signal-to-noise ratio in this case is given by the source strength divided by the number of background counts, not the source strength divided by the square root of the number of background counts. Since both the source and background counts scale as the area of the detector, one does not gain by using a larger detector. To overcome this limitation, one must somehow determine the local background. Since it is impractical to measure the background before one loses a source, the solution is to measure the background *in situ* through the use of imaging techniques. This allows one to return the problem to one dominated by counting statistics, greatly increasing the performance of the system. A cartoon model grid search is used to motivate the discussion that compares the performance of a coded aperture imager to that of an omnidirectional detector in this application.

THE MISSING SOURCE PROBLEM

A classic problem in radiation detection is to look for a source that may or may not be present. This same problem must be dealt with to varying degrees in fields as diverse as environmental remediation, where one seeks residual contamination, through halting the illegal traffic in smuggled nuclear materials. Two general problems exist, the first uses a fixed radiation detector to look at passing objects, the second uses a moving detector to scan a region. In many respects the problems are very similar and we will posit a cartoon search based on the second scenario.

In the cartoon, we are given an area Z, which is to be scanned for a source of strength S counts/sec in a minimum time. We assume that we have a detector sensitive to the source out to a distance δr, in a time interval δt, and use this information to define a scan path through the region as shown in Fig. 1. The spacing between tracks must be less than or equal to $2\delta r$ and the edge tracks must be within a distance δr of

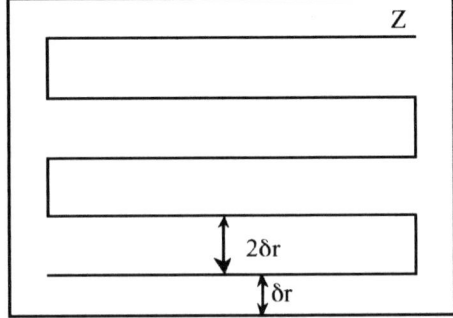

Fig. 1. Search of area Z with sensitivity to distance δr. The space between search paths is twice the sensitive range.

CP632, *Unattended Radiation Sensor Systems for Remote Applications*, edited by J. I. Trombka et al.
© 2002 American Institute of Physics 0-7354-0087-3/02/$19.00

the edge of the region. The speed with which the path can be traversed is approximated as V ~ δr/δt. The length of the path is given by the length of the region sides, which are assumed to be of equal length, z. Then the distance, D, that must be traversed is given by one long path of length z-2δr times the number of times we must go back and forth, i.e. (z-2δr)/2δr plus one more long path for the sides:

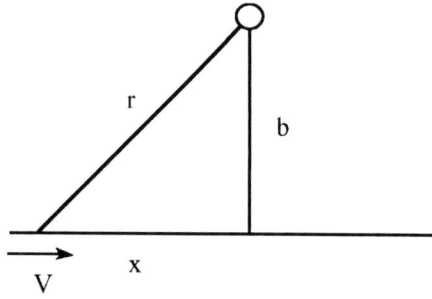

Fig 2. Trajectory past a point source with an impact parameter b.

$$D = (z - 2\delta r)\frac{z - 2\delta r}{2\delta r} + z - 2\delta r = (z - 2\delta r)\frac{z}{2\delta r} \approx \frac{Z}{2\delta r} \tag{1}$$

Where the last step assumes that δr << z. If we divide by the velocity, we arrive at the search time:

$$T = \frac{Z}{2\delta r^2}\delta t \tag{2}$$

Clearly this is a strong function of the, as yet undetermined, factor δr.

To identify δr, i.e. to what distance we can detect a source, we start by integrating the flux as we traverse near a point source with an impact parameter b as shown in Fig. 2. The differential flux, dI, is given by:

$$dI = \frac{A\varepsilon I_0 \cos\theta}{4\pi r^2}e^{-r/\lambda}dt = \frac{A\varepsilon I_0 b}{4\pi(b^2 + x^2)^{3/2}}e^{-r/\lambda}dt \tag{3}$$

Here A is the detector area, ε is the detector quantum efficiency, I_0 is the source strength, λ is the mean free path of the radiation in air and θ is the angle between r and the normal to the detector surface.

OMNIDIRECTIONAL DETECTOR MODEL

To add a framework to the cartoon, we posit a lost one milliCurie, ^{137}Cs source and search for it using a NaI detector with an area of 100 cm². For simplicity, we set ε to unity and momentarily ignore the background. The radiation rate in counts/sec seen by the detector as it traverses past the source at a velocity of a meter/second is shown for a number of different impact parameters in Fig. 3. The problem seems easily solved with detection out to many tens of meters. At this time we note that at smaller impact parameters one not only sees more counts, but the curve is also narrower.

Noise

In reality, there will be background radiation that we will take as one count/cm²/keV/min. With the NaI energy resolution of ~ 8% at 660 keV this comes out to ~ 100 counts/sec in our detector. Of course, counting statistics will cause variations in counts to the curves of Fig. 3 as shown for a few impact parameters in Fig. 4. The distance to which detection can be reliably determined is reduced, but can still be taken as several tens of meters.

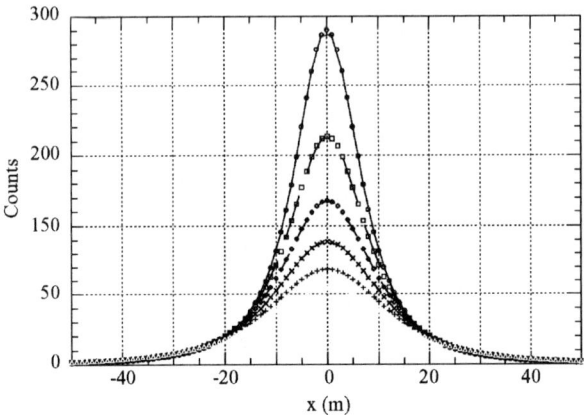

Fig. 3. Intensity seen by an ideal 100 cm² detector as it passes past a 1 mCi source at impact parameters of 12 m, 14 m, 16 m, 18 m and 20 m (respectively decreasing amplitude curves.)

To determine the impact parameter at the limit of sensitivity that defines δr, we can pick a signal-to-noise ratio (SNR) that defines a positive source detection. This should be high enough that there are no, or very few, false detections, since each false detection will add some time to the search while we investigate. The naive assumption is that we know the background rate as B counts/cm²/sec/keV and can then calculate:

$$SNR = \frac{I}{\sqrt{I + BA\varepsilon t}} \tag{4}$$

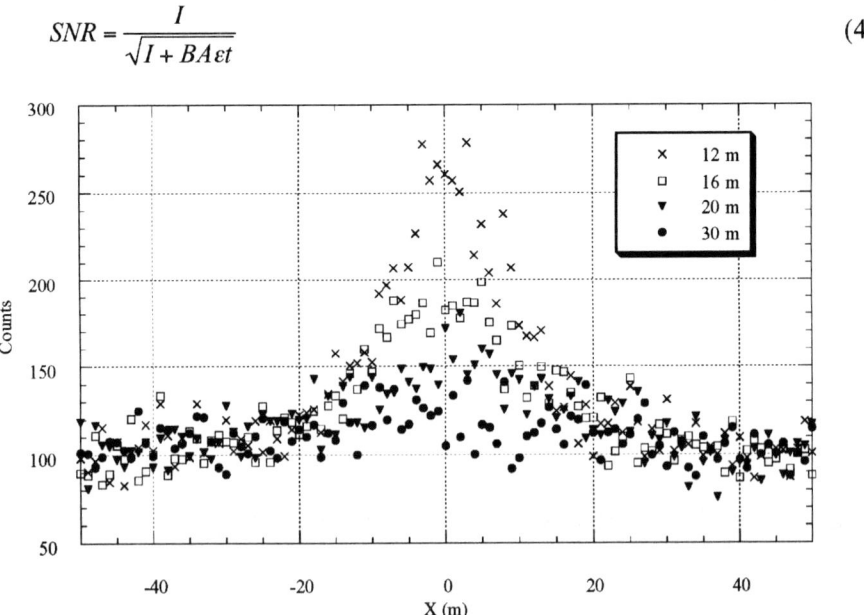

Fig. 4. Curves similar to Fig. 3 but with 100 counts/sec of background noise and counting statistics added to the simulation.

where I is the integral of (3) along the trajectory. We note that I is proportional to both the detector area and the dwell time (which is inversely related to the velocity.) So we might think that we could improve things by increasing either of these quantities.

Unfortunately, this simple view is not correct. It is based on the assumption that we know B. In fact, we find that B will vary many times itself due to the variation in local building and environmental materials.[1-3] This says that the denominator in the SNR expression (the amount of expected fluctuation) is not proportional to the square root of B, but rather to B itself:

$$SNR = \frac{I}{BA\varepsilon t} \propto \frac{A\varepsilon t}{A\varepsilon t} = Constant \qquad (5)$$

Both the numerator and denominator are proportional to Aεt and we do not win by slowing down or using a larger detector! We demonstrate this fact by revising our earlier point source model to include a simple, variable background term. At an impact parameter of 20 m we add a distributed source, 45 m long, which has a strength chosen to ~ double the background count rate when our detector is opposite it. This is shown in Fig. 5, where we run the simulation without counting statistics. In Fig. 6, we rerun the problem at source distances of 10 m, 15 m, 20 m and 25 m. While the source is clearly detected at 10 m, by the time it is 20 m away, it has merged with the extended background structure and cannot be reliably seen. Even if we increase the detector size two orders of magnitude, to 10,000 cm^2, the source still blends into the distributed structure by the time it is 20 m away (see Fig 7!)

The advantage of a larger detector, is that it increases the sensitivity to a source within the distance scale over which the background varies. We assume this is dominated by the variation in man-made materials and assign a distance scale of order 10 m. Hence, to push our detector to a size which is sensitive to more than ~ 10 m is counter productive since the detector cannot distinguish between a distant distributed source and a distant point source.

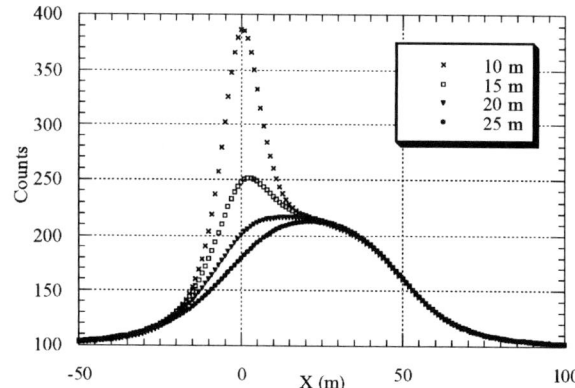

Fig. 5. Signal levels as an idealized detector tracks past a source and background structure. The location of the background is fixed while the impact parameter to the source is increased from 10 m to 25 m in steps of 5 m. The background structure is defined to approximately double the 100 counts/sec background rate.

Increased sensitivity through imaging

To increase δr, we can attempt to measure the background *in situ*. After all, this

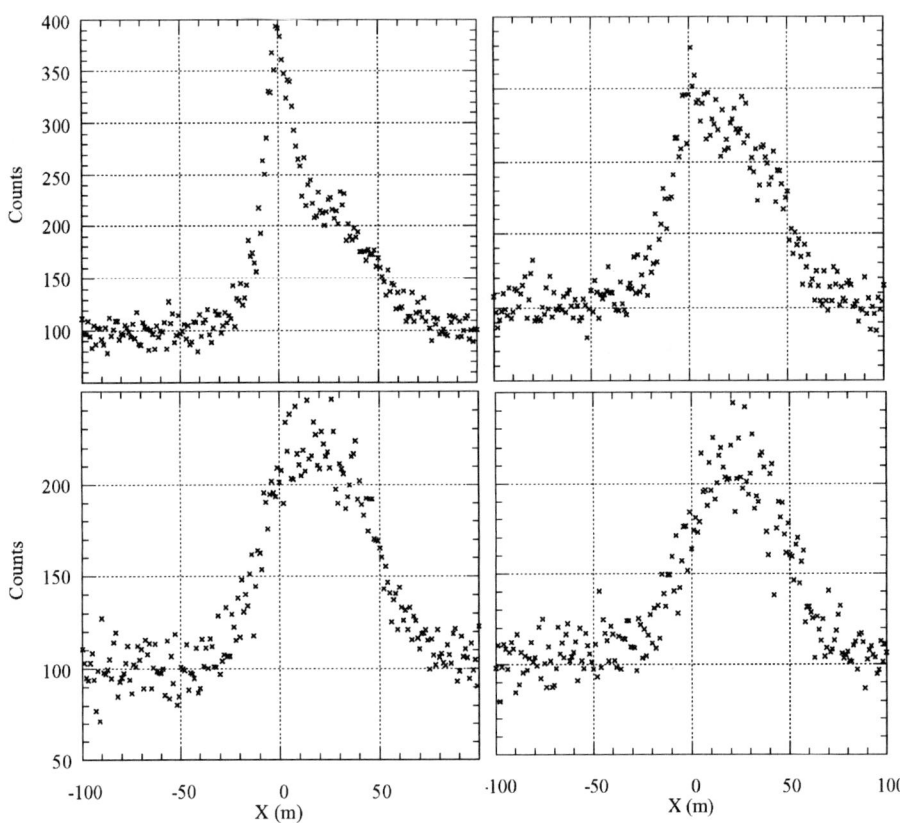

Fig. 6. Response of an omnidirectional detector to the same configuration used in Fig. 5 at source impact parameters of 10 m (top left,) 15 m (top right,) 20 m (bottom left) and 25 m (bottom right.) The source may be marginally visible at 15 m.

is what one really does as one traverses the track of Fig. 1. One starts by checking the count rate at the beginning of the track and then monitoring it as a function of time (position.) Two techniques can be employed to trip an alarm: 1) An absolute threshold which is so far above any expectations of background that it is almost always a sure sign of a detection, or 2) One can monitor the count rate and look for a local spike, i.e. generate a proximity image.

The first technique is severely limited in sensitivity and is not an optimum approach. The second technique, while practical, is limited by the spatial resolution of an omnidirectional detector. This can be estimated from the curves in Fig. 3 by equating the size resolution at a given impact parameter to the FWHM of the intensity curves (see Fig. 8.) At the point where this size approaches the length scale of local variability, one is unable to differentiate between a source and a background object. Once again we are limited to distances of order 10 m.

One can break through this detection distance barrier by the use of an advanced imaging technique. The simplest approach is to highly collimate the detector so that it

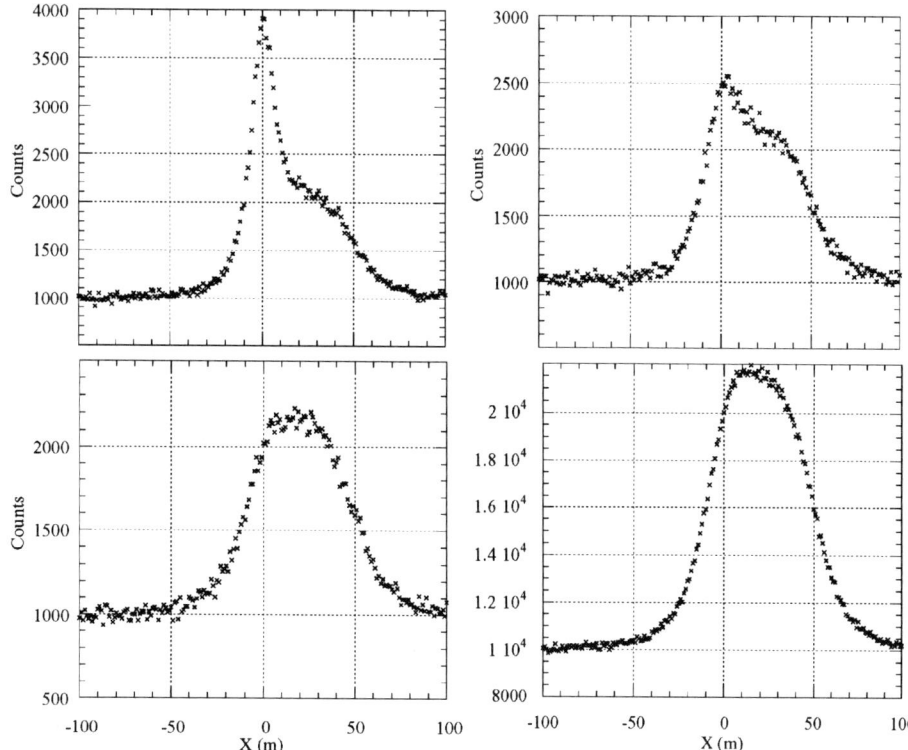

Fig. 7. Results on the source configuration with a 1000 cm^2 detector at 10 m (top left,) 15 m (top right) and 20 m (bottom left.) The source remains undetected at 20 m. Even with a 10,000 cm^2 detector the source is revealed at 20 m (bottom right) only as an asymmetry to the background object. It is arguable that no detection occurs.

only sees a few meters at a distance of δr. The problem is that this restricts the time that an object is in the field of view, limiting the radiation one can collect. A more appropriate approach is to use one of the indirect imaging techniques that have been developed by the astrophysical community. These techniques rely on a shadow mask to modulate the radiation seen by the detector as either a function of time (rotation modulation collimator[4]) or location (coded apertures[5], Fourier transform cameras[6].)

Coded Aperture Imager Model

To demonstrate the advantages of such imaging tools, we have extended our simple

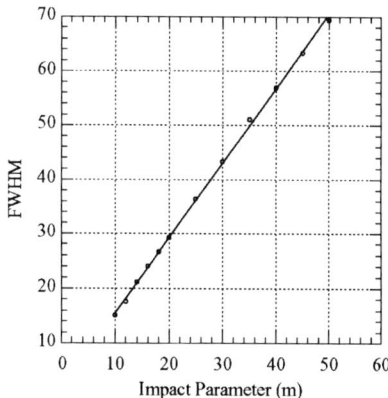

Fig. 8. Full width at half maximum of the curves in Fig. 3 set a first order estimate on the size resolution as a function of impact parameter.

detector model to simulate the response of a one-dimensional coded aperture system. The details of how such a system images is beyond the scope of the present paper. In short, an aperture system comprising a pixel pattern of pixels either open or opaque to the radiation of interest is placed a distance in front of a position-sensitive detector. The pattern is selected so that a far-field, point source casts a shadow pattern that is unique for each location in the field of view. The image is created using a cross-correlation technique that uses all of the data in the detector to recreate the counts in each pixel of the image. This knowledge is important to properly generate the counting statistics noise. The system has a simple angular resolution defined by the mask feature size divided by the spacing between the mask and the detector. By multiplying the angular resolution times the distance to the source, one obtains the resolution at the source. Finally, to avoid artifacts, we impose a field of view (e.g. with a slat collimator) that restricts each pixel in the detector to see no more than half of the mask (which is twice the width of the detector). The size of the field of view is fixed for the purposes of the model to be 25 pixels each of 2.8 degrees.

To determine the response of the imaging system, we impose a linear collimation function on equation (3) to obtain the flux seen by the detector from a given source at any time:

$$dI = \begin{cases} \dfrac{w/2 - |p|}{w/2} \dfrac{A \varepsilon b}{4\pi(b^2 + x^2)^{3/2}} e^{-r/\lambda} dt, & \text{for } |p| \le \dfrac{w}{2} \\[2em] 0 & ; \text{ for } |p| > \dfrac{w}{2} \end{cases} \tag{6}$$

Where p is the pixel number and w is the width of the detector in pixels. The detector is passed along the search trajectory in a series of steps and (6) is used to generate the number of counts seen in each pixel of the image based on a two-pass approach. On the first pass, at each location, the counts from all sources visible to the detector are summed to obtain the total counts in the detector. This number is used to calculate a random variation with a normal distribution based on the square root of the number of counts in the detector. This random number provides the base counts in each image pixel at this location. During the second pass, the number of counts from the sources visible to each image pixel, are added to that pixel. At the end of this step, a series

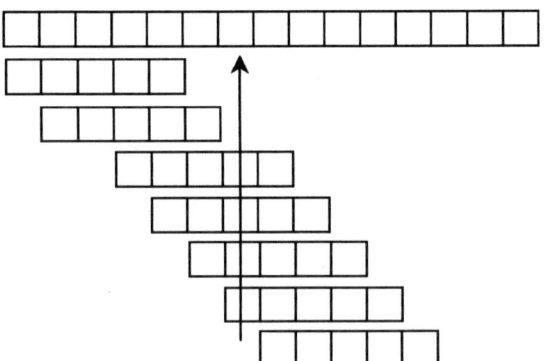

Fig. 9. Schematic of how a 5 pixel detector moves horizontally as a function of time (down the page). The data is collected into each detector position and then summed (as shown by the arrow) into a "world" view which is fixed in space (top row).

66

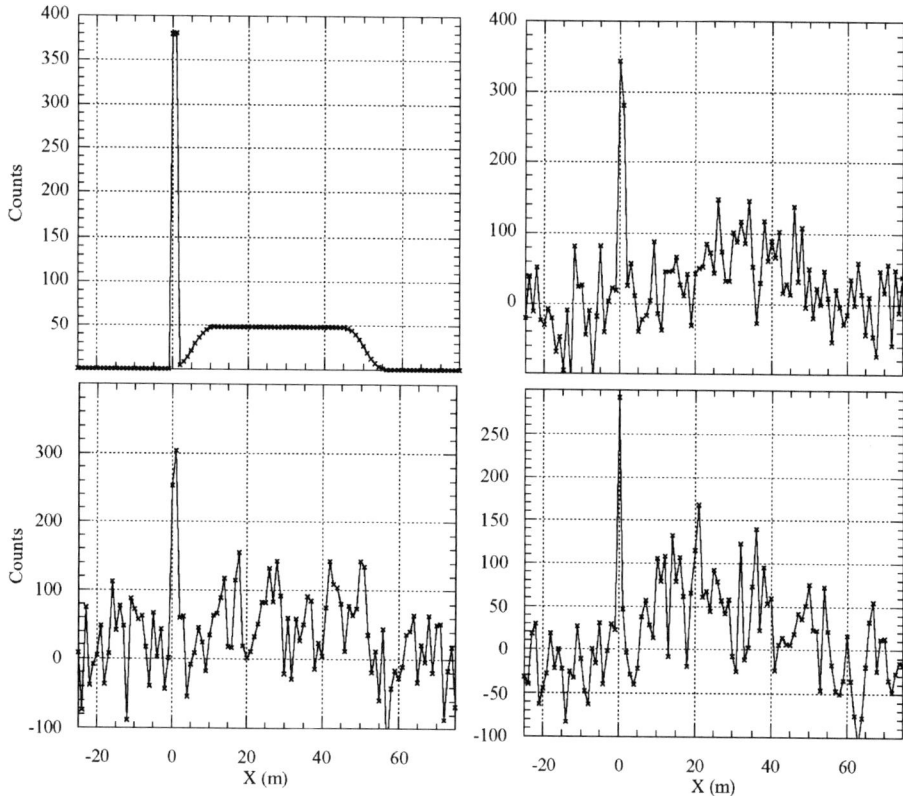

Fig. 10. The same source configuration as in Fig. 5 but data taken with an imaging detector. The ideal response with no counting statistics at 10 m (top left) and the more realistic response at 10 m (top right,) 15 m (bottom left) and 20 m (bottom right.) The source is still visible at 20 m.

of 25 pixel detectors, each offset by one pixel, exist. The data from the appropriate pixels are then summed into a new set of pixels based on a world coordinate system (see Fig. 9.) Note that the world coordinate system makes an implicit assumption about the distance to the source. If this distance is not correct, then the strength of a point source will be diluted across several world pixels due to parallax errors. The final world coordinate image approximates the SNR of a coded aperture imager without the additional complication of propagating photons through a coded aperture and recreating the image.

The results for the same source configuration used in Fig. 5 are shown with no random noise in the top left panel of Fig. 10. The first observation about this plot is that the integrated count rates are lower than in the omnidirectional detector. Two factors contribute: first, the mask obscures half of the detector area, and second, the collimators further reduce the counts. The plot also does not show an offset due to the non-varying background. However, these counts will contribute to the random fluctuations as shown in the rest of the Figure. The source now appears as a spike above the local background and no longer blends into the distributed structure as the impact parameter

67

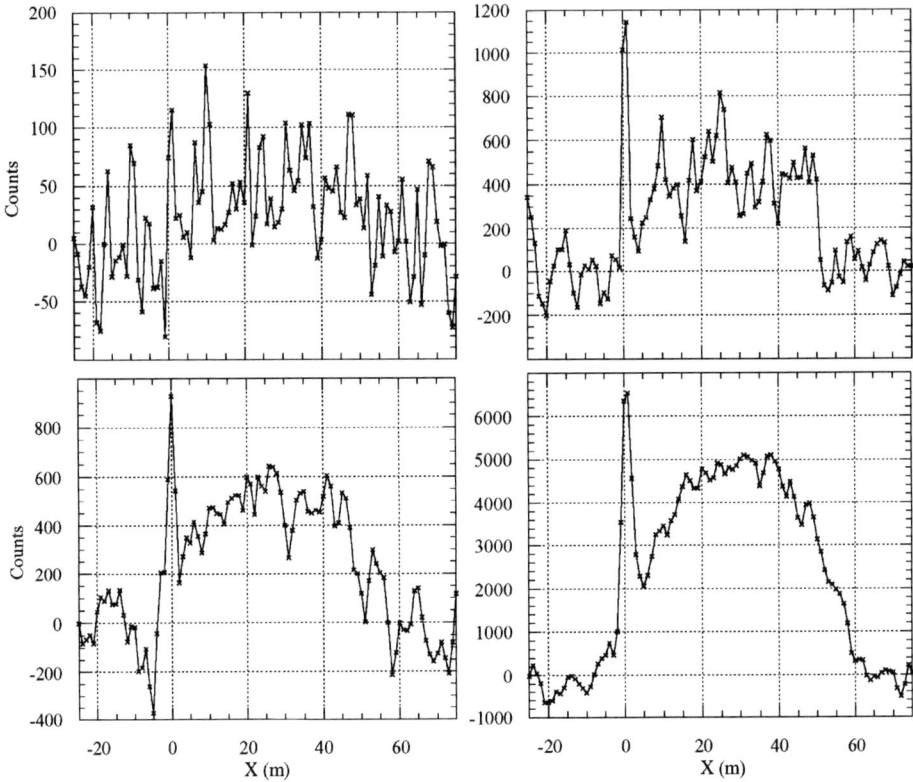

Fig. 11. The 100 cm^2 imager succumbs to statistical noise at ~ 20 m. By 25 m, the source is no longer visible (top left.) Increasing the detector area to 1,000 cm^2 provides sensitivity at 30 m (top right) and 40 m (bottom left.) The signal-to-noise ratio can be further improved if we use a 10,000 cm^2 imager which provides a clean detection at 50 m, bottom right.

is increased. However, the random fluctuations of other pixels show that the detection limit will occur at a few tens of meters as well (Fig. 11.)

The power of the imager is revealed if we increase the area of the detector. At 1,000 cm^2 the imager can clearly resolve the source out to 40 m as shown in Fig. 11. The omnidirectional detector was unable to achieve this even if the detector was increased 10 times greater than this (Fig. 7.) As a final demonstration of the promise of this technique, we have simulated a more complicated environment including background variations a factor of 6 above the nominal level. Even at 40 m, the source is still visible (see Fig. 12.)

Conclusion

The complexity of the natural background radiation makes solution of the simple cartoon model unrealistic for all but the most trivial search scenarios. However, the model does provide a framework in which to approach the problem in a systematic

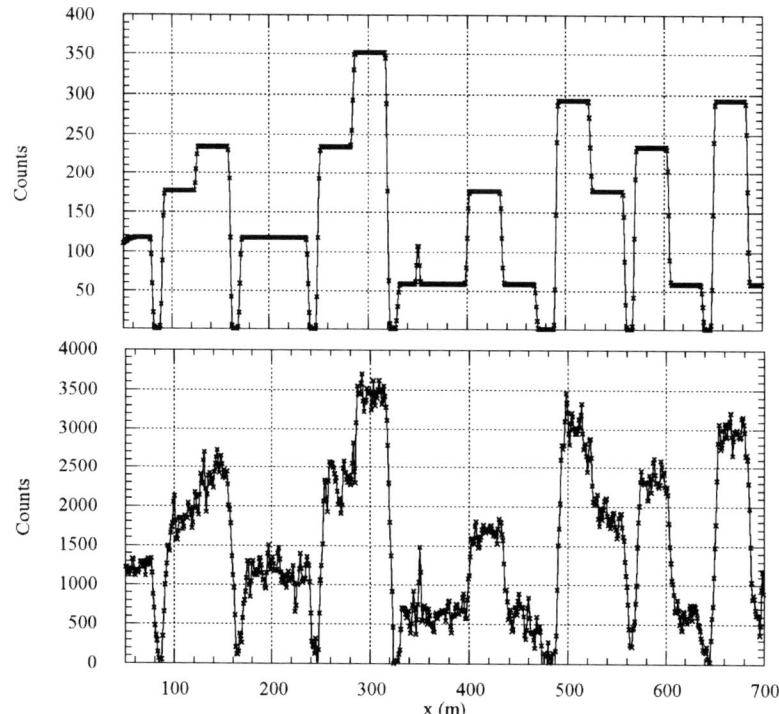

Fig. 12. Complicated background structure (top—50 m from the scan path of a 100 cm² detector, no statistical noise) does not hide the source (located at x = 350 m) scanned at an impact parameter of 40 m with a 1,000 cm² detector (bottom.) (The background is fixed at 50 m.)

fashion. It clearly reveals that increasing the sensitive range of a detector is an important goal—entering as the second power to reduce the time required to sweep an area. With this motivation, an investigation of the search properties of an omnidirectional detector show it is poorly suited to this task. The inability to resolve point and distributed sources at even modest distances, is a fatal flaw of this instrument which cannot be overcome by any means including extremely large detector areas. In contrast, a well designed imaging instrument does not suffer from this problem and can take full advantage of the extra sensitivity that increased detector area affords.

Acknowledgements

This work was performed under the auspices of the U.S. Department of Energy by Lawrence Livermore National Laboratory under Contract W-7405-Eng-48.

References

1. N.N. Jibiri, *Assessment of health risk levels associated with terrestrial gamma radiation dose rates in Nigeria*, Environment International, **27**, 21, 2001.
2. A.C. Paul, et al., *Population Exposure to Airborne Thorium at the High Natural Radiation Areas in India*, J. Environ. Radioactivity, **40**, 251, 1998.

3.   A.S. Mollah, et al., *Measurement of high natural background radiation levels by TLD at Cox's Bazar coastal areas in Bangladesh*, <u>Radiation Protection Dosimetry</u>, **18**, 39, 1987.
4.   H.W. Schnopper, R.I. Thompson, S. Watt, *Predicted Performance of a Rotating Modulation Collimator for Locating Celestial X-ray Sources*, <u>Space Science Reviews</u> **8**, 534, 1968.
5.   E.E. Fenimore and T.M. Cannon, *Coded aperture imaging with uniformly redundant arrays*, <u>Appl. Opt.</u> **17**. 337. 1978.
6.   D. Palmer, T.A. Prince, *A laboratory demonstration of high-resolution hard X-ray and gamma-ray imaging using Fourier-transform techniques*, <u>IEEE Trans. Nucl. Sci.</u> **NS-34**, 71, 1987.

Unattended Radiation Sensor Systems for Remote Terrestrial Applications and Nuclear Nonproliferation

Lodewijk van den Berg, Alan E. Proctor, Ken R. Pohl, Alex Bolozdynya
and Raymond DeVito

Constellation Technology Corporation,
Largo, Florida 33777

Abstract. The design of instrumentation for remote sensing presents special requirements in the areas of power consumption, long-term stability, and compactness. At the same time, the high sensitivity and resolution of the devices needs to be preserved. This paper will describe several instruments suitable for remote sensing developed under the sponsorship of the Defense Threat Reduction Agency (DTRA). The first is a system consisting of a mechanical cryocooler coupled with a high-purity germanium (HPGe) detector. The system is portable and can be operated for extended periods of time at remote locations without servicing. The second is a hand-held radiation intensity meter with high sensitivity that can operate for several months on two small batteries. Intensity signals above a set limit can be transmitted to a central monitoring station by cable or radio transmission. The third is a small module incorporating one or more high resolution mercuric iodide detectors and front end electronics. This unit can be operated using standard electronic systems, or it can be connected to a separately designed, pocket-size module that can provide power to any detector system and can process detector signals. It incorporates a shaping amplifier, a multichannel analyzer, and gated integrator electronics to process the slow signal pulses generated by room temperature solid state detectors. The fourth is a high pressure xenon (HPXe) ionization chamber filled with very pure xenon gas at high pressure, so that the efficiency and spectral resolution are increased above the normally available gas-filled tubes. The performance of these systems will be described and discussed.

PACS numbers: 29.40.-n,29.40.Wk,29.30.Kv.

INTRODUCTION

The necessity to perform observations and measurements of nuclear radiation at remote locations has led to the development of sensors and supporting instrumentation that can operate for months or years without maintenance or servicing. In addition, these systems often need to be compact and lightweight, have minimal power consumption, and withstand adverse environmental conditions.

This paper will describe several systems developed by Constellation Technology Corporation (Constellation) that conform to the requirements stipulated above. These systems can also be used advantageously for more normal, routine measurement

CP632, *Unattended Radiation Sensor Systems for Remote Applications,* edited by J. I. Trombka et al.
© 2002 American Institute of Physics 0-7354-0087-3/02/$19.00

applications, where they provide significant savings in manpower, measurement time, and logistical supplies.

ELECTROMECHANICALLY COOLED DETECTOR

The electromechanically cooled high-resolution detector (EMC) has been designed to replace a liquid nitrogen cooled equivalent detector. The EMC, shown in Figure 1, is man-portable and suitable for use in locations where liquid nitrogen is not available, or frequent attention is inconvenient and time-consuming. The "box" section contains a Stirling engine cryocooler and associated electronics. The detector section contains a conventional high-purity germanium (HPGe) detector, preamplifier, and temperature sensor to control the cooling system. Several types of detectors can be used interchangeably, which allows the EMC to serve multiple applications.

FIGURE 1. Photograph of electromechanically cooled detector system

The EMC is usually powered from a 12 VDC supply in the transport case, but any equivalent 12 VDC source, such as a vehicle battery, is acceptable. The power supply in the transport case can operate from a 100-240 VAC/50-60 Hz source and includes a 12 V rechargeable battery which can operate the cryocooler for 40 minutes during transport or external power interruption.

The performance of the EMC detector is comparable to that of liquid nitrogen cooled detectors, with minor degradation in resolution at low energies due to mechanical vibration and electrical noise. A typical example of a high-resolution energy spectrum is shown in Figure 2.

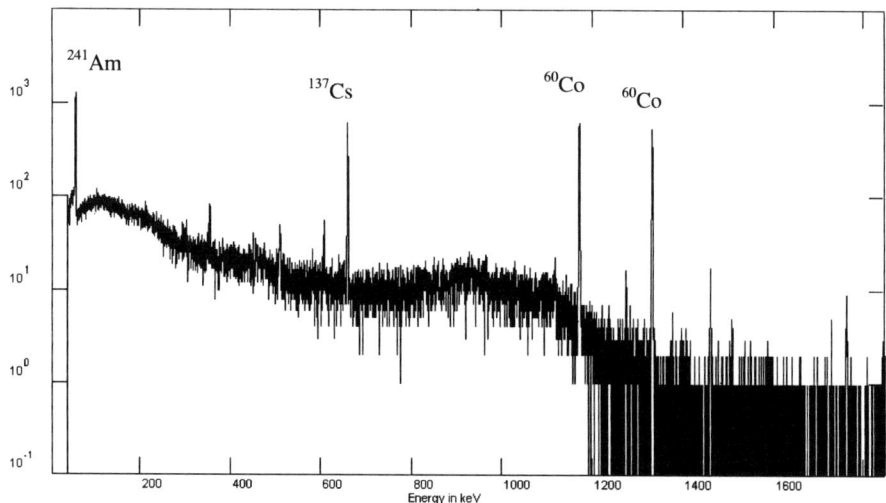

FIGURE 2. High energy spectrum acquired with mechanically cooled HPGe detector

The EMC was developed for arms control inspectors to use at sites where liquid nitrogen is not available. Inspectors would ship the EMC to their base location, cool the detector overnight, transport it to the measurement site while operating on battery power, and perform the inspection. While the EMC is more costly than liquid nitrogen cooled detectors or other commercially available mechanically cooling systems, no other detector combines the portability, operational capability, and freedom from cryogens.

RADIATION INTENSITY COUNTER

The radiation intensity counter is a small, self-contained pocket-size instrument that is suitable for safeguard, environmental, and non-proliferation applications. The sensing element consists of a 25 mm x 25 mm x 5 mm mercuric iodide detector that operates as a counter, although regions of interest can be defined. The unit provides a sensitivity of <20 microrad/hour at 662 keV, and can operate for several months on two small alkaline batteries. Figure 3 shows a photograph of this device.

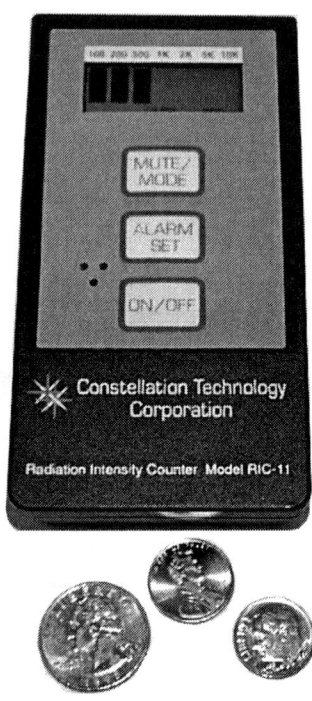

FIGURE 3. Photograph of radiation intensity counter (RIC)

The count rate is visually presented to the user in the form of a liquid crystal bar graph. Electronic or audible alarms will be generated when the radiation intensity exceeds a preset level. The instrument can be carried for personal protection, or can be permanently installed for monitoring or surveying applications. In that case, the electronic alarm signal can be transmitted to a central monitoring station.

MERCURY MODULE

The Mercury Module is a small metal enclosure with dimensions 90 mm x 50 mm x 50 mm. It contains a mercuric iodide detector with dimensions of up to 25 mm x 25 mm x 3 mm, a preamplifier, and other elements of the front-end electronics. Mercuric iodide detectors are very efficient because of the high density and the high atomic number of the material, and they can operate at ambient temperatures without cooling. Since the detectors are configured as modular units, they can be stacked on top of each other to increase the active detector volume to approximately 7.5 cm^3. Figure 4 shows a photograph of the Mercury Module.

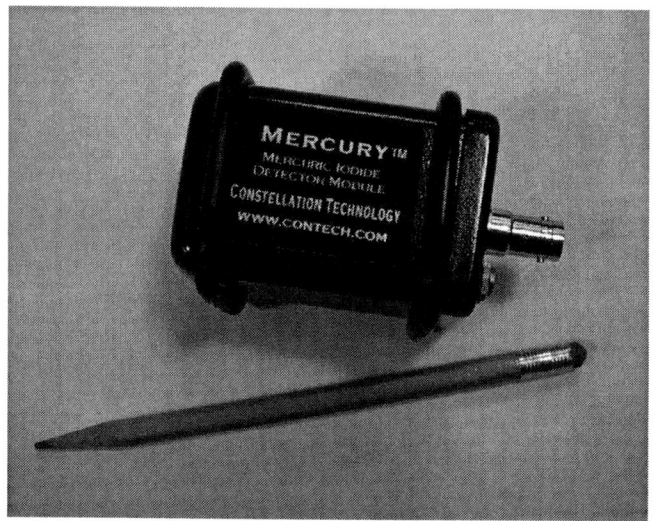

FIGURE 4. Photograph of Mercury Module

The base of the module contains the connections for the high voltage, the power supply for the preamplifier, and the detector signal output. The module can be operated using a standard NIM power supply and signal processing unit. An example of the energy spectrum measured using a mercury module is shown in Figure 5.

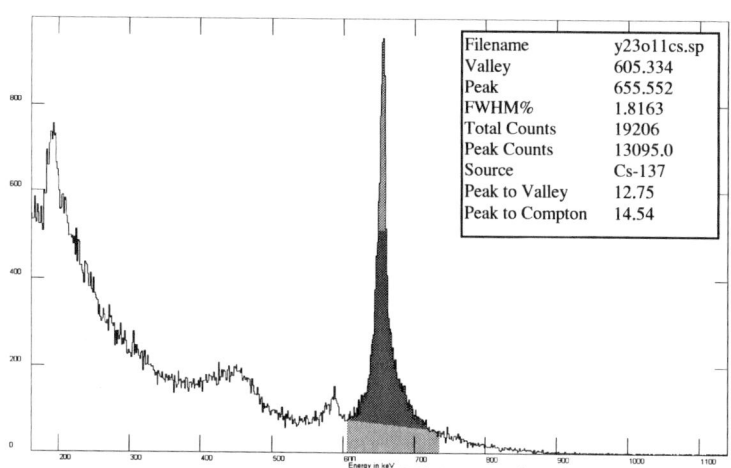

Filename	y23o11cs.sp
Valley	605.334
Peak	655.552
FWHM%	1.8163
Total Counts	19206
Peak Counts	13095.0
Source	Cs-137
Peak to Valley	12.75
Peak to Compton	14.54

FIGURE 5. Spectrum of ^{137}Cs radiation measured with a mercuric iodide detector

The compact size of the Mercury Module makes it well suited for measurements in confined spaces or for incorporation in hand-held systems. The modules can also be permanently installed in areas of high radiation density because of the high resistivity to radiation damage of the mercuric iodide detectors.

MICROMAX

The use of compact, light-weight portable sensor systems for remote applications requires that the power supply and signal processing system to support the sensors can equally be operated with a minimum of resources. These features are incorporated in a small, pocket-size unit called the MicroMax, which contains a surface-mount high voltage supply, a shaping amplifier with gated integrator, and a multichannel analyzer. Figure 6 shows a photograph of the MicroMax.

FIGURE 6. Photograph of MicroMax

The bias supply in this package can be adjusted up to 4000 V, so that the system can be used with a variety of detector systems. The shaping time is usually set at 2 microseconds, and the integrator times can be adjusted between 6 and 20 microseconds. The gated integrator system is especially useful when using detectors with slow risetime pulses, where the integrated charge is a better evaluation of the energy of a gamma ray photon than the height of the signal pulse. The number of channels in the analyzer can be set from 512 to 4096, so that spectra with different resolution can be adequately analyzed. The data from the analyzer can be displayed on and stored in a pocket-size PC attached to the MicroMax. The package consisting of sensor, MicroMax, and PC can be operated for approximately 48 hours on 2 AA batteries included in the MicroMax. This system is especially useful in combination with a Mercury Module since the mercuric iodide detector has a very low leakage current.

HP XENON DETECTORS

A portable xenon spectrometry system has been developed which combines the high spectral resolution of xenon detectors with PC based operating controls. The use of extremely high purity gas at high pressure results in improved resolution and increased efficiency. The detector contains 0.8 kg of gas and its dimensions are 12 cm diameter and 30 cm long. The sensitive area is therefore large in all directions. The system operates over a temperature range of 5-60°C and the detector is radiation hard. A photograph of the basic detector and PC arrangement is shown in Figure 7.

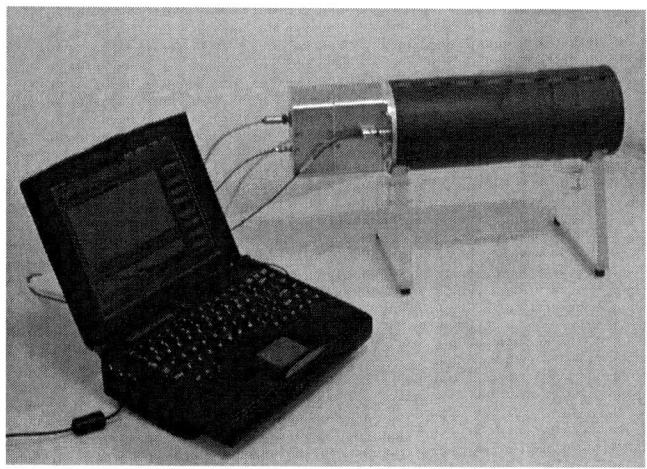

FIGURE 7. Photograph of xenon detector arrangement

The PC provides power to the high voltage system and the preamplifier and processes the signals from the detector using a PC/MCA board. This MCA provides up to 4096 channels at variable shaping times so that it can be used with different types of detectors.

This system can be used in airborne instrumentation or for the monitoring of nuclear facilities.

SUMMARY

Several systems developed for the unattended and remote sensing of nuclear radiation have been described. Critical parameters in the design of these systems were reliability, long-term stability, compactness, comfortable portability for one person, and low power consumption whenever possible. We have approached several of these requirements with the use of unusual detector configurations and newly developed efficient detector materials. This was combined with the application of advanced electronic systems to the detector power supply and signal processing. Several of these systems have general applications, for example in environmental x-ray fluorescence analysis, in neutron activation analysis, and for the measurement of ionizing radiation in space experiments.

ACKNOWLEDGMENTS

This work was supported by DTRA under contract no. DTRA01-99-C-0187.

Aerial Measuring System Sensor Modeling

Rebecca Detwiler

Bechtel Nevada
P.O. Box 98521, M/S RSL-24
Las Vegas, NV 89193-8521

Abstract. The AMS fixed-wing and rotary-wing systems are critical National Nuclear Security Administration (NNSA) Emergency Response assets. This project is principally focused on the characterization of the sensors utilized with these systems via radiation transport calculations. The Monte Carlo N-Particle code (MCNP) which has been developed at Los Alamos National Laboratory was used to model the detector response of the AMS fixed wing and helicopter systems. To validate the calculations, benchmark measurements were made for simple source-detector configurations. The fixed-wing system is an important tool in response to incidents involving the release of mixed fission products (a commercial power reactor release), the threat or actual explosion of a Radiological Dispersal Device, and the loss or theft of a large industrial source (a radiography source). Calculations modeled the spectral response for the sensors contained, a 3-element NaI detector pod and HpGe detector, in the relevant energy range of 50 keV to 3 MeV. NaI detector responses were simulated for both point and distributed surface sources as a function of gamma energy and flying altitude. For point sources, photo-peak efficiencies were calculated for a zero radial distance and an offset equal to the altitude. For distributed sources approximating infinite plane, gross count efficiencies were calculated and normalized to a uniform surface deposition of 1 $\mu Ci/m^2$.

The AMS rotary-wing (helicopter) system is often used to characterize the plutonium (Pu) ground contamination following a nuclear weapons accident or other Pu dispersal mechanism. Therefore, helicopter calculations modeled the transport of ^{241}Am as this is the "marker" isotope utilized by the system for Pu detection. The helicopter sensor array consists of 2 six-element NaI detector pods, and the NaI pod detector response was simulated for a distributed surface source of ^{241}Am as a function of altitude.

This project deals with the modeling the Aerial Measuring System (AMS) fixed-wing and rotary-wing sensor systems, which are critical U.S. Department of Energy's National Nuclear Security Administration (NNSA) Consequence Management assets. The fixed-wing system is critical in detecting lost or stolen radiography or medical sources, or mixed fission products as from a commercial power plant release at high flying altitudes. The helicopter is typically used at lower altitudes to determine ground contamination, such as in measuring americium from a plutonium ground dispersal during a cleanup. Since the sensitivity of these instruments as a function of altitude is crucial in estimating detection limits of various ground contaminations and necessary count times, a characterization of their sensitivity as a function of altitude and energy is needed.

CP632, *Unattended Radiation Sensor Systems for Remote Applications*, edited by J. I. Trombka et al.
© 2002 American Institute of Physics 0-7354-0087-3/02/$19.00

Experimental data at altitude as well as laboratory benchmarks is important to insure that the strong effects of air attenuation are modeled correctly. The modeling presented here is the first attempt at such a characterization of the equipment for flying altitudes.

The sodium iodide (NaI) sensors utilized with these systems were characterized using the Monte Carlo N-Particle code (MCNP) developed at Los Alamos National Laboratory.

For the fixed wing system, calculations modeled the spectral response for the 3-element NaI detector pod and High-Purity Germanium (HPGe) detector, in the relevant energy range of 50 keV to 3 MeV. NaI detector responses were simulated for both point and distributed surface sources as a function of gamma energy and flying altitude. For point sources, photopeak efficiencies were calculated for a zero radial distance and an offset equal to the altitude. For distributed sources approximating an infinite plane, gross count efficiencies were calculated and normalized to a uniform surface deposition of 1 $\mu Ci/m^2$.

The helicopter calculations modeled the transport of americium-241 (^{241}Am) as this is the "marker" isotope utilized by the system for Pu detection. The helicopter sensor array consists of 2 six-element NaI detector pods, and the NaI pod detector response was simulated for a distributed surface source of ^{241}Am as a function of altitude.

DESCRIPTION OF GAMMA SOURCES MODELED

Point Sources

Due to the large source-to-detector distances and the desire for a simulated detector response as a function of energy, the sources were directionally biased. This method was tested with calculations made with no directional biasing at an altitude of 100 meters (m), and photopeak results agreed within the statistical errors. As shown later in the paper, gross counts also agreed within errors with experimental data up to an altitude of 1000 feet (ft), although further work is needed to investigate discrepancies above that altitude. To minimize running time by not repeating close energies, a simulated gamma spectrum of spaced energies was used to express photopeak counts as a function of gamma energy. Energies ranged from 300 keV to 3 MeV for most fixed-wing altitudes. Point sources were modeled at two radial distances, one of zero radial distance with respect to the fixed-wing pod, and one at an offset radial distance equal to the altitude.

Distributed Ground Sources

The distributed sources were modeled after a uniform infinite plane surface distribution. In practice, a surface circular source of radius equal to the altitude was used for the fixed wing due to very low statistics from inadequate biasing ability for the distributed source. For fixed-wing distributed sources, a simulated gamma spectrum was also used, while only [241]Am was used for the helicopter. Helicopter distributed sources were also modeled initially with the radius equal to the altitude. Additional runs were made at the lowest two altitudes for distributed sources with larger radii. Directional biasing was limited to biasing in the upper hemisphere for distributed sources.

MODELING ENVIRONMENT

The fixed-wing and helicopter systems were modeled above a 200m layer of earth with composition given by ANSI 6.6.1-1987 and inside a hemisphere of air with a 1000m radius. A density of 1.25E-3 g/cc was used for the air, 3.67g/cc for NaI crystals, and 5.3234g/cc for the HPGe crystal. The body of the fixed-wing aircraft was simplified to an aluminum sphere containing the detector pods.

DETECTOR AND POD MODELING

The detector response as a function of energy was modeled for both the NaI and HPGe detectors, and photopeak counts were recorded from the generated spectra for the point sources. Although the Gaussian smoothing function added to the tally gives a more realistic detector energy response function, the results of counts in the energy bin containing the gamma photopeak energy with no Gaussian smoothing are identical to those obtained by summing the energy bins of the photopeak with Gaussian smoothing. Therefore, for photopeak calculations, spectra did not have the Gaussian function added, although the sample spectra shown later do have Gaussian smoothing.

Fixed Wing

Both a NaI pod and HPGe detector were modeled for the fixed-wing system. The NaI pod contained 3 NaI detectors. The detectors were housed in foam inside the fiberglass box, and were shielded with a thin layer of aluminum. The foam, fiberglass, and aluminum shielding were all modeled, using typical densities for the packing foam and fiberglass. The HPGe detector was modeled in foam inside a fiberglass case with a plastic cover.

Helicopter

The NaI pod, containing six NaI detectors aligned symmetrically about the center with three on each side, was modeled for the helicopter. Detection ends pointed toward the pod center and photomultipliers were at the opposite ends. The aluminum helicopter pod was simplified to a uniform layer. Again, the aluminum shielding around the NaI crystals and the packing foam were included in the modeling, as was a thin cadmium shield directly above the detectors.

BENCHMARK MEASUREMENTS

The benchmark measurements recorded spectra from a single detector for both the fixed-wing pod and helicopter pod and results were compared with MCNP calculations. For the fixed-wing pod, measurements were made for the larger NaI detector inside the pod, with the pod-pointing head on and at 90° from the source at 1 m, for ^{241}Am, cesium-137 (^{137}Cs), and cobalt-60 (^{60}Co) sources. The middle-sized detector was measured for one source, ^{137}Cs, at the 90° orientation only. Photopeak counts were compared to calculated values. The helicopter benchmark was made with the pod mounted on the helicopter, at 90° from a source centered with respect to the right outer NaI crystal, approximately 1m below the crystal used. Sources used were ^{241}Am, ^{137}Cs, and sodium-22 (^{22}Na), and both photopeak and gross counts were compared. Refer to Table 1 and Table 2.

Table 1. Benchmark Measurements and Calculations for Fixed Wing

| | | NaI Detector – Large | | | |
| | | 90° Orientation | | 0° Orientation | |
Isotope	Energy	Exp	MCNP	Exp	MCNP
^{241}Am	0.05963	2.81E-03	2.34E-03	7.10E-04	6.21E-04
^{137}Cs	0.662	1.18E-03	1.31E-03	2.14E-04	1.74E-04
^{60}Co	1.1173	7.69E-04	8.40E-04	1.56E-04	1.394E-04
	1.332	6.01E-04	7.44E-04	9.90E-05	1.31E-04
NaI Detector – Medium					
^{137}Cs	0.662	3.26E-04	3.22E-04		

Table 2. Benchmark Measurements and Calculations for Helicopter

Outer NaI Detector					
		Photo-Peak Efficiencies		Gross Count Efficiencies	
Isotope	Energy	Exp	MCNP	Exp	MCNP
^{241}Am	0.05963	2.5E-3	3.07E-03	2.70E-03	3.54E-03
^{137}Cs	0.662	1.53E-3	1.50E-03	5.32E-3	5.09E-03
^{22}Na	1.275	8.74E-04	8.76E-04	1.33E-02	1.42E-02

RESULTS AND ANALYSIS

The output of the MCNP detector response tally of counts per energy bin was given in counts per source gamma. These results were then normalized to appropriate measurable quantities. For the photopeak counts from the point source, the counts were normalized to a source strength of 1 μCi, giving a count rate in counts/s per 1 μCi. For the distributed sources, the gross counts were normalized to gross counts per second for the entire area of the surface deposition for a 1 μCi/m^2 deposition, or

$$\text{(Gross Counts / } \gamma) \text{ x } (1\gamma/d) \text{ x } ((3.7e4 \text{ d/s)/ } \mu\text{Ci)} \text{ x } (1 \text{ } \mu\text{Ci/m}^2) \text{ x Area(m}^2) =$$
$$\text{(Counts/s) / } (\mu\text{Ci/m}^2)$$

The gross count rate per 1 μCi/m^2 at altitude can then be normalized to the dose rate mRem/hr or exposure rate mR/hr at 1m above ground level (AGL) from the count rate-energy curves shown in Figures 10 – 13 and appropriate exposure rate or dose rate conversion factors given in Table 3.4 of the *FRMAC Assessment Manual* [1]. The manual gives dose rates and exposure rates for a 1 μCi/m^2 infinite plane surface deposition evaluated at 1m AGL for a particular isotope. As the results here are for a mono-energetic source, the count rates at the appropriate energies corresponding to the gamma lines of the isotopes must be multiplied by the branching ratios and then summed. The resulting total count rate for the isotope at the given altitude per μCi/m^2 deposition must then be divided by the isotope's conversion factor from Table 3.4 to give the gross count rate at altitude for that isotope per unit dose or exposure rate at 1m AGL.

The following example illustrates the above method for the ^{60}Co isotope at 1000 ft. At this altitude, the gross count rates at 1.173 MeV and 1.332 MeV are roughly 2.1E+1counts/s and 2.4E+1 counts/s, and the branching ratio for each is 1, giving a total count rate for 1 μCi/m^2 ^{60}Co deposition of (2.1E+1)x1 + (2.4E+1)x1 = 4.5E+1 (counts/s) / (μCi/m^2). The gross count rate at 1000 ft normalized to the EDE (effective dose equivalent) rate at 1m AGL would then be [4.5E+1(counts/s) / (μCi/m^2)] / [2.2E-2(mRem/hr) / (μCi/m^2)] = 2.05E+3 (counts/s) / (mRem/hr).

RESULTS FOR FIXED-WING POINT SOURCES

The point source modeling results for the fixed-wing aircraft are shown below. Photopeak efficiencies are shown separately for each of the three NaI detectors. Due to time constraints, many of the runs were not long enough to allow counts in the small 1"x1"D NaI detector. Figures 1–5 show results for point sources with zero radial distance from the center of the fixed-wing pod, while Figures 6–9 show results with a radial offset equal to the altitude. Figures 2-5 and 6-9 show the photopeak efficiencies for a point source strength of 1 μCi as a function of energy for the simulated gamma spectra used, while Figure 1 shows the efficiencies as a function of altitude for one point source, 137Cs. The results show the effects of air attenuation at distances larger than 100m in the drop off of efficiencies faster than the 1/r2 dependence. The energy curves also show attenuation effects in a greater efficiency at higher energies for distances larger than 100m.

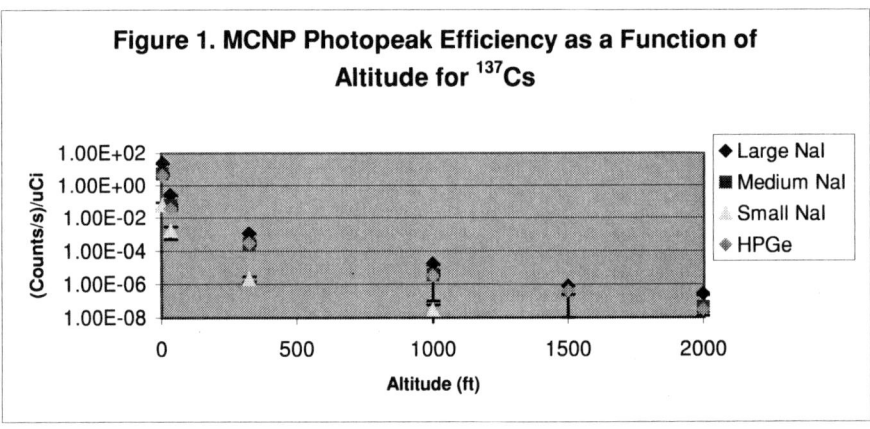

Figure 1. MCNP Photopeak Efficiency as a Function of Altitude for ^{137}Cs

Zero Radial Distances

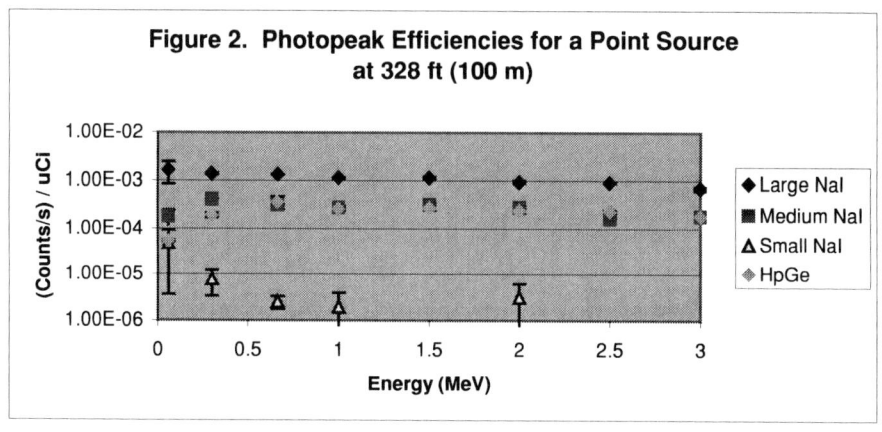

Figure 2. Photopeak Efficiencies for a Point Source at 328 ft (100 m)

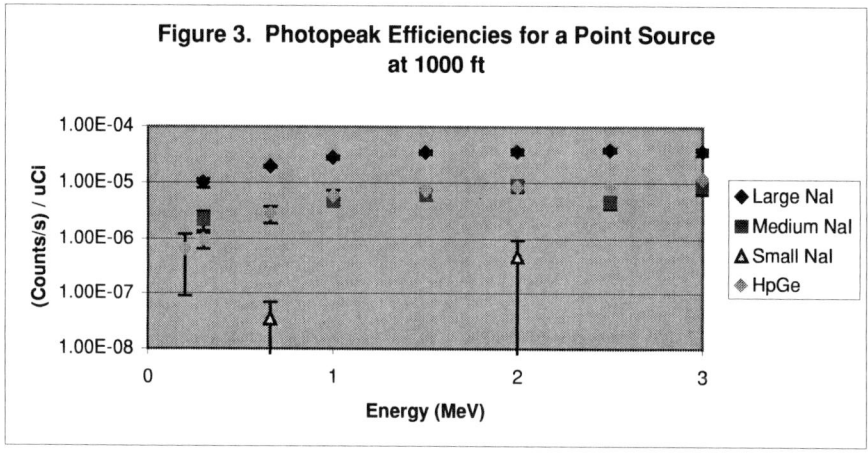

Figure 3. Photopeak Efficiencies for a Point Source at 1000 ft

Zero Radial Distances (continued)

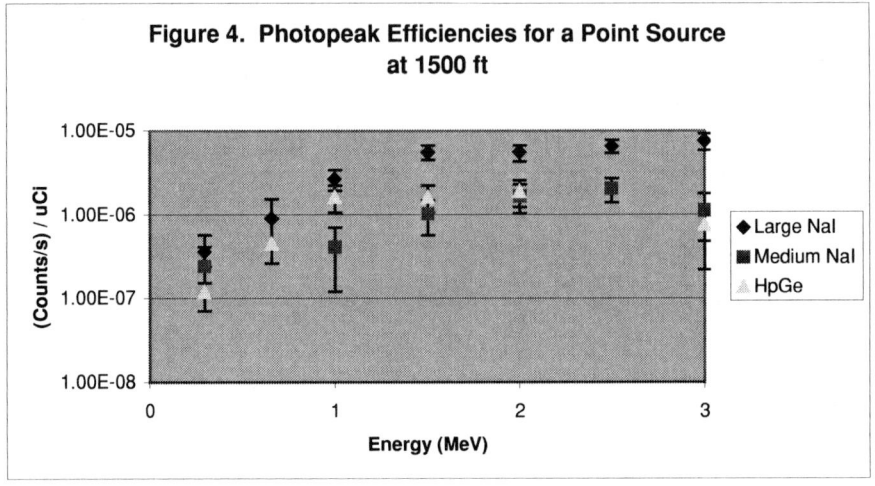

Figure 4. Photopeak Efficiencies for a Point Source at 1500 ft

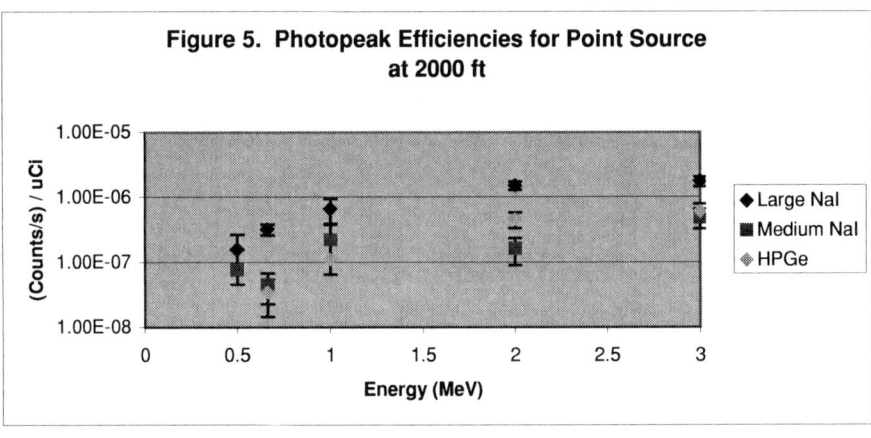

Figure 5. Photopeak Efficiencies for Point Source at 2000 ft

OFFSET RADIAL DISTANCES

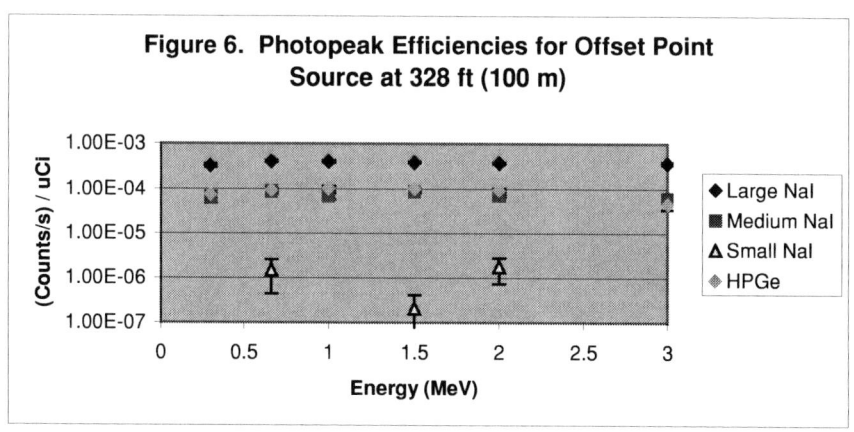

Figure 6. Photopeak Efficiencies for Offset Point Source at 328 ft (100 m)

OFFSET RADIAL DISTANCES (CONTINUED)

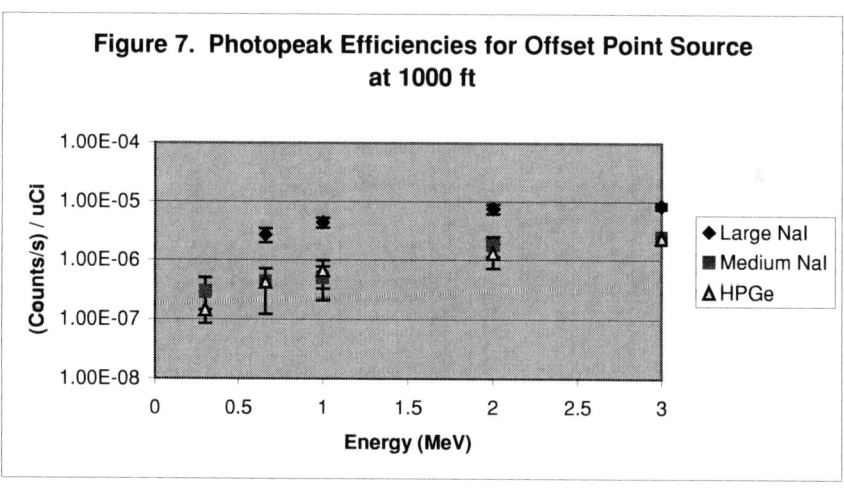

Figure 7. Photopeak Efficiencies for Offset Point Source at 1000 ft

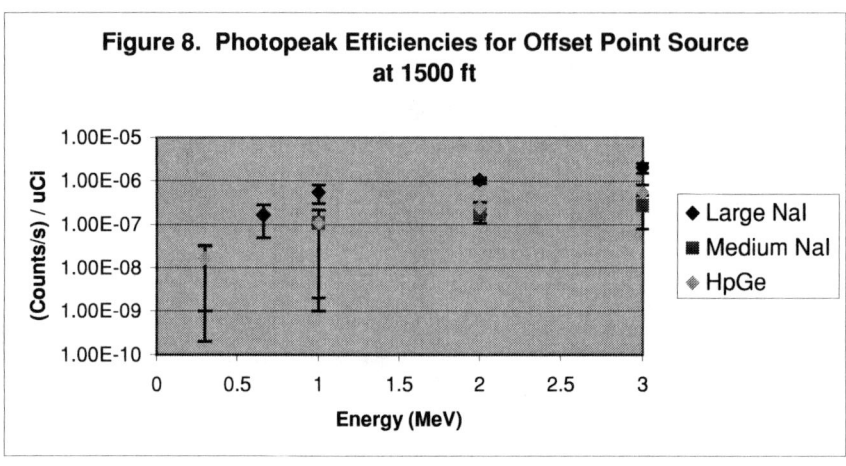

Figure 8. Photopeak Efficiencies for Offset Point Source at 1500 ft

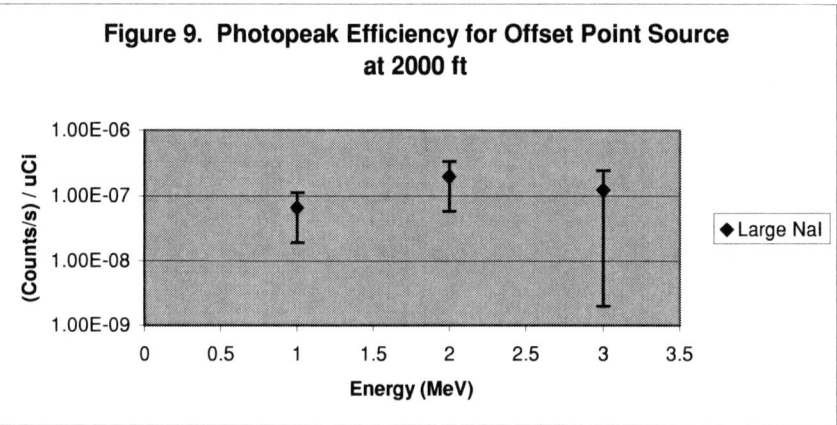

Figure 9. Photopeak Efficiency for Offset Point Source at 2000 ft

COMPARISON TO DATA

Figures 10 and 11 show a comparison of fixed-wing data to real data for an iridium-192 (^{192}Ir) source. Gross count rate for the large NaI detector is shown. The data and calculations at 500 and 1000 ft are not far off (30% and 10%), although the altitudes above 1000 ft show a drop off of calculations compared to the data. This is not explained by the expected air attenuation based air absorption at this energy, which with the distance change predicts a value at 1500 ft of roughly 30% that at 1000 ft. The sharper drop of Figure 10 as compared with Figure 11 above 1000 ft may be explained by lower representation of Compton scattering at smaller angles from the source due to stronger source directional biasing. However, reducing the biasing above that of the runs shown for Figure 11 does not seem to show an appreciable increase in gross count rate.

However, the photopeak efficiencies do follow to at least a factor of 2 or better the expected drops due to air attenuation and distance. Figure 12 shows the MCNP results for the 0.662 keV energy of ^{137}Cs compared to the calculated drop-offs from the 1m value due to the $1/r^2$ drop and air attenuation. A value of 0.0028/m was assumed for the air attenuation coefficient $(\mu)^2$.

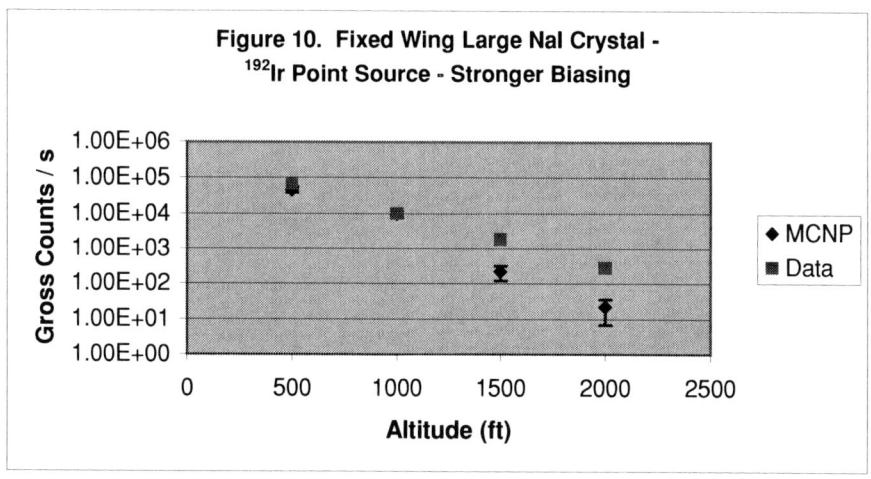

Figure 10. Fixed Wing Large NaI Crystal - ^{192}Ir Point Source - Stronger Biasing

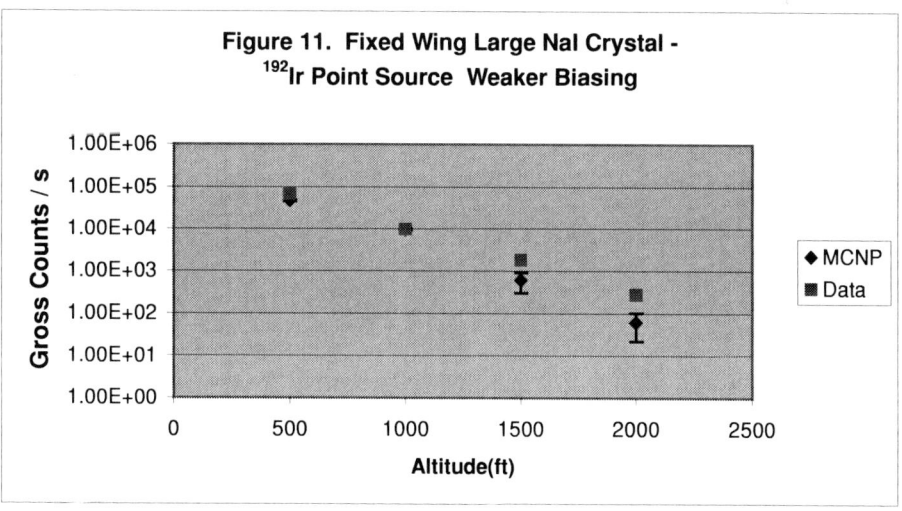

Figure 11. Fixed Wing Large NaI Crystal - ^{192}Ir Point Source Weaker Biasing

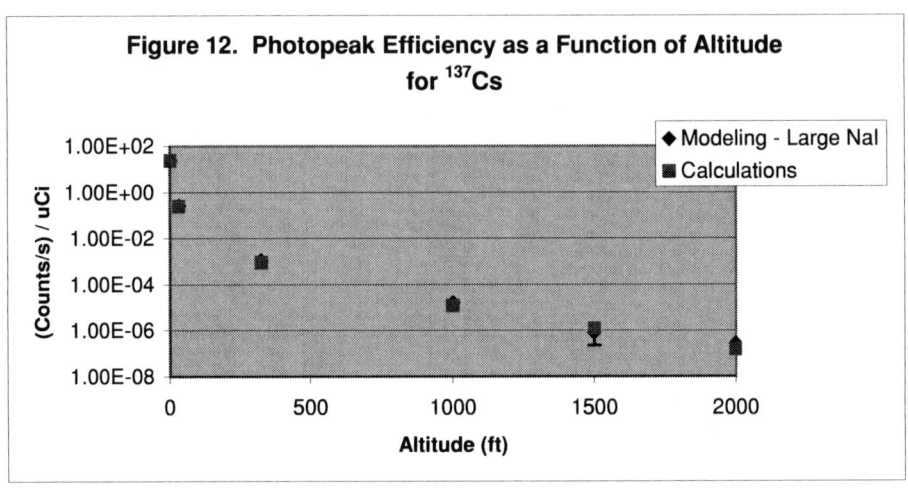

Figure 12. Photopeak Efficiency as a Function of Altitude for ^{137}Cs

RESULTS FOR FIXED-WING DISTRIBUTED SOURCES

The results for the fixed-wing distributed sources for are shown in Figures 13 – 16. As noted, they are uniform, circular surface sources approaching an infinite plane, with the radius equal to the altitude. Results of gross count rates are shown for the three sizes of NaI crystals, as well as for the HPGe detector. Counts were seen in the small circular 1"x1"D NaI detector only for the 100 m (328 ft) and 1000-ft altitudes. Error bars for the 1"x1"D NaI and HPGe detectors and other detectors at low energy points were cut off in several graphs to show the rest of the data more effectively. The count rates normalized to 1 μCi/m^2 are shown for four altitudes of 100 m (328 ft), 305 m (1000 ft), 457 m (1500 ft), and 610 m (2000 ft).

90

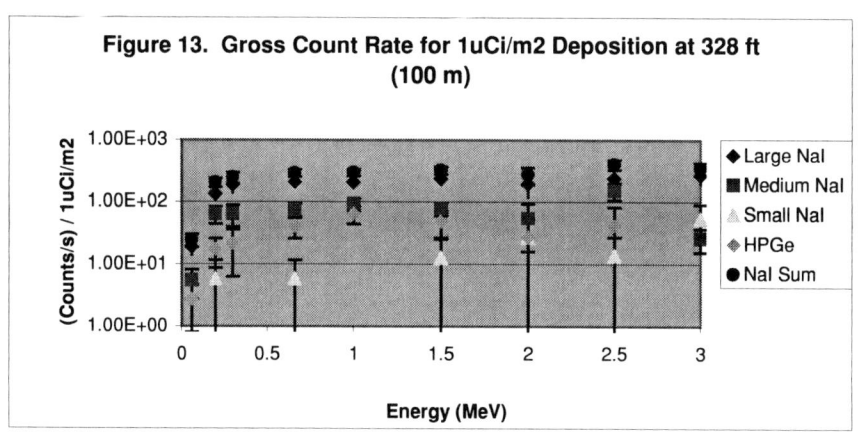

Figure 13. Gross Count Rate for 1uCi/m2 Deposition at 328 ft (100 m)

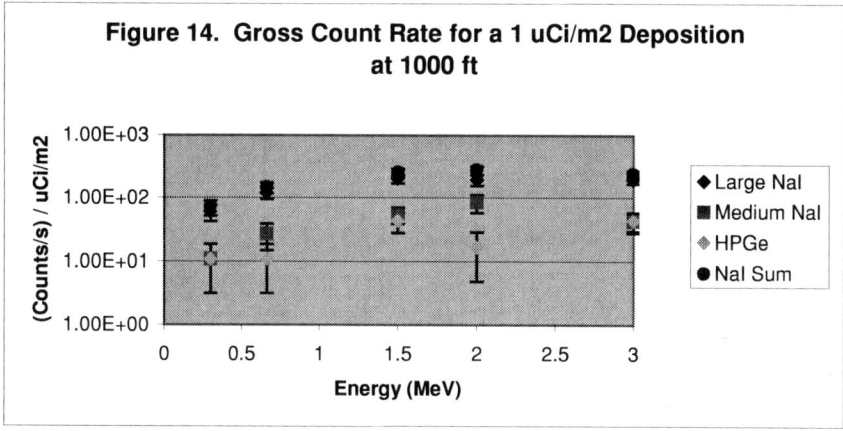

Figure 14. Gross Count Rate for a 1 uCi/m2 Deposition at 1000 ft

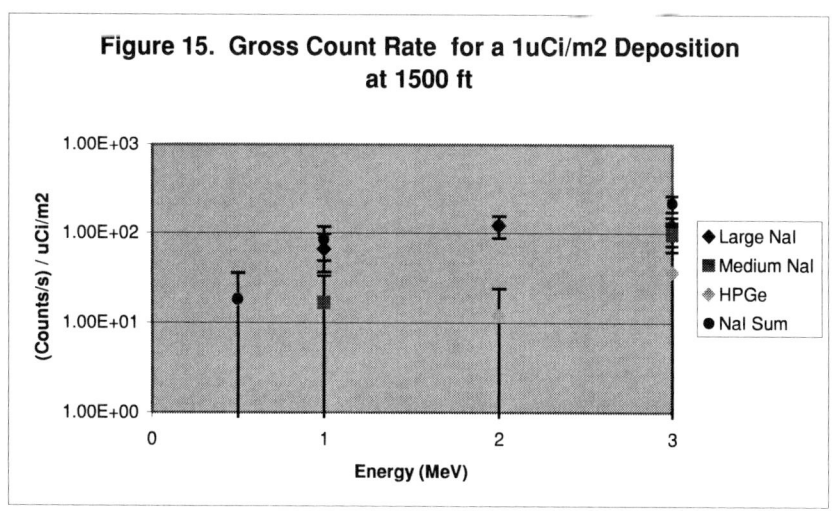

Figure 15. Gross Count Rate for a 1uCi/m2 Deposition at 1500 ft

Results for Distributed Sources (continued)

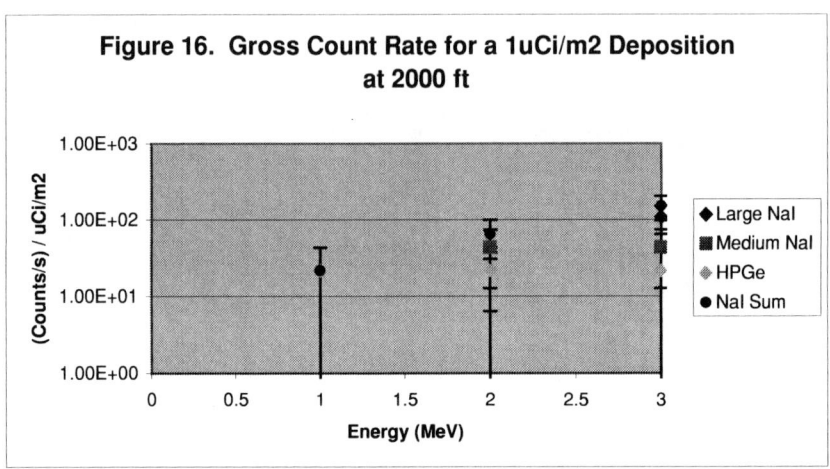

Figure 16. Gross Count Rate for a 1uCi/m2 Deposition at 2000 ft

RESULTS FOR HELICOPTER

The calculations for the helicopter, as previously stated, are made for a [241]Am distributed surface source and are shown as count rate for one NaI pod normalized to a deposition of 1 μCi/m^2 as with the fixed-wing distributed sources. The distributed source was circular as before, with the radius equal to the altitude for the first graph, Figure 17. Altitudes were 50 ft, 150 ft, 300 ft, and 500 ft. The final two graphs, Figures 18 and 19, show the count rates as a function of the radius of the distributed source for the lowest two altitudes of 50 ft and 150 ft, for a radius of 1x, 2x, 4x, and 8x the altitude. The results indicate that for an altitude of 50 ft, at a radius of 8x the altitude, the distributed source approaches an infinite plane, while at an altitude of 150 ft, a larger radius may be needed. Due to time constraints, similar runs were not made for the higher flying altitudes.

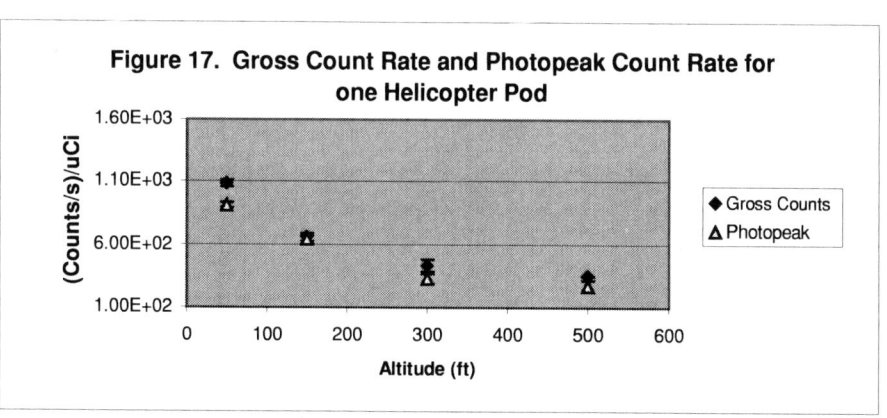

Figure 17. Gross Count Rate and Photopeak Count Rate for one Helicopter Pod

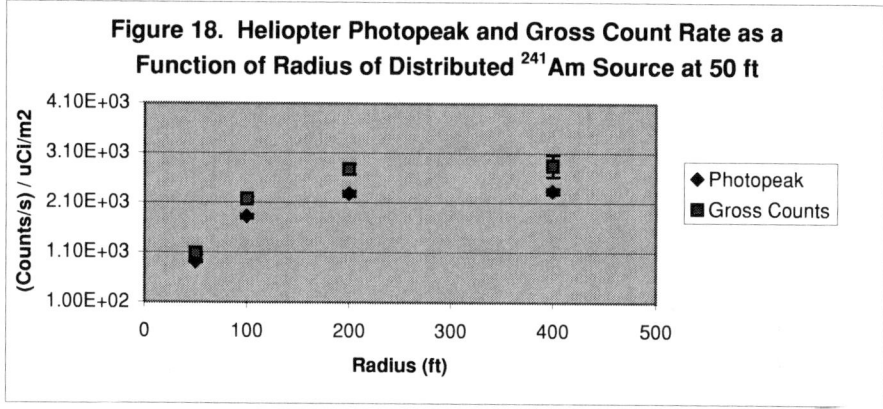

Figure 18. Heliopter Photopeak and Gross Count Rate as a Function of Radius of Distributed ^{241}Am Source at 50 ft

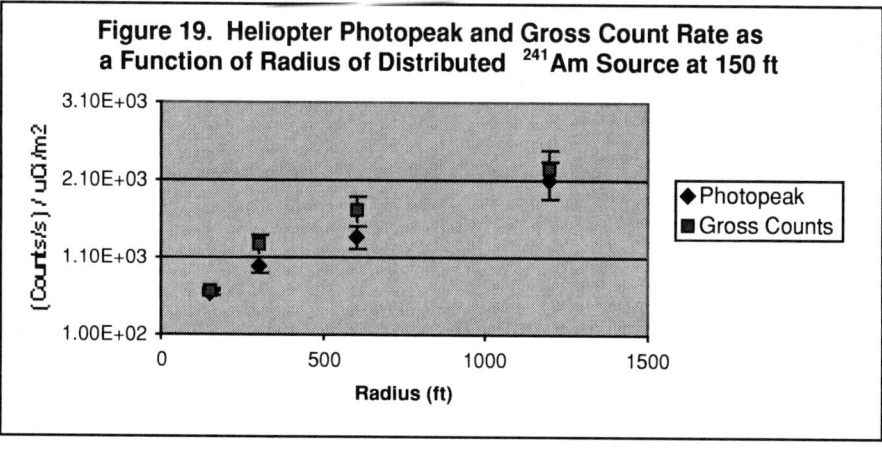

Figure 19. Heliopter Photopeak and Gross Count Rate as a Function of Radius of Distributed ^{241}Am Source at 150 ft

SAMPLE FIXED-WING SPECTRA

Figures 20 and 21 below show the spectra for HPGe and the largest NaI detector in the fixed-wing aircraft at an altitude of 328 ft (100 m) with no radial offset, while Figures 22 and 23 show the same spectra for an altitude of 1000 ft. Both have Gaussian broadening added to the tally, with 2.5keV Full Width Half Maximum (FWHM) for the HPGe spectra and 40 keV for the NaI spectra. The generation of spectra at altitudes of 1000 ft and greater was difficult due to low statistics in individual energy bins, as is shown in the spectra.

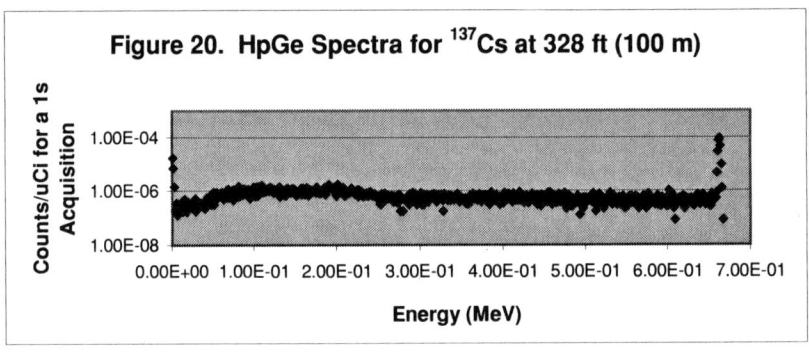

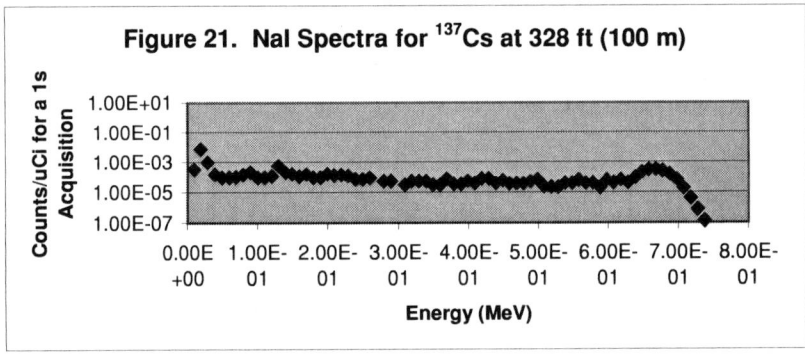

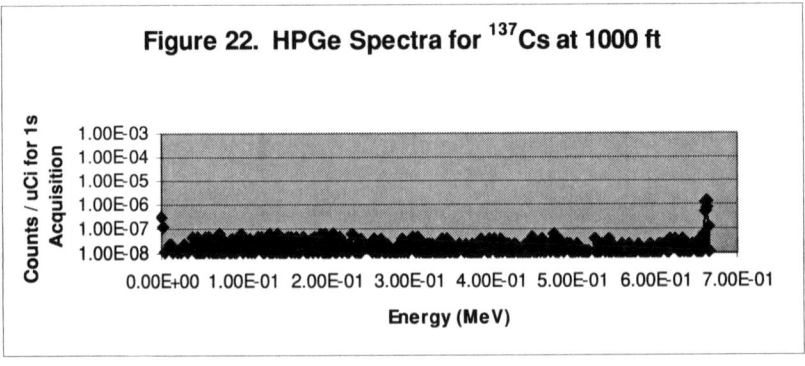

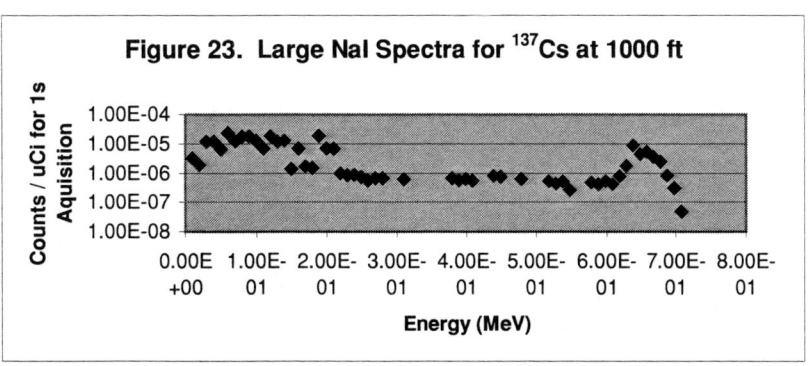

Figure 23. Large NaI Spectra for ^{137}Cs at 1000 ft

GEOMETRIES

Figures 24 – 29 show the geometries of the detector pods that were modeled for the fixed-wing and helicopter systems.

Figure 24. Fixed-wing NaI pod (blue) and HPGe pod (orange)- Vertical Cut

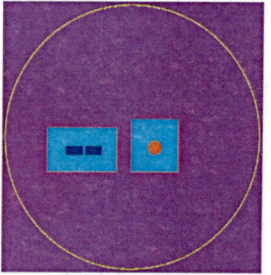

Figure 25. Fixed-wing - Horizontal Cut

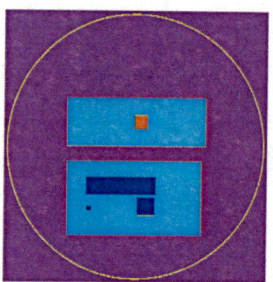

Figure 26. NaI Pod Showing all 3 NaI detectors - Vertical Cut

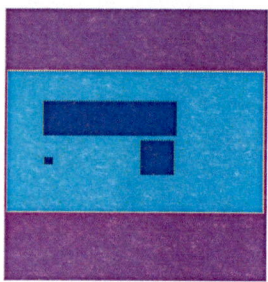

Figure 27. NaI Pod with large and small NaI – Horizontal Cut

Figure 28. Helicopter B200 Pod with 6 NaI detectors - Vertical Cut

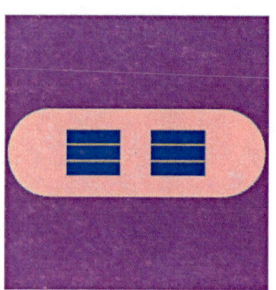

Figure 29. Helicopter Pod - Horizontal Cut

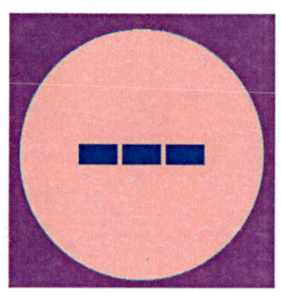

CONCLUSION

These results represent the first attempt to model the AMS systems at flying altitudes using MCNP. As the comparison to data shows, the gross count results appear to be reliable up to 1000 feet. However, a problem with the model at higher altitudes will require more investigation. Clearly more data points are needed for comparison of fixed-wing photopeak efficiencies, gross counts at other energies, as well as the helicopter modeling.

Another difficulty to be addressed in future work is the poor statistics of the modeling. This is both due to poor transport caused by the air attenuation at the large distances (especially for lower energies) and inability to generate enough photons in a reasonable running time. While the results for the lower altitudes for the fixed-wing and helicopter results show fairly good statistics and number of data points, the modeling data for higher altitudes, especially for the distributed sources, suffers from low statistics and few data points at lower energies. Future endeavors will include developing a method for directional biasing of a large distributed source, and investigating other methods to speed calculations while retaining accuracy for both photopeak and gross count rate.

Additionally, the modeling could be improved by better estimates of the effective thickness of the fixed-wing aircraft, as the detector pods are inside the aircraft rather than outside, as with the helicopter.

ACKNOWLEDGMENTS

This work was supported by the U.S. Department of Energy, National Nuclear Security Administration Nevada Operations Office, under Contract No. DE-AC08-96NV11718. **DOE/NV/11718—709.**

REFERENCES

1. FRMAC Assessment Manual, Volume 2, U.S. Department of Energy, Las Vegas, NV, 1996.
2. Radiological Health Handbook, edited by U.S. Bureau of Radiological Health, 1970.

ASTROPHYSICS APPLICATIONS

Miniature Neutron-Alpha Activation Spectrometer

Edgar Rhodes*, James Paul Holloway[†], Zhong He[†], and John Goldsten*

*Space Department, Johns Hopkins University Applied Physics Laboratory,
11100 Johns Hopkins Road, Laurel, MD 20723-6099

[†]Department of Nuclear Engineering & Radiological Sciences, University of Michigan,
1906 Cooley Bldg., 2355 Bonisteel Blvd., Ann Arbor, MI 48109-2104

Abstract. We are developing a miniature neutron-alpha activation spectrometer for in-situ analysis of chem-bio samples, including rocks, fines, ices, and drill cores, suitable for a lander or Rover platform for Mars or outer-planet missions. In the neutron-activation mode, penetrating analysis will be performed of the whole sample using a γ spectrometer and in the α-activation mode, the sample surface will be analyzed using Rutherford-backscatter and x-ray spectrometers. Novel in our approach is the development of a switchable radioactive neutron source and a small high-resolution γ detector. The detectors and electronics will benefit from remote unattended operation capabilities resulting from our NEAR XGRS heritage and recent development of a Ge γ detector for MESSENGER. Much of the technology used in this instrument can be adapted to portable or unattended terrestrial applications for detection of explosives, chemical toxins, nuclear weapons, and contraband.

INTRODUCTION

We are developing a miniature neutron-alpha activation spectrometer (MiNAAS) for in-situ analysis of chem-bio samples, including rocks, fines, ices, and drill cores, suitable for a lander or Rover platform for Mars or outer-planet missions. In the neutron-activation mode, penetrating analysis will be performed of the whole sample using a γ spectrometer and in the α-activation mode, the sample surface will be analyzed using Rutherford-backscatter and x-ray spectrometers. The instrument is expected to provide composition over a wide range of elements, including the rock-forming elements, rare earths, radioactive elements, and light elements present in water and biological materials. The detectors and electronics will benefit from remote unattended operation capabilities resulting from our NEAR XGRS heritage and recent development of a Ge γ detector for MESSENGER. Much of the technology used in this instrument can be adapted to portable or unattended terrestrial applications for detection of explosives, chemical toxins, nuclear weapons, and contraband.

Novel in our approach is the development of a switchable radioactive neutron source (SRNS) and a small high-resolution γ detector (SHGD). The SRNS is based on the separation of α-emitting radioisotope material and light-element material that has a large α cross section for generating neutrons. When the α-emitter material is placed

CP632, *Unattended Radiation Sensor Systems for Remote Applications*, edited by J. I. Trombka et al.
© 2002 American Institute of Physics 0-7354-0087-3/02/$19.00

in close juxtaposition with the light element material, the neutron source is switched on. When the materials are separated, the neutron source is switched off, preventing unwanted activation and radiation damage to spacecraft and instrument components when measurements are not being made, without any shield. The SRNS does not have the bulky ancillary equipment required by an accelerator-based neutron source, such as an ultra-high voltage power supply (which tends to be relatively unreliable). The SRNS can provide a reasonably high neutron flux but is small, requires very little power, and is well-suited to remote unattended operation.

The SHGD consists of a small HPGe or CZT γ detector inside a CsI(Tl) or BGO anticoincidence cup, depending on resolution requirements and environmental conditions, along with custom signal processing electronics. The cup provides suppression of cosmic rays and detector γ scattering, along with an escape coincidence mode that further suppresses background at energies above 1 MeV, and also supplemental detection of high-energy γ lines.

INSTRUMENT DESIGN CONCEPT

The instrument must be suitable for a lander or Rover platform that meets mass, power, and environmental constraints of planetary missions. For Mars missions, a total mass of ~ 2-5 kg (including electronics) and an input power ~ 2-10 W is envisioned. For missions to the outer planets, a total mass of ~ 1 kg and input power of ~ 1 W would probably be necessary. There will be two operational modes, neutron activation for choosing samples and for bulk element analysis of each sample, and alpha activation for surface element analysis of each sample.

In neutron activation mode, the switching neutron source is on. If Be is used as the target material, neutrons up to 11 MeV will penetrate 10 cm or so of rock and react with nearly every element (primarily by inelastic scatter, some absorption), producing penetrating gamma-rays with energies characteristic of the element. Depending on interaction cross-section and element composition, one can detect rock-forming elements (Na, Mg, Si, Fe, Ca, etc), rare earths (Sm, Eu, for example), radioactive elements (K, Th, U), light elements in waters, ices, and bio-materials (C, N, O, and H by capture if there is neutron moderator present).

A conceptual design of the MiNAAS instrument is shown in Fig. 1, operating in the neutron activation mode. An alpha source is located in the center position opposite the target surface. An on/off neutron source is accomplished by moving a thin Be foil between the alpha source and the target. A small stepper motor drives the foil holder mechanism. This allows one source to provide alpha particles for the Rutherford backscatter and X-ray fluorescence spectrometers, and neutrons for the prompt gamma spectrometer. While in the alpha-only position, the gamma ray spectrometer can make needed measurements of the gamma-ray background. The gamma-ray detector does not need to be located near the source and therefore is conveniently placed on the side of the sensor head. This allows room for the BGO or CsI anticoincidence cup.

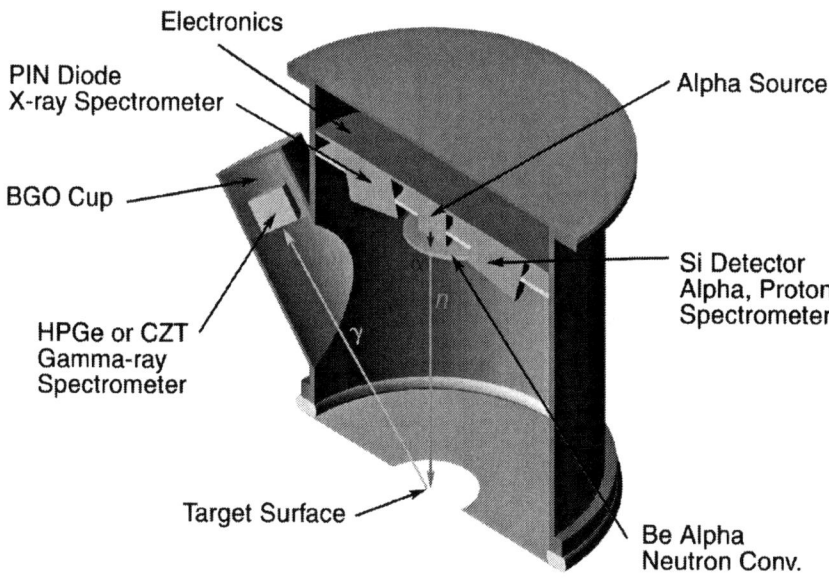

FIGURE 1. Conceptual design of MiNAAS instrument, shown in neutron activation mode.

Located close to the source is a small stack of surface-barrier solid-state detectors used for the backscatter and proton spectrometers, in the alpha activation mode. A thin front detector measures the energy of the backscattered alpha particles, allowing differentiation of light elements (C, N, and O). A thicker rear detector stops energetic protons that penetrate the front detector. In this way, the experiment can discriminate between alphas and protons. For protons, the coincident energies of the two detectors are combined to yield the total energy. The alpha/proton reaction can identify Na, Mg, Al, Si, and S.

For the X-ray spectrometer, a separate miniature Si PIN photodiode or drift detector measures characteristic x-rays in the 1-10 keV region. A thin Be window in front of the detector shields it from alphas and protons, yet is transparent to low-energy x-rays. X-rays in the sample excited by fluorescence from alphas and L-shell x-rays from the source allow identification of intermediate mass elements (Mg through Ni). Shown in Fig. 2 is the conceptual instrument design, operating in the alpha activation mode.

An initial penetrating scan of the planet surface by the gamma-ray spectrometer can examine soils, ices, loose materials, other objects, and major constituent elements, to choose samples from. Chosen samples can be analyzed in bulk and the source can then be switched to alpha activation mode for surface analysis of samples, eg. observing weathering layers, rinds, dust, or soil layers. The combination of the two complementary modes allows possible inferences into prebiotic conditions, petrology, planetary differentiation, igneous evolution, and weathering history.

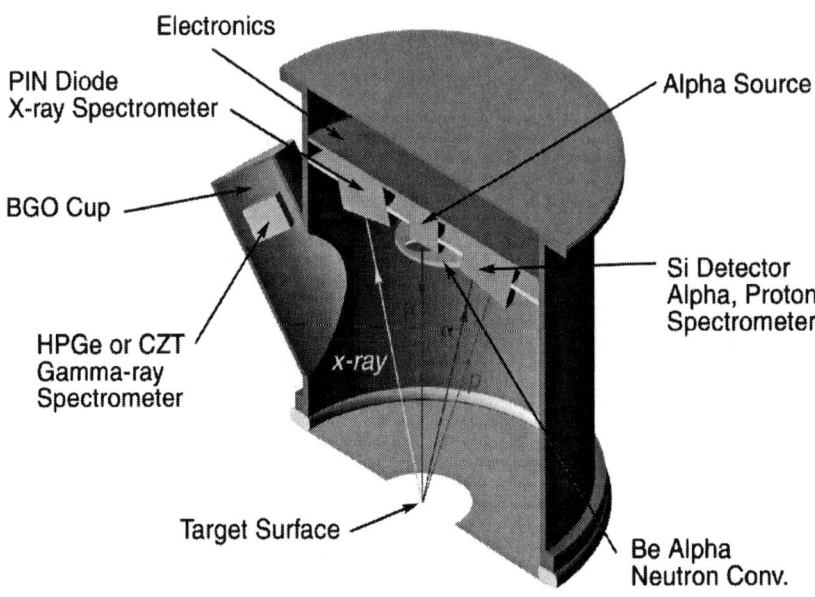

Electronics

PIN Diode
X-ray Spectrometer

Alpha Source

BGO Cup

Si Detector
Alpha, Proton
Spectrometer

HPGe or CZT
Gamma-ray
Spectrometer

x-ray

Target Surface

Be Alpha
Neutron Conv.

01-0208G-2

FIGURE 2. Conceptual design of MiNAAS instrument, shown in alpha activation mode.

SWITCHING RADIOACTIVE NEUTRON SOURCE

Exploration of planet composition and geology using neutron activation analysis requires a neutron source of adequate intensity in close proximity to the sample, that produces neutrons of sufficient energy to penetrate through the rock. Gamma rays characteristic of the elements will be provided by inelastic scattering, absorption, and moderation followed by capture of the neutrons inside the sample. However the source must be small in size and mass and require little power, in order to meet mission requirements for a Mars Rover or spacecraft exploring the outer planets. Also, the ability to switch the source on and off is highly attractive, to simplify pre-launch operations, avoid any activation of the spacecraft during flight, and allow sharing of an alpha source between neutron activation and APX spectrometries.

A switching neutron source satisfying these criteria can be developed, based on separation of alpha-emitting radioisotope material from target material consisting of specific isotopes of certain light elements, such as Be, B, and Li, that have high cross sections for alpha/neutron reactions. When these materials are separated so that the alpha particles do not strike the target, the source is switched off, and when these materials are brought together so that the alphas strike the target, the source is switched on. A possible geometrical concept for such a source is shown in Fig. 3.

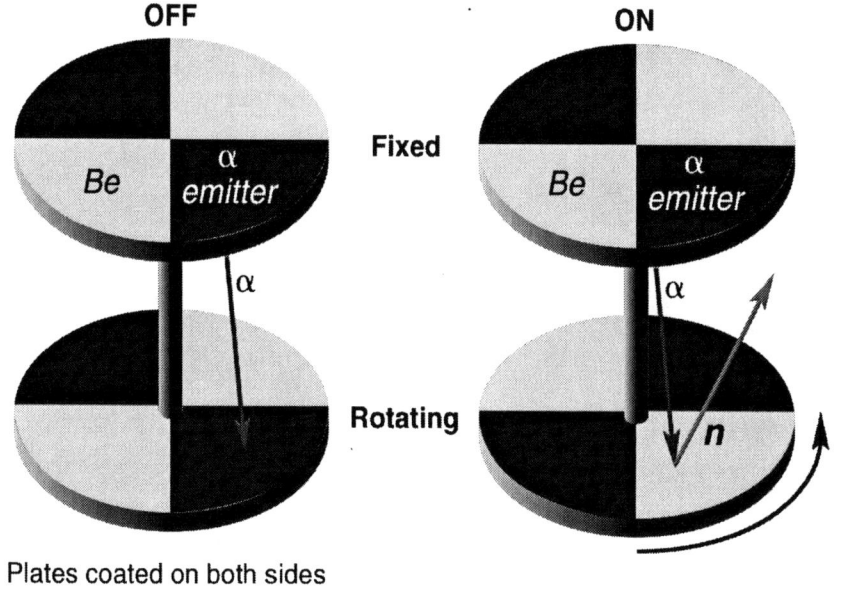

OFF   ON

Fixed

Be   α emitter

α

Rotating

Be   α emitter

α

n

Plates coated on both sides

01-0208G1

FIGURE 3. Possible geometrical concept for a switching radiosotope neutron source.

In an actual source, the plates would be located much closer together than shown in the figure, and raised ridges separating the segments would prevent alphas from crossing over to an adjacent plate. In Fig. 3, the alpha emitter is deposited to a depth equivalent to the alpha range; further depth would increase the background from reactions in the emitter material (such as gamma radiation, alpha/neutron reactions with impurities, and spontaneous fission) without a concurrent increase in the switchable neutron rate, lowering the on/off switching ratio. For a given alpha emitter and target, neutron intensity is increased by increasing the plate area or stacking pairs of plates.

Depending on the conversion rate of the alphas emitted in the layers to neutrons and the half-life of the alpha emitter, this Switching Radioisotope Neutron Source (SRNS) concept has the potential to provide a compact low-mass switching neutron source requiring only the minimal power to rotate the movable plates one segment, which can be done quite reliably and lends itself well to remote unattended operation. No massive shield is required, such as would be needed to "shut off" a conventional radioisotope source. The competing type of switching neutron source is an accelerator, which requires power for bulky ancillary equipment, as well as an ultra-high voltage supply (~ 100 kV for even a deuterium beam and tritium target), which has been shown to be unreliable in space applications.

The conversion rate of alpha particles to neutrons and the resultant neutron energy spectrum are basically determined by the choice of SRNS target material, Be, B, or Li as considered here and shown in Table 1 below. Be provides the highest neutron yield and the highest neutron energy, up to 12 MeV, but there is a 4.43 MeV gamma from de-excitation of a ^{12}C state that occurs with ~ 80% probability. B provides an intermediate yield and an energy spectrum similar to a fission spectrum, while Li provides a low yield, minimal gamma radiation, and a low-energy spectrum more suitable for moderation to thermal energies.

TABLE 1. Properties of Selected SRNS Target Materials

Target	Be	B	Li
Avg. N Energy, MeV	4.5	2.7	0.5
Max. N Energy, MeV	12	5	1.5
Neutron Yield, %Be	100	25 – 50	0.8 – 5
Gamma Energy, MeV	4.43	2.31	negl.
Gammas per Neutron	0.8	0.07	negl.

The choice of alpha emitter basically determines the overall alpha rate and corresponding neutron intensity, as well as the period of source usefulness, through its alpha half-life, and the background when the source is in its off position, through its spontaneous fission half-life, alpha/neutron reactions with impurities, and gamma decay products. A selection of alpha emitters is shown in Table 2 below. In rows three through six, the alpha emitter is assumed to be deposited uniformly to a depth equal to the range of the highest energy alpha emitted. In row five, a simple transport code was used to determine the neutron yield per alpha, based on the experimental alpha/neutron cross sections of Be, B, and Li.[1] In row six, the "on/off" switching ratio of neutron intensity is based on the results of row five and the ratio of the alpha emitter alpha half-life to its spontaneous fission half-life times the average number of neutrons per fission.

TABLE 2. Properties of Selected SRNS Alpha/Neutron Sources

alpha emitter		Th-228	Cm-242	Po-210	Ac-227	Pu-238	Am-241
half-life		1.91 y	163 d	138 d	21.6 y	87.8 y	432 y
energy, keV		80-2600	44	800	50-870	44-766	60
gamma dose 1m, mr/hr-cm^2		17,000	0.73	0.45	230	0.003	0.63
neutrons per cm^2 × 10^4 "on"	Be	2100	1900	1200	190	5.9	1.1
	B	600	610	470	55	2.2	0.4
	Li	100	10	10	8.4	0.064	0.012
neutron "on/off" ratio	Be	high	41	high	high	1250	high
	B	high	14	high	high	460	high
	Li	high	2	high	high	15	high

In Table 2, "high" in row six means that the switching ratio is greater than 10^5 based on the calculation. Such a high ratio will not normally be attained due to alpha/neutron reactions and spontaneous fissions caused by the presence of impurity isotopes inside the alpha emitter layer. The neutron yield in column six for Pu-238 and a Be target has been verified by radiochemistry glove box experiments.[1]

Although Th-228 and Ac-227 can provide high neutron yields for usable half-lives and high switching ratios, Table 2 shows that the gamma radiation from their daughters makes these isotopes undesirable for most purposes. Cm-242 and Po-210 can provide high neutron yields with low gamma dose, but their half-lives are too short for most purposes, and the Cm-242 switching ratio is not very high. Although their neutron yields are only moderate, Pu-238 and Am-241 have long half-lives and can provide relatively high switching ratios with low gamma dose. Preliminary design considerations indicate that an SRNS of intensity greater than 10^6 n/s can be fabricated from Pu-238 or Am-241 and Be having a total mass that is a relatively small fraction of a kilogram, leaving plenty of mass allocation for a gamma-ray detection system.

MINIATURE HIGH-RESOLUTION GAMMA-RAY DETECTOR

An HPGe detector would provide outstanding energy resolution and efficiency, but would have to be cooled to ~ 90K to keep radiation damage from cosmic rays under control and avoid frequent annealing. The Applied Physics Lab experience in developing an HPGe spectrometer for the MESSENGER mission to Mercury, cooled by a minicryocooler and thermally isolated by a Kevlar string suspension and nested low-emissivity shields, indicates that a similar configuration may be feasible in the cold Mars environment, for a relatively large rover that would allow the added mass and power required. This configuration is probably not feasible for a small rover or a mission to the outer planets. A room temperature high-resolution semiconductor gamma-ray detector would be more amenable to these power and mass constraints.

Prompt neutron activation produces a multitude of characteristic gamma rays in the 0.1-10 MeV region, but the mean free path length for high-energy gammas in a moderately dense material is ~ 5 cm, a value inconsistent with the size of current room temperature high-resolution detectors. However, by focusing on the low-energy portion of the gamma ray spectrum (0.1-1.5 MeV), smaller, realizable detector volumes can be considered that achieve sufficient energy resolution to provide significant science return for most of the key elements (Na, K, Al, Ca, Mg, Mn, Si, Ti, Fe, Cl, Th, Nd, U, Sm, Eu, Gd).

Cadmium Zinc Telluride (CZT), a wide-bandgap semiconductor that requires no cooling, appears to be a good candidate detector material. A typical energy resolution of ~ 3% FWHM for ^{60}Co at 300 K is readily available. One of our co-investigators, Zhong He of University of Michigan, has obtained better than 2% FWHM for ^{137}Cs for a 1 cm^3 crystal [2] and has been getting similar results for crystals up to 1.5 cm x 1.5 cm x 1.0 cm [3], using a coplanar electrode to eliminate low-energy tailing and a correction method for electron trapping. Although the experimental results with the reference detector indicate that the presently attainable CZT energy resolution is adequate, the presence of closely spaced lines from other elements makes further improvement in energy resolution strongly desirable. If radiation damage to CZT from cosmic rays turns out to be a serious problem, a future candidate room temperature high-resolution semiconductor detector could be HgI_2, which is much more robust against radiation damage.

Zhong He's unique depth-sensing technique can be used to form multiple coplanar-grid anodes on the anode surface of a single crystal to form larger detectors without sacrificing energy resolution. A 3 cm x 3 cm x 1 cm coplanar grid CZT detector may be feasible; it would be the world's largest CZT detector. Our proposed single detector could, in principle, be replaced by a three-dimensional array of CZT detectors in an effort to increase detector efficiency and to extend the resolution to a higher energy range, but this would require a substantial amount of processing electronics. Such an approach is considered too complex for a rover-based instrument and may be too massive, and is presently not being considered.

The signal-to-background ratio is dominated by the gamma ray continuum produced through Compton scattering in the detector, particularly for small detectors. An active cup surrounding the CZT would work in anti-coincidence to suppress the scattering, as well as help suppress cosmic ray background, help provide directionality to the CZT view field, and help protect the CZT from radiation damage and activation. Additionally, the high density and moderate energy resolution of the cup itself will provide sensitivity to some of the stronger high-energy gamma lines such as H (2.2 MeV), C (4.4 MeV), and O (6.1 MeV), that lie beyond the energy range of the CZT detector. A CsI(Tl) or BGO scintillator cup coupled to a PIN photodiode would eliminate a photomultiplier tube and its high voltage power supply.

RADIATION TRANSPORT COMPUTATIONS

We have used the MCNPX coupled neutron-gamma transport code to compute the gamma-ray flux per source neutron induced by an alpha-Be neutron source placed close to a rock of the "soil-free" composition of a Martian andesite, roughly 15 cm in radius, sitting atop Martian soil (composition from site A-2 of the Pathfinder mission [4]). The neutron source and gamma detector are both 15 cm above the soil, and ~1cm from the rock. The resulting gamma spectrum from the rock sample at the detector is shown in Fig. 4 for the low-energy region appropriate for a CZT detector. Multiple lines are seen from which to identify and quantify most of the major elements in the rock sample.

The gamma flux at the detector from the nearby Martian soil induced by the neutron source was found to be 50 to 70 times smaller than that from the sample, depending on the energy. But of further concern is the magnitude of the sample gamma flux compared to that induced naturally on Mars by cosmic ray bombardment. This latter gamma flux has been estimated using the Lahet and MCNP codes [5]. Assuming a neutron source of 10^6 n/s, the weakest neutron source we contemplate, the relative peak fluxes from the neutron source and from cosmic rays have been estimated and are shown in Table 3, which indicates that we will have signal peaks from 2 to 3 orders of magnitude larger than peaks produced by natural background.

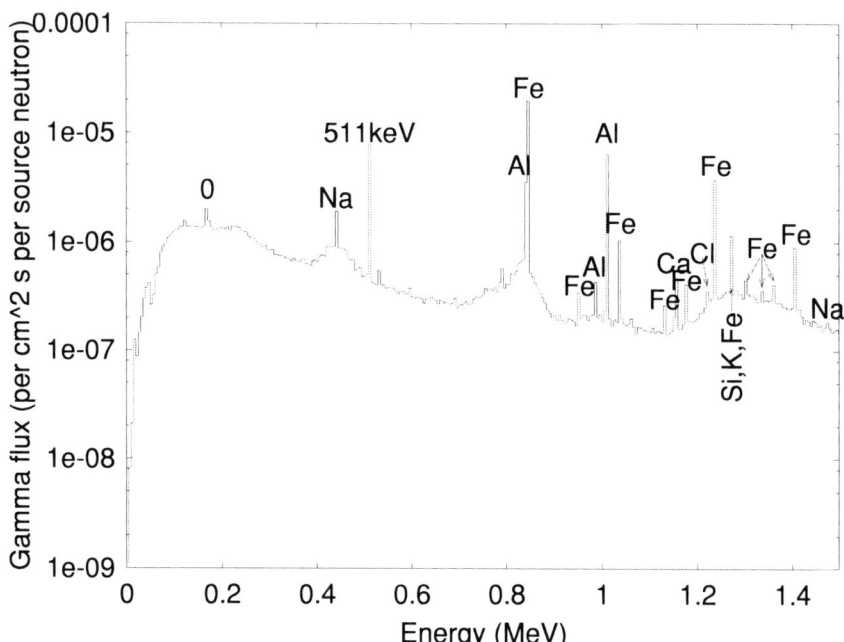

FIGURE 4. Computed neutron activation gamma flux spectrum of Mars adesite rock.

TABLE 3. Peak Gamma Counts in CZT Detector

target	w/o	Energy keV	GCR flux	α-Be flux	peak effic.	counts/s	hrs for 10^4 cnts
Na-23	1.93	439.9	0.0036	1.47	0.22	2.9	1.0
Al-27	5.61	1014	0.0055	6.11	0.06	3.3	0.8
Si-28	28.98	1779	0.0694	74.49	0.024	16.1	0.2
Ti-48	0.42	1037		0.949	0.058	0.5	5.6
Fe-56	8.57	846.7	0.0421	19.6	0.079	13.9	0.2

Also shown in Table 3 are the peak efficiency of a 3cm×3cm×1cm deep CZT detector, the resultant counts per second, and the number of hours to attain 1% Poisson count statistics, for key isotopes in the rock sample. An hour or less is required to accumulate 1% statistics for all isotopes shown except for Ti-48, which has a very low fractional composition. These computations do not consider background from the rover, gamma scattering inside the detector, or scattered gammas from cosmic rays. (In Table 3, flux is per cm^2/s).

The ultimate goal of our work is to determine composition from the uncollided prompt gammas, which will be detected as photopeaks and escape peaks in our gamma detector system. One of us, James Holloway, has recently studied the mathematics of the inverse problem of extracting composition information from large sample prompt gamma measurements [6] and has shown that the problem can be solved even in the presence of the significant complicating factors of neutron slowing down and thermalization in the unknown sample, and in the presence of nearby materials. An iterative solution to the problem is based on repeated neutron-gamma transport computations in materials of estimated composition and known geometry, and results on the convergence of the iteration are in hand.

Recently [7] we have undertaken prompt gamma measurements of large samples of simple binary and ternary mixtures, irradiated with a 5Ci AmBe source. From these measurements, the composition of the mixtures was successfully reconstructed using the computation driven iteration, providing us with both experience and a proof-of-principle in large sample prompt gamma analysis measurements. Sample composition was successfully determined even in the presence of known interferences from other materials in the measurement setup.

REMOTE OPERATION, CALIBRATION, AND SAMPLING

The Applied Physics Laboratory has considerable experience with remote operation, calibration, and sampling of instrumentation through development and operation of many spacecraft and spacecraft instruments, including deep space missions and a number of nuclear spectrometers. The miniature neutron-alpha activation spectrometer will receive the benefits of this experience.

Critical parameters affecting operation and calibration of the instrument must be internally monitored, such as the temperature and voltage of power supplies, the temperature of preamplifiers, amplifiers, ADC's, and detectors, the leakage current of semiconductor detectors, and detector total counts. Based on values of the critical parameters, autonomous safing and recovery routines must be employed, as it is not feasible to perform these operations expeditiously from the ground, due to delays in communication with the spacecraft and slow telemetry rates over long distances.

Variable telemetry rates to accommodate mission objectives are quite useful, in conjunction with data accumulation internal to the instrument or spacecraft for indefinite periods. Disparities between sampling periods and strategies and telemetry delays and rates must be accommodated. Minimal data and housekeeping information need to be transmitted periodically, and autonomous event detection algorithms may be needed to record specific data (such as gamma-ray burst timing). Instrument software should be divided into "core" code and "application" code. "Core" code, that needs to be robust and unchangeable, includes diagnostics, error checking, communication, and boot routines. The remaining, "application" code should be uploadable and flexible, with new functionality easily added.

It is quite useful to have a complete terrestrial model that duplicates the flight instrument as closely as possible (usually the "engineering" model). This allows testing effects of new operations and software uploads, as well as post-launch calibration of many parameters of the flight instrument. Of course, the flight instrument must receive an adequate pre-launch calibration of parameters unique to it (temperature, INL, DNL, gain and discriminator settings, etc.). Sources available for energy calibration include natural radioactive sources in space (and on planets), on-board radioactive sources, and on-board electronic pulsers (though it is difficult to eliminate pulser temperature dependence). By choosing electronic, detector, and

110

mechanical components and operating conditions for long life with minimal aging, the instrument will remain in calibration for a relatively long time period.

REFERENCES

1. Bowers, D. L., Rhodes, E. A., and Dickerman, C. E., "A switchable radioactive neutron source: Proof-of-principle",*J. Radioanalyt. and Nucl. Chem.* **233**, 161-165 (1998).
2. He, Z., Knoll, G. F., Wehe, D. K., and Miyamoto, J., "Position Sensitive Single Carrier CdZnTe Detectors", *Nucl. Instr. & Meth. A* **388**, 180-185 (1997).
3. Perez, J. M., He, Z., and Wehe, D. K., "Stability and Characteristics of Large Volume CZT Coplanar Electrode Detectors," IEEE Trans. Nucl. Sci. **48(3)**, 272-277 (2001).
4. Rieder, R., Economou, T., Wanke, H., Turkevich, A., Crisp, J., Bruckner, J., Dreibus, G. and McSween, H., "The chemical composition of Martian soil and rocks returned by the mobile alpha proton X-ray spectrometer: Preliminary results from the X-ray mode," *Science* **278**, 1771-1774 (1997).
5. Masarik, Jozef, and Reedy, Robert, "Gamma ray production and transport in Mars," *J. Geophys. Res.* **101**, 18,891-18,912 (1996).
6. Holloway, James Paul, and Akkurt, Hatice ,"Some Aspects of the Mathematical Modeling of Prompt Gamma Neutron Activation Analysis," in *Advances in Reactor Physics, Mathematics and Computation into the Next Millennium (Physor 2000)*, Amer. Nucl. Soc. CDRom ISBN: 0-89448-655-1 paper no. 80), Pittsburgh, PA, May 7-11, 2000.
7. Akkurt, Hatice, Holloway, James Paul, and Smith, L. Eric, "Testing an iteration for prompt gamma composition analysis," Trans. Amer. Nucl. Soc. **85**, 438-440 (2001).

Large Volume HgI$_2$ Gamma-Ray Spectrometers

Zhong He, James Baciak

Nuclear Engineering and Radiological Sciences Department
The University of Michigan, Ann Arbor, Michigan 48109-2104

Abstract. This paper demonstrates the enhanced capability of single polarity charge sensing, the 3-dimensional position sensing technique, developed at the University of Michigan and previously successfully demonstrated on CdZnTe detectors, to improve the spectroscopic performance of HgI$_2$ and to extend its range for spectrometry to an unprecedented thickness of 10 mm. Energy resolutions of close to 1% FWHM at 662 keV gamma-ray energy were obtained from individual depth locations underneath pixel anodes, and 1.4-2.0% FWHM energy resolutions from 5 out of 6 tested pixel anodes on two 10 mm thick detectors.

Introduction

HgI$_2$ has properties of high atomic number (Z=80-53), high density (ρ= 6.3 g/cm^3), wide band-gap (2.13 eV) and high bulk resistance (10^{12}-10^{13} Ω). These properties make it a very attractive material for efficient gamma-ray detectors capable of room-temperature operation. However, problems of charge trapping, material non-uniformity and temporary change of its properties result in poor spectral performance and limited thickness (not more than 3 mm) of a conventional detector using planar-electrodes.

The 3-dimensional position-sensitive single-polarity charge sensing technique developed at the University of Michigan [1] should be able to eliminate the problem of trapping of holes, to correct for the trapping of electrons, and to mitigate the material non-uniformity to the scale of the position resolution (~1 mm in 3-dimensions) of the detector system. Figure 1 illustrates the principle of the 3-dimensional position sensing technique. A two dimensional array of anode pixels is fabricated on the anode surface and all pixels are biased at the same voltage potential. A large number of electron-hole pairs proportional to gamma-ray energy deposition are generated from the gamma-ray interaction. The electrons move towards the anode and are collected by one of the pixel anode directly located above the location of gamma-ray interaction. The induced signal E' on the pixel anode is dominated by the number of electrons collected, and has a slight dependence on the depth of interaction as shown in the left-bottom of Figure 1. The lateral position (x, y) of the gamma-ray interaction is identified from the location of the pixel anode, and the depth (z) of interaction can be obtained from the ratio of the cathode signal to the signal of the pixel anode [2]. The actual energy deposition E_0 can then be deduced from the signal of the pixel anode E' and the depth (z) of interaction.

CP632, *Unattended Radiation Sensor Systems for Remote Applications*, edited by J. I. Trombka et al.
© 2002 American Institute of Physics 0-7354-0087-3/02/$19.00

Because of the promising results obtained on 5 mm thick HgI_2 spectrometers [3,4], the 3-dimensional position-sensitive single polarity charge sensing technique has been applied to 10 mm thick detectors. Several prototype detectors have been fabricated by Constellation Technology Corporation [5], each detector has four pixel anodes with dimensions about 1×1 mm surrounded by a large anode [3]. Signals from each pixel anode and the cathode are readout using Amptek A250 preamplifiers and shaped using standard Canberra 243 amplifiers. A sample prototype HgI_2 detector with dimensions of 1×1×1 cm is shown in Figure 2.

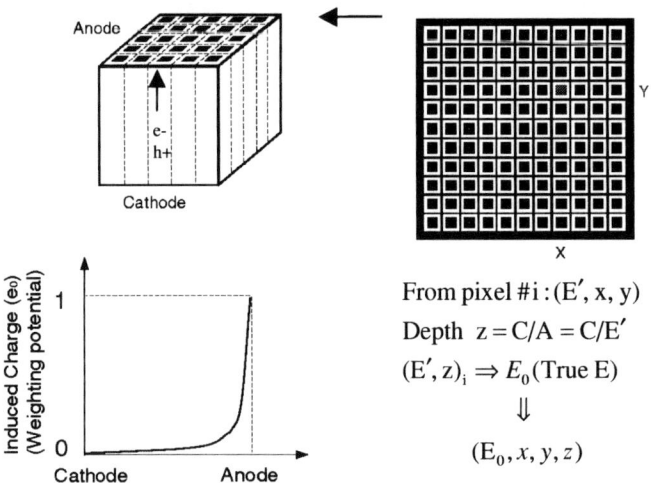

FIGURE 1. Illustration of the 3-dimensional position-sensitive technique.

FIGURE 2. A sample 3-dimensional position-sensitive prototype HgI_2 gamma-ray spectrometer.

Detector performance

Six pixel anodes, three on each 10 mm thick HgI_2 detector, were tested. Energy spectra of 662 keV gamma rays obtained at three different interaction depths from pixel anode #2 of detector #93203N92 are shown in Figure 3. The cathode was biased at –2500 V. The shaping times on the pixel anode and the cathode were 16 and 8 μs respectively. A 10 μCi ^{137}Cs source was located a few cm from the cathode surface. It can be seen that close to 1.1% FWHM energy resolution was achieved in the middle region between the cathode and the anode underneath pixel #2, and the mercury K X-ray escape peaks can be clearly identified due to the small volume of each voxel contributing to each energy spectrum.

It should also be noticed that the Compton continuums are quite flat down to about 20 keV region as predicted by Monte-Carlo simulations. This agreement between the observed and simulated spectra supports that all detection volume underneath the corresponding pixel anode is active. This is a distinctive feature compared with some of previous published results based on other techniques [6,7], in which sharp rising continuums are common towards the lower energy region of the spectra. If only a fraction of the detector volume contributes to the photopeak efficiency, and the other part records signals at lower amplitudes than the true energy deposition in the detector, a rising continuum towards the lower energy region would appear from the convolution of the incident gamma-ray spectrum with the detector response.

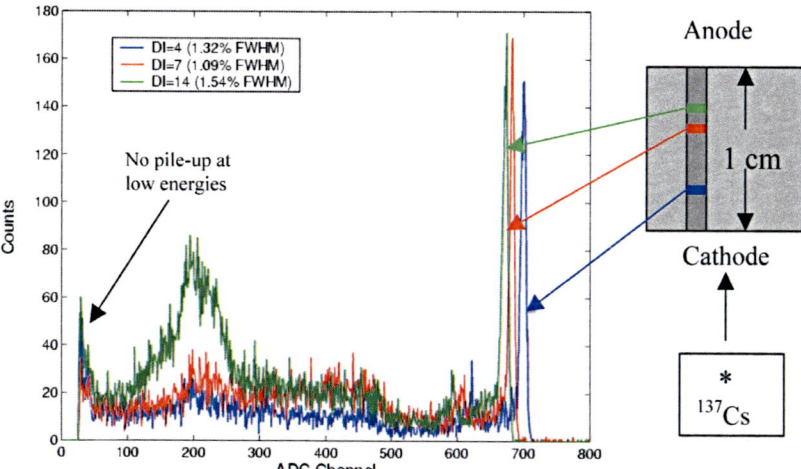

FIGURE 3. Energy spectra of ^{137}Cs at three gamma-ray interaction depths underneath pixel #2 obtained from detector #93203N92.

A digital depth correction method was employed to align the photo-peaks at different interaction depths so that the pulses corresponding to the same energy deposition appear at the same ADC channel. This technique can correct for the variation of electron trapping at various depth locations, and therefore mitigate the problems of material non-uniformity. Energy spectra from all events recorded on pixel #2 of detector #93203N91 are shown in Figure 4. The cathode bias was also –2500 V. The anode and cathode shaping times were both 8 μs in this case. If we only take

115

advantage of single polarity charge sensing technique based on the small pixel effect, an energy resolution of about 3% FWHM is obtained. After applying the depth correction, the energy resolution is improved to 1.4% FWHM. This result demonstrates that the 3-dimensional position sensing technique is superior to the simple single polarity charge sensing technique using the small pixel effect. Notice that the spectra shown in Figure 4 were obtained from a different detector than that shown in Figure 3, so that representative spectra from both tested detectors are reviewed.

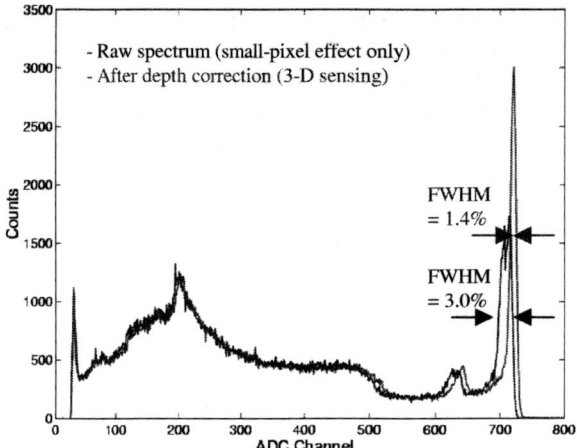

FIGURE 4. Energy spectra of ^{137}Cs from all depths underneath pixel #2 obtained from detector #93203N91. Both the raw and depth corrected spectra are shown in comparison.

Conclusions

After depth correction, 5 out of 6 tested pixel anodes on two detectors showed energy resolutions at 662 keV gamma-ray energy in the range between 1.4% to 2% FWHM. These results clearly demonstrate that 3-dimensional position sensitive single polarity charge sensing technique is promising on HgI_2 gamma-ray spectrometers with thickness of 1 cm. All tests were performed at a very low cathode bias of −2500V due to the breakdown in the circuitry at higher voltages. Efforts are underway to increase the bias voltage in our test setup so that detector performance at higher bias voltages can be studied.

ACKNOWLEDGMENTS

We want to acknowledge the support of Constellation Technology Corporation on fabricating the detectors, and Dr. Raymond P. Devito for his support on this project.

REFERENCES

1. He, Z., et al. *Nuclear Instruments and Methods in Physics Research A* 422 (1999) 173-178.
2. He, Z., et al. *Nuclear Instruments and Methods in Physics Research A* 388 (1997) 180-185.
3. He, Z., and Vigil, R.D., "Investigation of HgI$_2$ Gamma-Ray Spectrometers," to appear in: *Nuclear Instruments and Methods in Physics Research A*, 2002.
4. Baciak, J. E., He, Z and Devito, R.P., "Electron Trapping Variations in Single-Crystal Pixellated HgI$_2$ Gamma-Ray Spectrometers," to appear in: *IEEE Transactions on Nuclear Science*, 2002.
5. Constellation Technology Corporation, 7887 Bryan Dairy Road, Suite 100, Largo, Florida 33777.
6. Berg, Lodewijk V.D., "Recent Advances in the Development of Mercuric Iodide Radiation Detectors," presented during 12th International Workshop on Room-Temperature Semiconductor X and Gamma-Ray Detectors, Nov.4-10, San Diego, CA 2001.
7. Patt, B.E., et al., *Nuclear Instruments and Methods in Physics Research A* 380 (1996) 276-281.

A Germanium Orthogonal Strip Detector System for Gamma-Ray Imaging

Ethan L. Hull [*a], Morgan T. Burks [a], Chris P. Cork [a], William Craig [b], Del Eckels [b], Lorenzo Fabris [a], Anthony Lavietes [b], Paul N. Luke [a], Norman W. Madden [a], Richard H. Pehl [a], Klaus P. Ziock [**b]

a Lawrence Berkeley National Laboratory, b Lawrence Livermore National Laboratory

Abstract. A coded aperture, germanium-detector based gamma-ray imaging system has been designed, fabricated, and tested. The detector, cryostat, and signal processing electronics are discussed in this paper. The latest version orthogonal strip planar detector is 11-millimeters thick, having 38x38 strips of 2-millimeter pitch. The planar detector was fabricated using amorphous germanium contacts. The strips on each face of the detector lie in a chorded-circular pattern to more efficiently utilize the area of the 10-cm diameter germanium crystal. The detector is held in a mount that allows convenient installation and removal of the detector, lending itself to eventual tiling of such detectors into large arrays. The cryostat includes provisions to install a large volume coaxial germanium detector immediately behind the planar detector in the same cryostat. Many gamma rays Compton scatter from the planar detector into the coaxial detector. The energies of these coincident interactions are summed to increase the gamma-ray detection efficiency for higher energy gamma rays (> 200 keV). This hybrid detector configuration recovers many of the gamma rays that would otherwise scatter out of the planar detector. Each strip is read out with a compact, low noise, external FET preamplifier specially designed for this instrument. A bank of shaping amplifiers, fast amplifiers, and constant-fraction discriminators provide readout information that includes 3-dimensional position information on multiple simultaneous photon interaction positions in the detector at the 2 mm level. The energy deposition at each location is also read out with germanium spectroscopy grade resolution. This information is sent to a computer where the image is formed. The excellent energy resolution of the germanium detector system provides isotopic imaging.

INTRODUCTION

A planar detector with orthogonal strips on opposite sides of the detector plane can serve as a position sensitive detector. These detectors utilize a timing and energy coincidence between strips on opposite sides of the detector to determine the x-y position of the gamma-ray interaction on the detector plane. The holding, cooling, fabrication, signal extraction, preamplifiers, pulse-processing electronics, and readout electronics make these detector systems complicated and expensive. Recent years have seen considerable evolution in the miniaturization of preamplifiers [1] and pulse

* elhull@lbl.gov; phone 1 510 495 2312; fax 1 510 486 7557 Lawrence Berkeley National Laboratory; 1 Cyclotron Road, Berkeley, CA 94720; **ziock1@llnl.gov; phone 1 925 423 4082; fax 1 925 423 5998; Lawrence Livermore National Laboratory; P.O. Box 808, Livermore, CA 94550

CP632, *Unattended Radiation Sensor Systems for Remote Applications,* edited by J. I. Trombka et al.
© 2002 American Institute of Physics 0-7354-0087-3/02/$19.00

processing electronics [2, 3]. Furthermore, the segmentation of germanium detectors with amorphous contact technology enables the accurate positioning of fine detector contact structures [4, 5]. These advances make such detectors more viable today.

A uniformly redundant array (URA) coded aperture imaging technique has been successfully demonstrated for isotopic imaging with germanium-quality spectroscopy resolution. Imaging sources with a single pinhole collimator in front of a position sensitive detector provides a simple means of forming a direct image. A properly chosen URA generates a shadow pattern on the face of a detector that can be reconstructed by a computer program to give an indirect image. Such an image can achieve a significantly higher signal-to-noise ratio in a shorter time interval than the simple pinhole image. A URA is ideal for imaging point sources at a great distance, such as astronomical sources. Recently, it has been demonstrated that URA imaging is also viable in the modest-near field, ~ 40 cm to ~2 meters [6-8].

Gamma-ray detection efficiency is of the utmost importance for any planar imaging technique, whether it be a URA, pinhole collimator, or multiple parallel-hole collimator system. In germanium, the photoelectric cross section is rather small for gamma rays of a few hundred keV. The vast majority of such gamma rays interacting in a pixel of the detector will Compton scatter out of that pixel and, in many cases, completely out of the detector. This incomplete energy deposition is not useful for imaging specific isotopes based on spectral peaks. To enhance the efficiency of the segmented planar detector, a large volume germanium coaxial detector can be placed immediately behind the segmented planar detector. Many of the incident gamma rays Compton scatter at forward angles from the planar detector into the coaxial detector, depositing their remaining energy in the coaxial detector. Since both detectors have germanium-quality energy resolution, the sum of energy signals from these detectors also has excellent energy resolution. Simulations reported earlier and more recent measurements indicate that the addition of the coaxial detector increases the gamma-ray efficiency of the planar detector by a factor of four at 356-keV [9].

INSTRUMENT DEVELOPMENT

The Lawrence Berkeley Laboratory Measurement Sciences Group, in collaboration with Lawrence Livermore National Laboratory, has developed many fundamental enabling technologies making fully instrumented germanium orthogonal strip detector systems for use with URA imaging techniques more viable. A discussion of the germanium detector, cryostat, and electronics development follows. It must be emphasized that all three of these components are important to produce a working imaging system. By carefully developing each component of the system and integrating these components, from the detector to the formation of images, we have incrementally advanced the appropriate technologies. We have now successfully developed two versions of germanium orthogonal strip detector systems.

In simplest form, both systems consist of a germanium orthogonal strip planar detector, a commercial coaxial germanium detector, electronics to read out the detectors and a collimator in front of the planar detector. A basic outline of the instrument is shown in the block diagram in figure 1. Photons are incident from the source through a collimator, here a pinhole collimator. Many of the interacting photons will scatter from a pixel in the orthogonal strip detector into the much larger volume coaxial detector. The signals from both the planar and coaxial detectors are amplified and shaped by the pulse processing electronics. Fast coincidence and spectroscopy signals are sent to the digital read out electronics where the data are converted into digital words and sent to a computer where an actual image is formed.

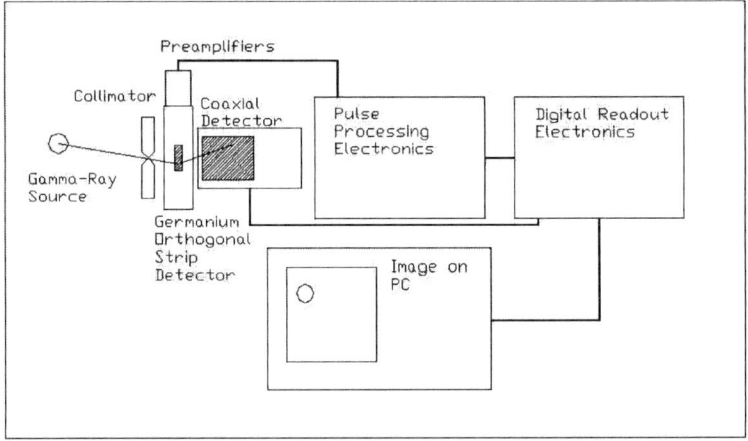

FIGURE 1 A schematic of the detector systems.

Orthogonal strip detectors and cryostats

The first and second-version detectors were fabricated from 7 cm and 10 cm diameter germanium wafers 1.1 cm thick, respectively. The detector crystals have net electrically active impurity concentrations of approximately $2x10^9$ /cm^3 n-type and $4x10^9$ /cm^3 p-type, sufficiently low to provide uniform electric field distribution and full depletion without excessively high bias voltage. The germanium wafers were ground into "hat" type structures. The brim of the hat structure allowed for handling during fabrication and holding in the cold plate. The detectors were fabricated by sputtering ~1000 Å of amorphous germanium on each side. Then shadow-mask aluminum evaporations deposited 19 strips on each side of the first-version detector and 38 strips on each side of the second-version detector. The strips on one side are orthogonal to those on the other side. The pitch of the strips is 2 mm with 0.5 mm gaps between the strips. A guard ring surrounds the strips on each side. Amorphous germanium contact technology lends itself to these rather complicated contact structures with simple shadow-mask evaporation. The bias voltages used with these detectors have been chosen very conservatively low to avoid any possible breakdown problems. The first-version detector is operated at –400 V, depleting at –200 V. The second-version detector is operated as high as –700 V, depleting at –400 V.

The germanium orthogonal strip detectors are held in aluminum cryostats that maintain a vacuum environment so the detector can be cooled to near liquid nitrogen temperatures; a photograph of the first-version cryostat with the detector in place is shown in figure 2. A copper dipstick cold finger, immersed in a separate dewar of liquid nitrogen cools the detector. The amorphous germanium contacts provide a barrier height of approximately half the germanium band gap rather than the entire band gap as with more conventional n+ and p+ contacts usually formed by lithium thermal diffusion and boron ion implantation. The lower barrier height requires that the detector be operated at temperatures fairly close to liquid nitrogen temperature to maintain low leakage current. The equilibrium temperature of the detector in the cryostat is approximately 85 K. The detector is held in an aluminum cold plate clamped to the end of the cold finger. The cold plate must hold the detector firmly enough to prevent it from moving but not so tightly that the crystal is fractured by mechanical stress due to thermal contraction when the system is cooled. The detector is electrically isolated from the cold plate by thin sheets of beryllium oxide.

FIGURE 2 The first (left) and second (right) version germanium orthogonal strip detectors. The second-version detector has a chorded-circular pattern of strips making more effective use the area of the germanium wafer.

The second-version detector, shown on the right side of figure 2, was held in a somewhat different looking cryostat with all the same basic necessities satisfied. The temperature of this system is approximately 82 K. The detector is held in an aluminum cold plate that bolts to an aluminum cold finger welded directly through the vacuum wall into an upright 7-liter dewar. This makes the overall size of the instrument smaller with the slight disadvantage of more frequent liquid nitrogen filling. The second-version detector has twice the number of strips, requiring tighter vacuum feedthrough and preamplifier packing. The strips on the faces of the detector are arranged in a chorded-circular pattern that allows much more efficient use of a circular wafer of germanium than the square array in the first-version. Consequently, the strips are not all the same length. The larger crystal diameter and the chorded-circular pattern of strips make the active area of the second-version detector a factor of 3.5 larger than the first-version detector.

In both instruments, electrical contact is made to the strips on the detector with small spring loaded "pogo" pins soldered into a ceramic composite circuit board. On the DC coupled, low voltage, side of the detector, stainless steel wires carry the charge generated by gamma-ray interactions in the detector from contacts on the ceramic board to vacuum electrical feedthroughs on the sidewall of the cryostat. On the AC coupled, high voltage, side of the detector, the signal leads go from contacts on the ceramic board to a bank of AC coupling capacitors inside the cryostat. From the low voltage side of the coupling capacitors, stainless steel wires carry the signals to the vacuum feedthroughs on a sidewall of the cryostat, allowing the signal feedthroughs to be low voltage parts rather than larger high voltage feedthroughs. The signal leads from each side of the detector extend in the direction parallel to the long side of the strips. The signal leads only extend in two directions leaving the other two sides free from mechanical encumbrances. This design lends itself to tiling these detectors into groups of four, possibly in a single cryostat. In the second-version system, the cold plate/detector assembly is intentionally made as modular as possible with custom made socket boards on the sides that mate to a custom plug wired to the vacuum feedthroughs. This makes connecting the detector strips to the feedthroughs and preamplifiers a simple operation.

Electronics

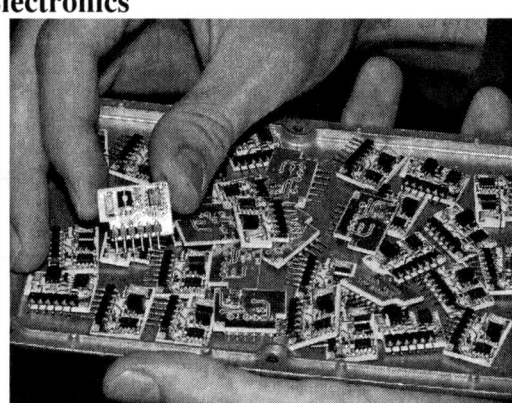

FIGURE 3 The first (left) and second (right) version preamplifiers (right) show the evolution of the miniaturized front-end electronics. The 11-mm packing density of the first-version preamplifiers was cut down to 6 mm in the second-version.

Miniature, high gain band-width product, low noise preamplifiers are housed on the outside of the cryostat just over the vacuum feedthroughs. Two different layouts of the preamplifiers used here are shown in figure 3. The first-version preamplifiers are laid out on boards aligned horizontally side by side. The second-version preamplifiers are laid out on boards that stand vertically side by side, allowing greater packing density. These charge sensitive preamplifiers utilize a low noise room temperature N-channel JFET as the first amplification element. Charge restoration is accomplished by resistive feedback. In both systems, the size of the feedback resistor is 10 GΩ and 1 GΩ on the DC and AC coupled preamplifiers, respectively. The lower value resistor, on the AC side, was selected to better accommodate the transient caused by either incrementing or decrementing the bias. The value of the charge integrating feedback capacitor is 0.5 pF for both AC and DC coupled versions of the preamplifiers.

The first-version pulse processing system reads the x-y position and energy of a single gamma-ray interaction in the detector. The energy signal is read from the DC side of the detector. This is done with a bank of shaping amplifiers, delay-line clipped fast amplifiers, and fast leading-edge discriminators. The top of figure 4 shows both versions of the detector and the pulse processing electronics. Signals from the fast discriminators, only one X and one Y for each event, and the analog energy signal from a single DC channel are sent to the digital read-out electronics. Here the analog signals are converted to digital words and sent to a personal computer to form the image and perform analysis. The first-version of the readout electronics does not accommodate multiple Compton events within the orthogonal strip detector.

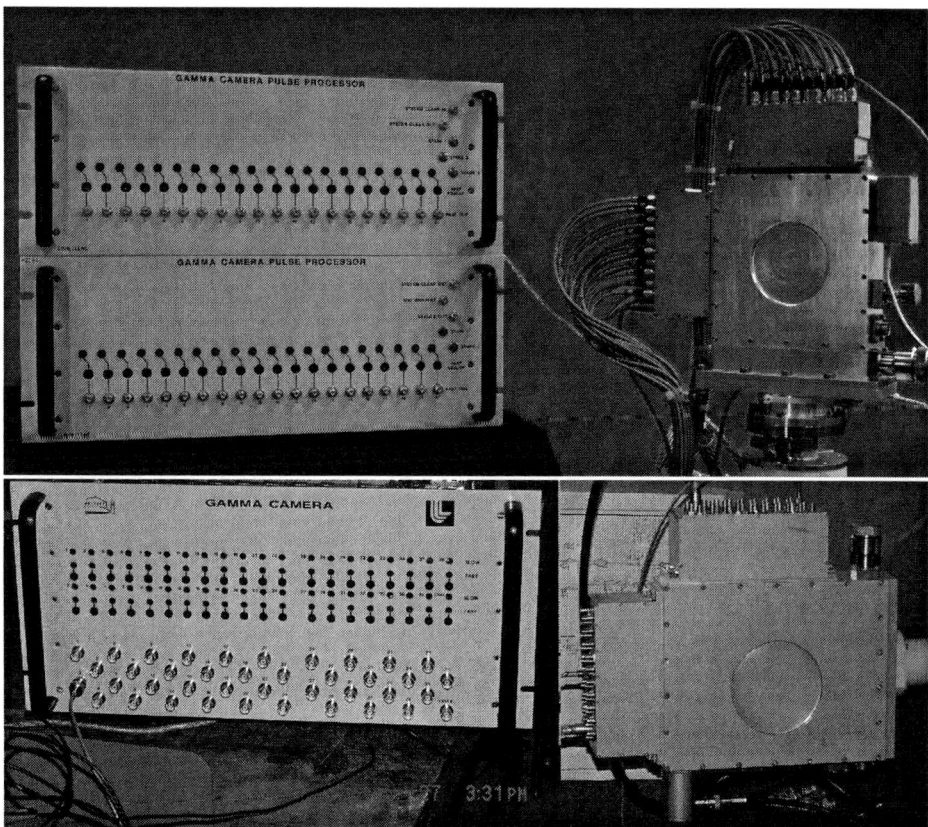

FIGURE 4 The first-version (19x19) system (top) reads the x-y position of a single gamma-ray interaction and energy. The second-version (38x38) system (bottom) reads out x-y-z positions of multiple gamma-ray interactions and their energies.

Measurements made on the first-version system had a significant impact on the design of the pulse-processing electronics for the second-version system. The charge collection physics in the strip detector presents some interesting effects. When charge is collected on a given strip, say channel 10, there is a transient signal induced on channels 9 and 11. In fact, there is a transient signal of lesser pulse height induced on all the other channels as well. These transient signals, from the preamplifiers, rise just as the signal on the collecting electrode but then go to zero as the charge carriers are actually collected by the collecting electrode. The transient signals from the preamplifiers on channels 9 and 11 can be as high as 25 percent of the full pulse height of the charge signal collected on channel 10. This presents a problem for the use of a fast leading-edge discriminator. Charge signals on channel 10 can be accompanied by transient signals on channels 9 and 11 of sufficient height to trigger their respective discriminators. This causes many of the single-site interactions to be misinterpreted as multiple-site events. This problem is addressed in the second-version electronics by reading the energy from every channel when any x and y channel are triggered. The transient signals are reduced to a negligible pulse height by the long integration time used in the spectroscopy channel, allowing discrimination between transient and real charge signals.

The second-version electronics includes a circuit to determine the relative arrival time of the electrons and holes on opposite sides of the detector. This information provides the z-position or depth of gamma-ray interactions in the detector which is important for use in both coded aperture and Compton imaging. A delay-line technique is used to determine the time of the half-height of the charge signals independent of their total signal height. The difference in these times for signals from opposite sides of the detector is recorded. This time difference is directly proportional to the z-position of the gamma-ray interaction from an electrode.

As well as accommodating transient recognition and z-position information, the second-version readout system handles 38 x 38 channels and multiple coincident gamma-ray interactions in the detector.

PERFORMANCE

Spectroscopy

Measurements were performed to characterize the properties of the germanium orthogonal strip detectors. Figure 5 shows the noise and spectroscopy performance of representative strips on the second-version detector. The capacitance of these strips is about 25 pF, dominated by the capacitance to the nearest neighbor strips. Given the capacitance and the noise figures of the FETs, the noise plots match the predicted series noise at short peaking times. The different value feedback resistors account for the degradation in the noise of the AC coupled channels over that of the DC coupled channels. A ^{133}Ba spectrum shows this detector is performing well, preserving germanium quality energy resolution.

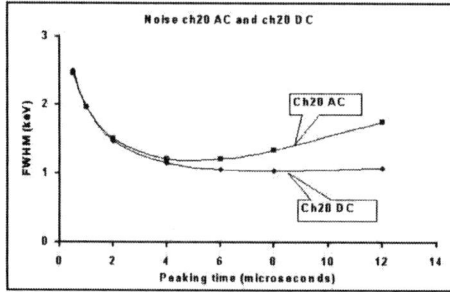
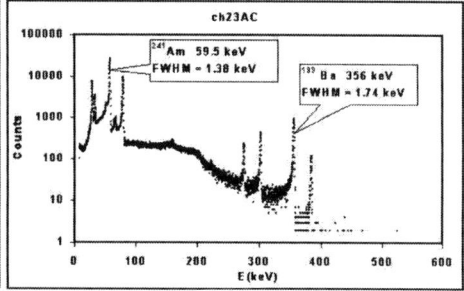

FIGURE 5 The noise (left) and spectroscopy (right) performance of strips from the second-version detector.

Gamma-Ray Imaging

The position sensitive performance of the first-version planar detector system is demonstrated by a shadow image of a key shown figure 6. The image was made with an ^{241}Am source approximately 30 cm over the DC side of the detector while a brass key was held against the entrance window of the cryostat. The outline of the key shows clearly in the image.

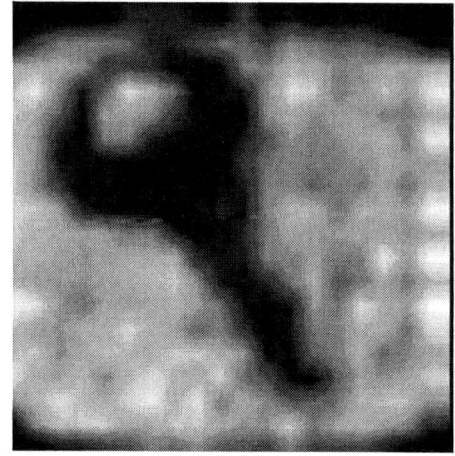

FIGURE 6 The smoothed shadow image of a key reflects the position sensitivity of the germanium orthogonal strip detector.

Pinhole images have also been collected with the first-version system. The collimator is made of a 2.5-cm thick lead sheet with a 2-mm diameter pinhole having 45-degree conical counter bores 1.15 cm deep on either side of the hole to provide a reasonable field of view. A simple image collected with this collimator between the detector and several sources in the field of view is shown in figure 7. The image was acquired in one hour and includes the following sources (from left to right) at the different distances from the DC side of the detector: 52 µCi ^{57}Co at 44 cm, 51 µCi ^{241}Am at 36 cm, and 14 µCi ^{133}Ba at 27 cm. The source distances were adjusted to account for the differing source strengths. The distance between the ^{133}Ba and the ^{241}Am is 6.4 cm. The distance between the ^{241}Am and the ^{57}Co is 10 cm. Each source is 6-8 mm in diameter. The top left image in figure 7 shows all counts regardless of energy. A cut through the center spot shows the profile or counts per pixel along the two-cursor axis. Clearly there is a great deal of background in the image, for this middle spot the peak pixel is only about 4 times the background. However, if we cut on the 59.5 keV gamma-ray peak from ^{241}Am in the energy spectrum from each pixel we see a clear spot that is about 100 times the background as shown in the bottom left image. The top and bottom right images are cuts on the 122-keV gamma-ray peak from ^{57}Co and the 356-keV gamma-ray peak from ^{133}Ba. With these two images the contrast is again very high. The energy and position resolution of the germanium detector provide unambiguous identification and separation of isotopes in the image. It is important to note that the sources were carefully arranged to show up in the top left image. If the ^{241}Am source were twice as far away from the detector it would have been indistinguishable from the background from the other sources. However, an energy cut on the 59.5-keV peak would bring its image out of the background with excellent contrast. The image was made with the orthogonal strip detector alone; the coaxial detector was not used here. Although simple pinhole images are rather inefficient, this example shows that the system functions properly. Much better signal-to-noise images from the URA in front of the detector can be found in [8].

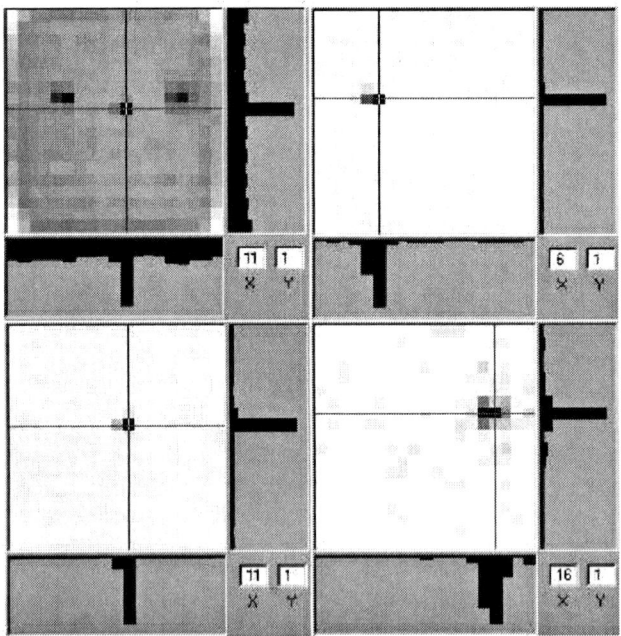

FIGURE 7 Images of three gamma-ray sources, from left to right ^{57}Co, ^{241}Am, and ^{133}Ba sources. The sources individually stand out quite dramatically when energy restrictions are placed on the gamma-ray energies included in the image.

The addition of the coaxial detector to the system enhances the efficiency of the higher energy photons greatly. At low energies, like 59.5 keV, the coaxial detector adds nothing to the efficiency of the system, as one would expect because there is very little Compton scattering in germanium at these energies and even if the gamma ray scatters, the residual gamma ray is unlikely to reach the coaxial detector. However, the efficiency improvement for the higher energy gamma rays is substantial. The addition of the coaxial detector increases the efficiency of the 356 keV peak by a factor of 4 over the planar detector alone [8].

FUTURE IMPROVEMENTS

There are several improvements planned to increase the efficiency and improve the resolution of these detector systems. With more efficient use of the germanium crystal, the active area of the planar detector can be increases by another factor of 1.4 over the area of the second-version detector. Mechanical cooling will be implemented to eliminate the need for bulky liquid nitrogen dewars. The planar and coaxial detectors will be placed inside the same vacuum cryostat, allowing them to be much closer. This should increase the efficiency enhancement from the coaxial detector from a factor of 4 to 5.3 at 356-keV.

The next version preamplifiers will be much lower noise, our current noise figure of ~1.3 keV, at $T_P = 4$ μs, will be reduced to 0.7 keV. This improves the spectroscopy performance of the detector. Furthermore, the lower noise allows a lower discriminator threshold in the fast electronics providing better z-position resolution for lower energy depositions. These small energy depositions correspond to small angle Compton scattered gamma rays in the planar, the most likely to scatter into the coaxial detector. The next version readout electronics will be made much more compact. Work will be done to properly recover complicated multiple Compton scattering events in the detectors.

ACKNOWLEDGMENTS

Special thanks are extended to Thomas Raudorf of Ortec for his assistance in fabrication of the second-version germanium orthogonal strip detector. Portions of this work were supported by the director, the Office of Nonproliferation and National Security, and the Office of Nonproliferation and Verification Research and Development of the U.S. Department of Energy under contract number DE-AC-03-76SF00098. Portions of this work were performed under the auspices of the U.S. Department of Energy by Lawrence Livermore National Laboratory under contract W-7405-ENG-48.

REFERENCES

1. L. Fabris, N. W. Madden, H. Yaver, Nuclear Instruments and Methods in Physics Research **A 424** 545 (1999).

2. E. L. Hull, M. T. Burks, C. P. Cork, W. Craig, D. Eckels, L. Fabris, A. Lavietes, P. N. Luke, N. W. Madden, R. H. Pehl, K. P. Ziock, "A germanium orthogonal strip detector system for gamma-ray imaging," To be published in Proceedings of the 46th Annual International Symposium on Optical Science and Technology, July 29-August 3, 2001.

3. M. T. Burks, M. Amman, S. E. Boggs, E. L. Hull, P. N. Luke, N. W. Madden, V. J. Riot, K. P. Ziock, "A Germanium Gamma Ray Imager with 3-D Position Sensitivity," Proceedings of IEEE NSS, November 2001, San Diego, CA.

4. P. N. Luke, C. P. Cork, N. W. Madden, C. S. Rossington, M. F. Wesela. IEEE Trans. Nucl. Sci. **39** No.4, 590 (1992).

5. M. Amman and P. N. Luke, "Position-Sensitive Germanium Detectors for Gamma-Ray Imaging and Spectroscopy," Proc. SPIE **4141**, 144 (2000).

6. K.P. Ziock, L. Nakae, "A Large-area PSPMT Based Gamma-ray Imager with Edge Reclamation," Proc. Of IEEE NSS, Nov. 2000, Lyon, France and to appear in IEEE Trans. Nucl. Sci,. June 2002.

7. R. Accorsi, F. Gasparini, R. C. Lanza, IEEE Trans. Nucl. Sci., **48** No. 6, 2411 (2001).

8. K. P. Ziock, N. Madden, E. Hull, W. Craig, T. Lavietes, C. Cork, "A Germanium-Based, Coded Aperture Imager," Proceedings of IEEE NSS, November 2001, San Diego, CA. and to appear in IEEE Trans. Nucl. Sci., 2002.

9. K.P. Ziock, et al, *A Germanium Based Coded Aperture Imager*, Proceedings of the 41st Annual INMM Meeting, New Orleans, LA, July 2000.

Employing thin HPGe detectors for gamma-ray imaging

K. Vetter[a], L. Mihailescu[a], K. Ziock[a],
M. Burks[b], C. Cork[b], L. Fabris[b], E. Hull[b], N. Madden[b], R. Pehl[b]

[a]Lawrence Livermore National Laboratory, 7000 East Avenue, Livermore, CA 94550, USA
[b]Lawrence Berkeley National Laboratory, 1 Cyclotron Road, Berkeley, CA 94720, USA

Abstract.

We have evaluated a collimator-less gamma-ray imaging system, which is based on thin layers of double-sided strip HPGe detectors. The positions of individual gamma-ray interactions will be deduced by the strip addresses and the Ge layers which fired. Therefore, high bandwidth pulse processing is not required as in thick Ge detectors. While the drawback of such a device is the increased number of electronics channels to be read out and processed, there are several advantages, which are particularly important for remote applications: the operational voltage can be greatly reduced to fully deplete the detector and no high bandwidth signal processing electronics is required to determine positions. Only a charge sensitive preamplifier, a slow pulse shaping amplifier, and a fast discriminator are required on a per channel basis in order to determine photon energy and interaction position in three dimensions. Therefore, the power consumption and circuit board real estate can be minimized. More importantly, since the high bandwidth signal shapes are not used to determine the depth position, lower energy signals can be processed. The processing of these lower energy signals increases the efficiency for the recovery of small angle scattering. Currently, we are studying systems consisting of up to ten 2mm thick Ge layers with 2mm pitch size. The required electronics of the few hundred channels can be integrated to reduce space and power. We envision applications in nuclear non-proliferation and gamma-ray astronomy where ease of operation and low power consumption, and reliability, are crucial.

INTRODUCTION

The ability to image and characterize known as well as unknown gamma-ray sources finds a variety of applications in national security, such as nuclear nonproliferation and stockpile stewardship programs, in nuclear waste control and monitoring, in nuclear medicine for cancer diagnosis and therapy as well as in astrophysics. Besides these applications, which are based on the natural gamma-ray emission of radioactive material of interest, gamma-ray imaging can be employed in alternate modes, such as particle induced gamma-ray emission, scattering or transmission modes.

Recent advances in the manufacture of two-dimensional segmented semi-conductor detectors enable to determine energies and three-dimensional positions of individual

CP632, *Unattended Radiation Sensor Systems for Remote Applications*, edited by J. I. Trombka et al.
© 2002 American Institute of Physics 0-7354-0087-3/02/$19.00

gamma-ray interactions in the detector [1-4]. Using tracking algorithms, which are based in the underlying interaction processes such as the Compton scattering and the photo-electrical effect, it is possible to determine the time sequence of the interactions [5]. Knowing the energies and positions of the first interactions allows to determine the incident angle of the incoming gamma ray without the use of a collimator which will significantly increase the efficiency a gamma-ray imaging system. However, to realize the full potential of gamma-ray tracking and Compton-tracking based gamma-ray imaging, particularly for low gamma-ray energies, the excellent position and energy resolution of a semi-conductor detector is needed.

In this study we are focusing on the use of thin high-purity Ge (HPGe) detectors in planar geometry, which are equipped with orthogonal strips on each side (so called double-sided strip detector (DSSD)). However, in the following we will briefly explore characteristics of the two possible geometries of HPGe detectors: the planar and the coaxial configuration.

Coaxial vs. planar HPGe detector

Generally, two-dimensionally segmented HPGe detectors can be built in two different geometries: planar and coaxial.

Coaxial detectors can be built up to diameters of 8cm and lengths of 14cm. The Boron contact at the outside can be segmented longitudinally and azimuthally as shown on the left side of figure 1. The advantage of coaxial detectors is their compactness and therefore their high efficiency in detecting gamma radiation and the large field-of-view (FOV).

Planar detectors can be built up to 20mm thick and with an area of up to 90x90mm2. They can be segmented two-dimensionally either in pixels on one side or in orthogonal strips on each side. Recent advances in detector processing enable to use reliable contacts on both sides of the crystal. To increase efficiency in particular for higher gamma-ray energies several of these planar DSSD detectors can be arranged in a stack.as shown in figure 1.

However, for imaging applications the high efficiency of large volume detectors is not the primary goal but the sensitivity to measure a weak signal out of background radiation. Only the capability to provide a spatial resolution as with an imaging device can effectively increase the signal-to-background ratio by distributing the background throughout the solid angle while maintaining the gamma-ray flux of interest locally in one point. Only increasing the efficiency will increase both, the signal and the background.

Ultimately, only an imaging capability is able to provide geometric dimensions and extensions of materials of interest.

To obtain energy and three-dimensional position information of individual interactions as a prerequisite for gamma-ray tracking in a two-dimensionally segmented coaxial detector, in coaxial HPGe detectors the pulse shapes of charge collecting as well as adjacent segments have to be analyzed which requires complex processing of the signals. However, in this way not only the radius can be determined but also the complementary longitudinal and azimuthal positions can be determined to much higher accuracy than the segment size.

The planar geometry is much simpler and allows to obtain positions and energies with higher accuracy than the segmentation size and the crystal thickness with much reduced signal processing. It is even possible to build a stack array of thin planar DSSD detectors, which don't require any complex processing to obtain energy and position of individual interactions. If the detector is thin enough and the pitch size of the strips is sufficiently small to minimize multiple scattering processes in one of these "voxels" and to provide sufficient position resolution just by the dimensions of these "voxels", then just the strip and crystal IDs and their measured energies can be used for tracking.

Features and advantages of thin DSSD HPGe detectors

In thin planar DSSD HPGe detectors no complex and high-bandwidth related pulse processing is required since only the energy has to be determined. The thickness of the planar crystals and the strip sizes on each side define a three-dimensional "voxel" which is dimensioned to provide sufficient position resolution and to reduce the number of multiple interactions in one "voxel". As indicated above, if an energy is measured in one "voxel" then the position of the interaction is known to the accuracy of this voxel just by it's ID number.

We envision crystal thicknesses of 2mm and pitch sizes of 2mm and therefore "voxel" sizes of $2mm^3$

Assuming a low count-rate environment in which neither timing nor position determination requires a bandwidth in the order of 10MHz the bandwidth can be lowered to about 100kHz to provide the necessary shaping to optimize the energy resolution for each "voxel". The costs for the processing electronics can be reduced due the reduced number of parts required and the lower costs of the parts. In addition, the electronics can be built more compact, which is very important for eventually integrating the hundreds of channels.

Since thin HPGe detectors can be depleted with a few tens instead of a few hundreds or even thousands of volts the high voltage support can be simplified. In particular, the dimensions of the high-voltage filter can be reduced as well as the Z of the material can be lowered to reduce scattering sources for gamma rays outside the detector.

Since no high-power demanding high-bandwidth analog or digital electronics is required the electronics power consumption can be minimized.

FET's and all parts of the electronics can be moved outside the cryostat and can be thermally decoupled from it to minimize necessary cooling of the Ge detector.

By using thin HPGe detectors the requirements for material quality, for example in terms of the concentration and distribution of impurities, can be reduced. Otherwise disturbing effects caused by charge trapping or by material and field inhomogeneities are minimized.

Lowering the requirements for the crystal quality lowers the costs in manufacturing these detectors since Ge material can be used which otherwise would have to be reprocessed to achieve sufficient quality for other detectors.

Since the requirements to obtain a three-dimensional position of interactions is reduced to filtering a signal out of the noise background which can be done very effectively with slow shaping amplifiers much smaller signals can be accepted for the position determination. Since small signals are associated with small energies which themselves are related with small scattering angles, smaller scattering angles can be accepted.

Finally, since the distances between the HPGe wafers can be changed, systems can be optimized for efficiency (small distances) or spatial resolution (large distances).

Figure 1 shows three possible arrangements of two-dimensionally segmented HPGe detectors and a table to compare basic properties of these systems.

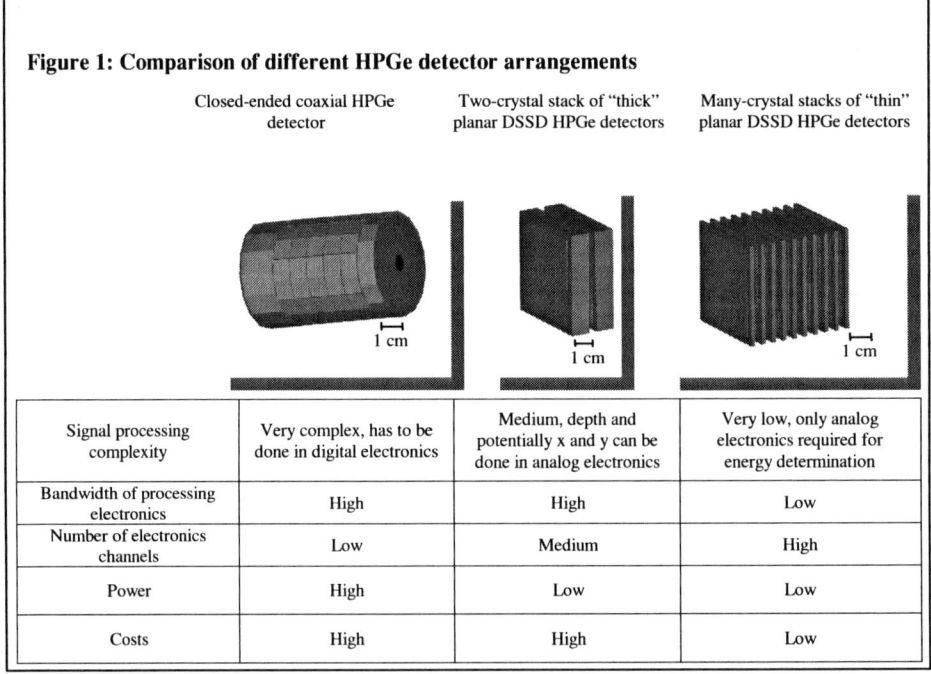

Figure 1: Comparison of different HPGe detector arrangements

	Closed-ended coaxial HPGe detector	Two-crystal stack of "thick" planar DSSD HPGe detectors	Many-crystal stacks of "thin" planar DSSD HPGe detectors
Signal processing complexity	Very complex, has to be done in digital electronics	Medium, depth and potentially x and y can be done in analog electronics	Very low, only analog electronics required for energy determination
Bandwidth of processing electronics	High	High	Low
Number of electronics channels	Low	Medium	High
Power	High	Low	Low
Costs	High	High	Low

Drawbacks of thin DSSD HPGe detectors

One drawback in using thin DSSD HPGe detectors to obtain energies and positions of individual gamma-ray interactions is the number of required electronics channels. Instead of using in the order of 40 as in a coaxial system or 100 in a mixed planar DSSD system, here we need to equip potentially more than 1000 channels with shaping amplifiers and digitizers. However, since the requirements for processing are low, integrated electronics including amplifiers and digitizers can be built and employed to minimize space requirements. In addition amplifier and digitizer systems can be envisioned in which each channel only requires 20-30mW per channel resulting in about 30W for a system of 100 channels.

Using thin detectors increases the probability that electrons, which are scattered out of the atom from the interacting gamma ray are escaping the Ge detector or at least the sensitive part of the detector without leaving the full energy. These events reduce the efficiency in detecting the full energy of the incident gamma ray and cause an increased background in the gamma-ray spectrum.

Figure 2 shows calculated losses in efficiencies for different thicknesses and numbers of planar HPGe detectors for three different gamma-ray energies: 186keV (^{235}U), 375keV (^{239}Pu) and 646keV (^{239}Pu). For example, in our proposed design of 10 layers of 2mm thick HPGe detectors we expect to loose about 15% of full-energy events at a gamma-ray energy of 646keV.

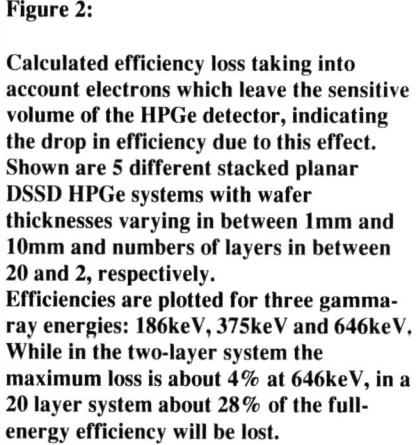

Figure 2:

Calculated efficiency loss taking into account electrons which leave the sensitive volume of the HPGe detector, indicating the drop in efficiency due to this effect. Shown are 5 different stacked planar DSSD HPGe systems with wafer thicknesses varying in between 1mm and 10mm and numbers of layers in between 20 and 2, respectively.
Efficiencies are plotted for three gamma-ray energies: 186keV, 375keV and 646keV. While in the two-layer system the maximum loss is about 4% at 646keV, in a 20 layer system about 28% of the full-energy efficiency will be lost.

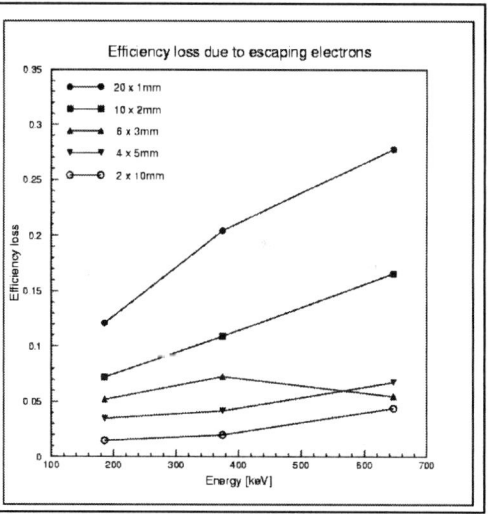

The advantage of the excellent energy resolution of Ge is associated with the requirement to cool the Ge crystal close to liquid nitrogen temperatures to reduce the thermal noise. However, recent developments in cooling technologies allow now to cool Ge detectors by electromechanical means and avoiding the many problems of liquid nitrogen, particularly for remote detection applications. The cryostat, wiring and feedthroughs have to be designed to minimize the thermal power consumption. We envision that a system can be designed and built with a heat load, which requires less than 30W electrical power.

CONCLUSIONS

We have studied an array of thin DSSD HPGe detectors to be used as gamma-ray imager. Three-dimensional positions of gamma-ray interactions are measured by the strip and crystal ID. Therefore, no complex signal processing is required to obtain positions. Only low bandwidth analog electronics is necessary to obtain energy for each "voxel" fired. Positions and energies from such an array can be used to track the interactions of the gamma ray through the detector and to determine the incident angle of the incoming gamma ray, which enables collimator-less gamma-ray imaging.

Low bandwidth electronics as well as thin planar HPGe detectors can be manufactured with very low cost and can be operated very reliably due to the reduced sensitivity to detector and electronics quality. Since only a signal, which is above the noise is required to determine the position of the scattering process, events with smaller scattering angles can be accepted.

In summary, the proposed array of thin two-dimensionally segmented HPGe detectors represents an appealing approach for remote gamma-ray imaging applications which realizes the full potential of gamma-ray tracking by minimizing costs, power and signal processing complexity.

ACKNOWLEDGMENTS

The work was performed under the auspices of the U.S. Department of Energy by University of California Lawrence Livermore National Laboratory under contract No. W-7405-Eng-48.

REFERENCES

[1] B.F Phlips, S.E. Inderhees, R.A.Kroeger, W.N. Johnson, R.L. Kinzer, J.D.Kurfess, B.L.Graham, N. Gehrels, IEEE Transactions on Nuclear Science, Vo. 43, No. 3 (1996) 1472

[2] M. Amman, P.N. Luke, Nuclear Instrumentation and Methods in Physics Research A 452 (2000) 155

[3] Y.F. Du, Z. He, G.F. Knoll, D.K. Wehe, W. Li, Nuclear Instrumentation and Methods in Physics Research A 457 (2001) 203

[4] K. Vetter, A. Kuhn, M.A. Deleplanque, I.Y. Lee, F.S. Stephens, G.J. Schmid, D.A. Beckedahl, J.E. Kammeraad, J.J. Blair, R.M. Clark, M. Cromaz, R.M. Diamond, P. Fallon, G.J. Lane, A.O. Macchiavelli, C.E. Svensson, Nuclear Instrumentation and Methods in Physics Research A 452 (2000) 223

[5] G.J. Schmid, M.A. Deleplanque, I.Y. Lee, F.S. Stephens, K. Vetter, R.M. Clark, M. Cromaz, R.M. Diamond, P. Fallon, A.O. Macchiavelli, Nuclear Instrumentation and Methods in Physics Research A 430 (1999) 69

The Analog System for the Gamma Ray Spectrometer on Mars Odyssey

Samuel R. Floyd, David A. Sheppard, James L. Odom, Andrew A. Dantzler, and Scott D. Murphy

NASA Goddard Space Flight Center Code 691, Greenbelt, MD 20771USA

Abstract. The Gamma Ray Spectrometer instrument on board the Mars Odyssey spacecraft was powered up for the second time since launch in April 2001. High voltage was applied to the Ge detector in February 2002. The system is performing well, but the detector is showing resolution degradation from solar particles and cosmic ray exposure during the long cruise. The analog system and high voltage part of the system were designed and built at NASA's Goddard Space Flight Center. This portion of the instrument system will be presented. Of the many design challenges, three stand out on this mission. The prohibited materials and elemental mass budget for everything used in the construction was defined and constrained. This type of constraint was unique and challenging. Signal overload recovery from cosmic ray and high-energy solar proton hits was a challenge because the environment could not be well defined and varies throughout the Sun cycle. The design effort to prevent microphonic problems in the very sensitive front-end electronics was a challenge because the building of the instrument preceded the spacecraft. The mechanical noise environment from reaction wheels and other mechanisms on the yet to be built and tested spacecraft was based on judgement and analysis. The analog system design and performance data will be discussed.

Background

Remote sensing gamma ray spectroscopy was successfully used to determine surface chemistry on the Apollo missions to the Moon and most recently on the NEAR mission to the asteroid Eros. There are similar high expectations for our current Mars mission, Mars Odyssey.

The Mars Odyssey Gamma Ray Spectrometer (GRS) instrument build was a cooperative effort to which the Goddard Space Flight Center team provided the germanium (Ge) detector analog electronics, the analog-to-digital conversion board, the high voltage bias supply and the analog electronics controller board. The principal investigator's institution, the University of Arizona's Lunar and Planetary Laboratory (LPL), oversaw the development of the GRS and delivered the full-up tested instrument under the project management of NASA's Jet Propulsion Laboratory (JPL). The germanium detector was built by Eurisys of France. In the science mapping configuration at Mars, the detector and analog electronics sit at the end of a 6 meter boom with the remaining instrument hardware on the spacecraft main body. The Ge detector is radiatively cooled to approximately 90K.

CP632, *Unattended Radiation Sensor Systems for Remote Applications,* edited by J. I. Trombka et al.
2002 American Institute of Physics 0-7354-0087-3

In space, the detector and all surrounding materials are subjected to solar particles and extragalactic cosmic rays. When these high-energy particles interact with the detector, they produce a very high signal which will saturate the preamplifier. These signals are called overloads and must be considered in the electronics base line restoration. In the early days of the GRS design, the overload event rate was not well understood and the design adequately handled about 100 per second. Much later, very late in the design effort, a concern developed that the overload event rate could be several hundred per second and an active base line restoration circuit was quickly implemented by LPL. In flight, cruise data has since shown that the cosmic ray event rate is around 115 per second. Because Mars is a large planet and will occult part of the cosmos, this number is expected to drop to about 70 per second in the mapping orbit.

Materials Budge

Cosmic rays can also have nuclear interactions in the detector as well as in the surrounding structural materials. The radiation produced by these events is referred to as induced activity and is an unwanted background in the signal from the planet, i.e. if you are looking for iron on Mars, you don't want iron near your detector. Therefore, the GRS design had a very restrictive materials budget. Table 1 shows the elemental mass budget for the construction of the GRS instrument where m is the elemental mass (in grams) and r is the distance from the center of the detector (in centimeters).

The GRS Front End was built with non-restricted materials such as Titanium and Magnesium, and non-metallics such as Ultem™, Vespel™, Teflon™.

Element	m/r^2 Limit (g/cm^2)
Na	0.09
Al	0.62
S	0.09
Cl	0.003
K	0.0008
Fe	0.08
Ca	0.06
Cr	0.014
Mn	0.1
Ni	0.0025
Th	5.0E-7
U	5.0E-8

Table 1. The amounts for Thorium and Uranium apply to the equivalent amounts of the major gamma-ray emitting daughter activities in equilibrium with the parent nuclide.[1]

System performance

The Gamma-Ray Spectrometer front-end electronics is a single detection channel. The designed energy range of the system is 200 keV to 16 MeV gamma ray photons. The assembly consisted of a cryogenically-cooled front-end with a charge-sensitive preamplifier, a shaping amplifier, and a 14 bit analog-to-digital converter. These electronics are interfaced to an ultra-pure n-type germanium detector operated at a nominal bias of 3000V, which is provided by a custom-designed high voltage supply. During the mission, cooling of the 215 cm^3 detector to 90K is accomplished via a two stage passive radiative cooler, with the inner stage operating at 90K and the outer stage operating at approximately 170K. For testing in the laboratory, a dewar with liquid nitrogen and a custom cryostat are used. When tested in the laboratory the end-to-end full up system energy resolution measured 2.65 keV FWHM at 1.332 MeV.

While in space on cruise to Mars, solar particles and cosmic rays have damaged the Mars Odyssey single crystal Ge detector structure. This damage is manifested as a degradation in resolution. The design incorporates a heater so that the detector may be annealed in-flight at up to 120 °C to repair this radiation damage. The Mars Odyssey mission and operations activities can be seen at the following web site; http://mars.jpl.nasa.gov/odyssey

High Voltage Setting

To determine the proper high voltage bias for the Ge detector it was thoroughly characterized to measure depletion as a function of voltage. This tends to be a labor intensive effort and there is an easy way and a hard way to get the job done. The detector arrived from Eurisys with their first stage preamplifier electronics mounted. The technique used for the test was to observe the MCA channel location of the 1332 keV photopeak (Co-60) as a function of applied bias voltage. This worked very well and when the photopeak stopped moving the detector was depleted, cleared of free charge, and its capacitance was at a minimum. Further voltage increase does not make any changes in depletion or capacitance, so the photopeak remains fixed. When this method was tried on the detector with flight electronics, it did not work very well.

Another method for determining the optimum bias voltage is to plot the area under the photopeak versus bias voltage. When the area is maximized the detector is depleted. This method is a bit more time consuming because accurate areas require good statistics which take longer integration times for each voltage data step.

The detector and electronics supplied by Eurisys is illustrated in figure 1a. Its capacitor equivalent is shown in figure 1b where the capacitor with the arrow represents the detector. Eurisys takes the signal from the high voltage side of the grounded detector through a 100 pf coupling capacitor to the JFET gate. The feedback capacitor is between the coupling capacitor and the JFET. In this design the gate is virtual ground. This maybe viewed as the detector capacitor in parallel with the coupling capacitor. Therefore, charge generated on the detector capacitor is shared with the coupling capacitor in the ratio of their values. As the bias voltage increases the detector capacitor decreases, until depletion. Therefore, more charge develops on the coupling capacitor. Since it is the charge on the coupling capacitor that is sensed and amplified, it appears on the MCA display as if the overall system gain has increased and the photopeak moves to a higher channel number.

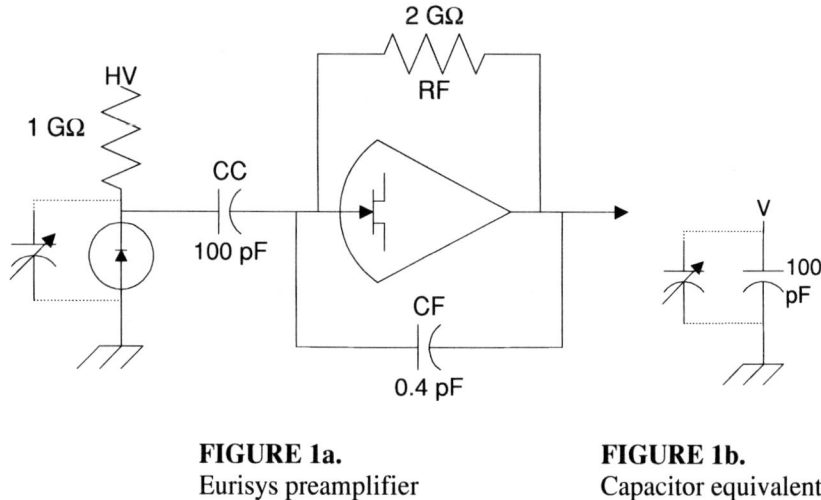

FIGURE 1a.
Eurisys preamplifier

FIGURE 1b.
Capacitor equivalent

The Goddard flight design has the detector signal going directly into a coupling capacitor, which is in series with the JFET gate, figure 2a. This design can be viewed as a detector capacitor in series with the coupling capacitor in series with the JFET gate, figure 2b. In this case, essentially all of the charge generated in the detector is sensed by the JFET gate (the very small capacitor), regardless of the bias voltage. The gain of the system hardly changes, but detector efficiency will change with bias voltage. Therefore, we used the efficiency versus bias voltage method to determine the proper high voltage bias for depletion.

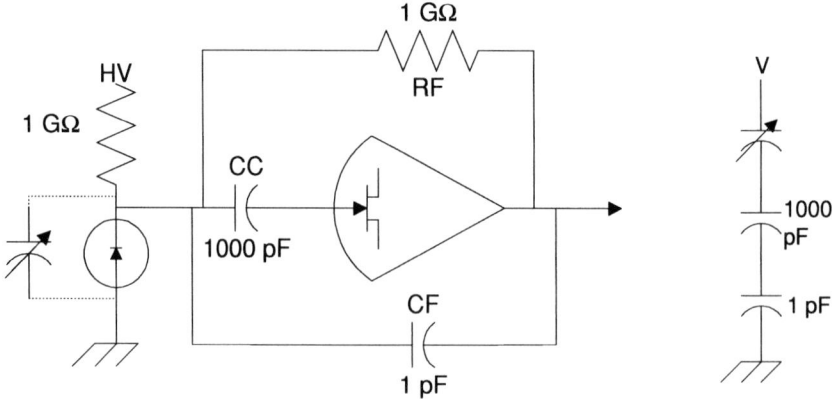

FIGURE 2a. Goddard preamplifier   **FIGURE 2b.** Capacitor equivalent

Preamplifier Design and Microphonic Suppression

The charge sensitive preamplifier consists of two separate sections. The first contains the AC-coupled front-end JFET and feedback components. The second is the Gamma Pulse Amplifier (GPA) that reads out the signal generated by the 170K Card and detector. The 170K Card assembly is mounted to the outer stage of the radiative cooler and is cooled to 170K during operation. A photo of the flight unit prior to installation in the cooler is shown in figure 3. Component selection and mechanical design was a key concern for this instrument due to the extreme sensitivity of the front end to noise contamination from electrical (leakage currents, e-field distortion, etc.) and mechanical (microphonics) sources. The mechanical design also addressed the fairly large loads generated by the amplification of Delta II's launch loads by the Gamma front end's mounting structure. As it turned out the 3 sigma design loads were almost 100 g's.

FIGURE 3. Gamma system Front End Assembly (also called the 170K Card) shown at above left. The 170K Card and Gamma Pulse Amplifier, above right.

Material selection further constrained the mechanical design so as to remain within the aforementioned materials budget. The front end's base is a gold plated (low emissivity) titanium plate which is bolted to the outer stage of the cooler. A single non-metallic (polyetherimide, Ultem 1000^{TM}) piece was machined to provide support for the polyimide printed circuit board and dielectric isolation from the ground-potential plane for the high voltage components.

The electrical components are secured by specially designed holders made from Vespel^{TM}. The holders were designed to mechanically constrain the lead soldered components and minimize any impact on the development of the high electric fields that surround the components during high voltage (~3000V) operation. The holders use essentially point contact restraints that leave a large fraction of the component's outer surface surrounded by vacuum. This minimizes any dielectric breakdown paths and electron migration paths. A solid piece of Teflon was used to secure the signal transistor input node. This served to both reduce the microphonics and minimize the electron conduction path at the point in the system that is the most sensitive. The signal transistor is lead solder mounted to the underside of the PC board so that the transistor gate lead passes through the PC board (without touching the board) and joined to the input node secured by the Teflon part.

FIGURE 4. Side view of Front End assembly. Note Nylon collar (white) on CTE joint holding Signal FET directly underneath the Signal Node (Teflon part on topside of board).

The FETS (both signal and monitor) were mechanically and thermally anchored by CTE joints. These innovative joints were designed to operate at cryogenic temperatures by using the large clamping forces generated by differential Coefficient of Thermal Expansion (CTE) between the copper joint receptacle and the steel FET canister. The CTE joints allow for assembly with no insertion force as the FET body is blind mated into the split copper CTE joint receptacle. A slip fit nylon collar surrounds the receptacle and FET can. When the system cools to ~170K the nylon shrinks on the copper receptacle generating large clamping forces between the copper receptacle and the FET can.

A Magnesium faraday shield in the center of the assembly helps to attenuate microphonics for the Anode biasing wire. The Anode biasing wire is a 1 mil diameter flying wire that travels from the Anode cross bar (shown in the center of the assembly in Figure 4) to the detector that is about 3.5" below the PC board. The Faraday shield is biased to the same voltage as the Anode wire (~3000 V) in order to minimize microphonics induced by a vibrating wire in a large gradient e-field.

All fastening hardware is made of non-metallics such as Ultem (polyetherimide) in order to reduce susceptibility to high-voltage arcing and adhere to materials budget constraints.

140

Summary

The GRS instrument has been evaluated during the long cruise and during the early part of the mapping phase of its mission. It appears that the three important design challenges were met. There is no evidence of a microphonic signal, the induced activity background is as low as practical and the cosmic ray overload event rate is manageable.

References

1. Excerpted from Mars Surveyor Program, Mars 2001 Orbiter Mission document "Functional Requirements Document" Rev v84, September 22, (1998).

Schottky Barrier CdTe(Cl) Detectors for Planetary Missions

Yosef Eisen[1] and Samuel Floyd[2]

[1] Department of Physics, The Catholic University of America, Washington DC 20064
[2] NASA/Goddard Space Flight Center, Greenbelt, MD 20771

Abstract. Schottky barrier cadmium telluride (CdTe) radiation detectors of dimensions 2mm x 2mm x 1mm and segmented monolithic 3cm x 3 cm x 1mm are under study at GSFC for future NASA planetary instruments. These instruments will perform x-ray fluorescence spectrometry of the surface and monitor the solar x-ray flux spectrum, the excitation source for the characteristic x-rays emitted from the planetary body. The Near Earth Asteroid Rendezvous (NEAR) mission is the most recent example of such a remote sensing technique. Its x-ray fluorescence detectors were gas proportional counters with a back up Si PIN solar monitor. Analysis of NEAR data has shown the necessity to develop a solar x-ray detector with efficiency extending to 30keV. Proportional counters and Si diodes have low sensitivity above 9keV. Our 2mm x 2mm x 1mm CdTe operating at -30^0C possesses an energy resolution of 250eV FWHM for ^{55}Fe with unit efficiency to up to 30keV. This is an excellent candidate for a solar monitor. Another ramification of the NEAR data is a need to develop a large area detector system, 20-30 cm^2, with cosmic ray charged particle rejection, for measuring the characteristic radiation. A 3cm x 3cm x 1mm Schottky CdTe segmented monolithic detector is under investigation for this purpose. A tiling of 2-3 such detectors will result in the desired area.

The favorable characteristics of Schottky CdTe detectors, the system design complexities when using CdTe and its adaptation to future missions will be discussed.

BACKGROUND

Remote sensing x-ray fluorescence spectroscopy from a spacecraft in orbit near a planetary body has been used to obtain elemental composition maps of the Moon[1] and the asteroid Eros[2]. This is a passive remote sensing of characteristic x-ray lines from the surface elements, where the excitation source is the Sun. High energy x-rays emitted from the Sun excites elements of the planetary surface, which produce the characteristic x-rays. To determine elemental composition, both the excitation source and the emitted characteristic x-rays must be measured. Measurements made during quiet Sun periods can detect the lower Z elements Mg (1.25 keV), Al (1.49 keV) and Si (1.74 keV). The higher Z elements, S (2.31 keV), Ca (3.69) and especially Fe (6.40 keV) and Ni (7.48 keV) can best be excited during solar flare periods. The planetary surface-observing detector must be able to resolve elements, such as Mg, Al and Si, with characteristic spectral lines less than 300 eV apart. The solar monitor must perform well at the solar flare conditions of high energy and high flux. Both detectors must operate for long periods in the radiation environment of space.

CP632, *Unattended Radiation Sensor Systems for Remote Applications*, edited by J. I. Trombka et al.
© 2002 American Institute of Physics 0-7354-0087-3/02/$19.00

The Sun is very active and its spectrum changes dramatically with solar flare activity (Sunspots). As the activity increases the flare temperatures increase and the solar X-ray flux occurs at higher energies. At higher energies the slope of the solar spectrum becomes less steep, the overall magnitude of the flux increases. Also, discrete K_α X-ray line emissions, which are very important in the excitation process, can be seen across the spectrum. A positive consequence of flares, is that the higher activity yields better fluorescence statistics at shorter integration times and compositional information on elements heavier than Si can be achieved. Knowledge of the incident solar spectrum up to 30 keV is desirable in order to deduce surface elemental concentration ratios and abundance.

THE DETECTOR

Schottky barrier cadmium telluride (CdTe) radiation detectors of dimensions 2mm x 2mm x 1mm and segmented monolithic 3cm x 3 cm x 1mm are under study at GSFC for future NASA planetary instruments. One single 2mm x 2mm x 1mm solar monitor detector is probable sufficient for the foreseeable missions. Six Schottky barrier CdTe 2mm x 2mm x 1mm detectors were procured from Acrorad Inc. and evaluated in our laboratory. The Schottky contact is made of indium. One of these detectors was subjected to a radiation damage study. The six detectors had virtually identical spectroscopic characteristics. The resolution when operated at temperatures around -30^{0}C was 260 eV FWHM for Fe-55, with a lower energy noise cutoff of 700 eV (Figure 1). The observed homogeneity of the CdTe material offers the potential for tiling many small chips to make a large area X-ray detector. However, a couple hundred chips would be necessary to make one 3cm x 3cm detector. Funding is now in place to study the larger area 3cm x 3cm segmented electrode pad monolithic detector.

To evaluate the 2mm x 2mm x 1mm solar monitor candidate, the NEAR mission environment was used to set some of the test conditions such as event rates. A solar flare is characterized by high fluxes of x-rays. If this detector had been on the NEAR mission, its event rate would have been about 10^4 interactions/sec. The detector was irradiated for 72 hours with a flux of 3×10^4 events per second using an ^{241}Am source. No degradation of performance was observed.

A "polarization" effect is observed with CdTe detectors operated at room temperature. The polarization effect may be viewed as a reduction in the effective electric field and a reduction in depletion depth. It is observed as a shift in the spectral peaks to lower energies. These detectors recover quickly with bias voltage power cycling. The polarization is due to charge trapping and detrapping. Detrapping is decreased considerably with a decrease in temperature. Trapping also decreases with increased bias voltage. These two controllable parameters will decrease the polarization. Our tests show that polarization is nearly unperceived, for observed periods up to two weeks, when the detector was operated at temperatures around -30^{0}C.

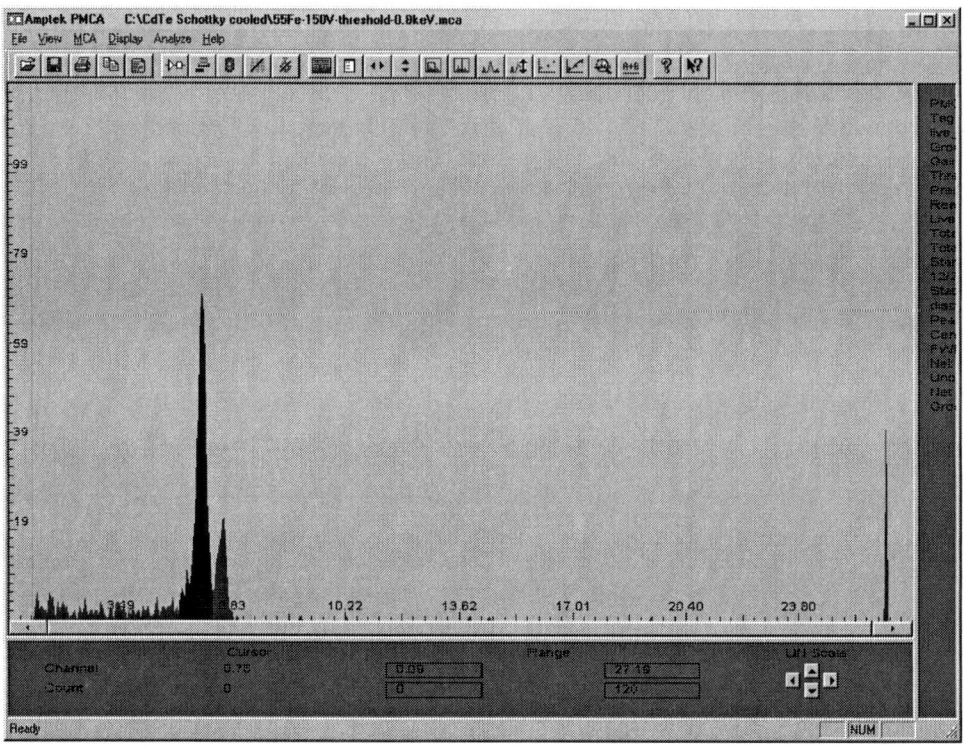

FIGURE 1. ^{59}Fe spectrum of CdTe detector prior to proton irradiation at -30^{0}C.

RADIATION DAMAGE STUDY

Our detector was irradiated with 200 MeV protons to study space radiation effects. The 200 MeV proton irradiation conditions were chosen because that is the test environment for the up coming MESSENGER mission to Mercury. Our detector was an easy add-on to their test.

Our Schottky CdTe detector of dimensions 2 mm x 2mm x 1mm was irradiated with 200MeV protons at the University of Indiana Cyclotron Facility. The rate of exposure was 10^9 p/s/cm^2. The detector was irradiated in three steps: 10^{10}p/cm^2, 5x10^{10}/cm^2 and 10^{11}p/cm^2. During irradiation, the detector was under bias of 300V and operated at room temperature. The detector was connected to an eV-5093 preamplifier and then to a 671 Ortec amplifier with a shaping time of 1µs. After each irradiation step the detector was exposed to an ^{241}Am calibration source. Since at room temperature the detector shows polarization as a function of time, prior to each measurement with ^{241}Am, the bias was turned off and then on again. Measurements were carried out at 300V and 500V. Figure 2 displays the ^{241}Am spectrum at 500V before and right after irradiation to 10^{10}p/cm^2. After an irradiation of 10^{10}p/cm^2, the 59.6keV peak shifted to 55.5keV and 56.7keV for 300V and 500V, respectively.

144

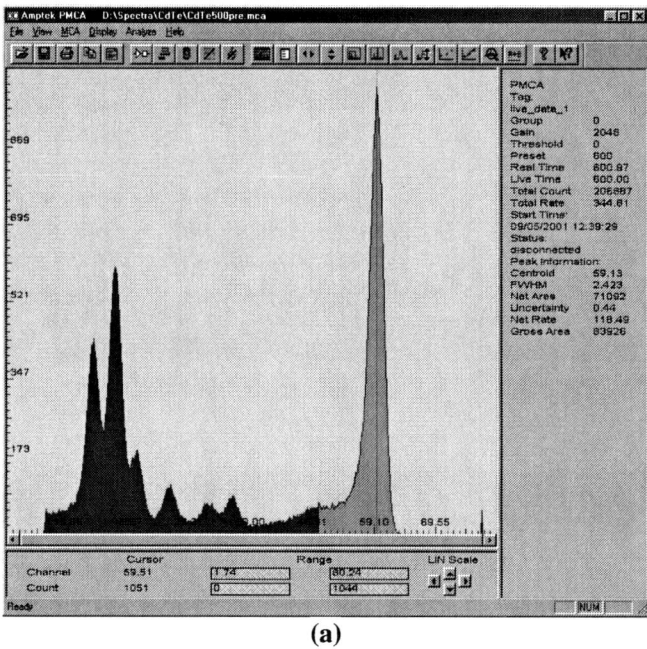

(a)

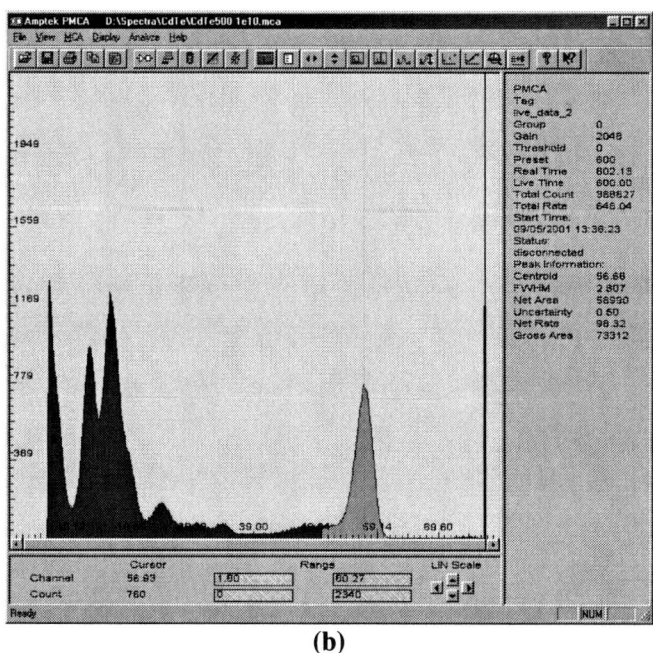

(b)

FIGURE 2. [241]Am spectrum at room temperature. a) Prior to proton irradiation. b) After proton irradiation to a dose of 10^{10}p/cm^2.

The energy resolution degraded from 2.6keV to 3.8keV FWHM at 300V and from 2.4keV to 2.8keV FWHM at 500V. At 5 x 10^{10} p/cm^2 the 59.6keV peak was degraded completely. The degradation in peak shape can also be related to an increased polarization effect during the measurement period. Radioactive background lines due to spallation reactions with elements composing the CdTe(Cl) detector were also observed in the spectrum at energies below 45keV. These lines were primarily of short half-life and did not interfere with the measurements a day after the irradiation. At 10^{11}p/cm^2 the peaks were severely broadened and no spectroscopic information could be obtained. In order to recover the damage caused by an exposure of 10^{11}p/cm^2, the detector was annealed at 65^{0}C for 40 hours. We then tested the detector at 300V, 500V and 600V both at room temperature and at –30^{0}C. The following was observed; a) at room temperature and at 500V, the 13.9keV and 17.8keV peaks are resolved. The peak at 59.6keV is still broad. At 600V the 20keV line is starting to be resolved. We also irradiated the detector with 5.9keV ^{55}Fe x-rays. As shown in Figure 3, at 600V the 5.9keV peak is well resolved from noise and has a resolution of 1.9keV FWHM. b) At –30^{0}C, the detector shows good energy resolution and peak to valley ratios for both the low and high energies of ^{241}Am as shown in Figure 4 for a voltage of 600V.

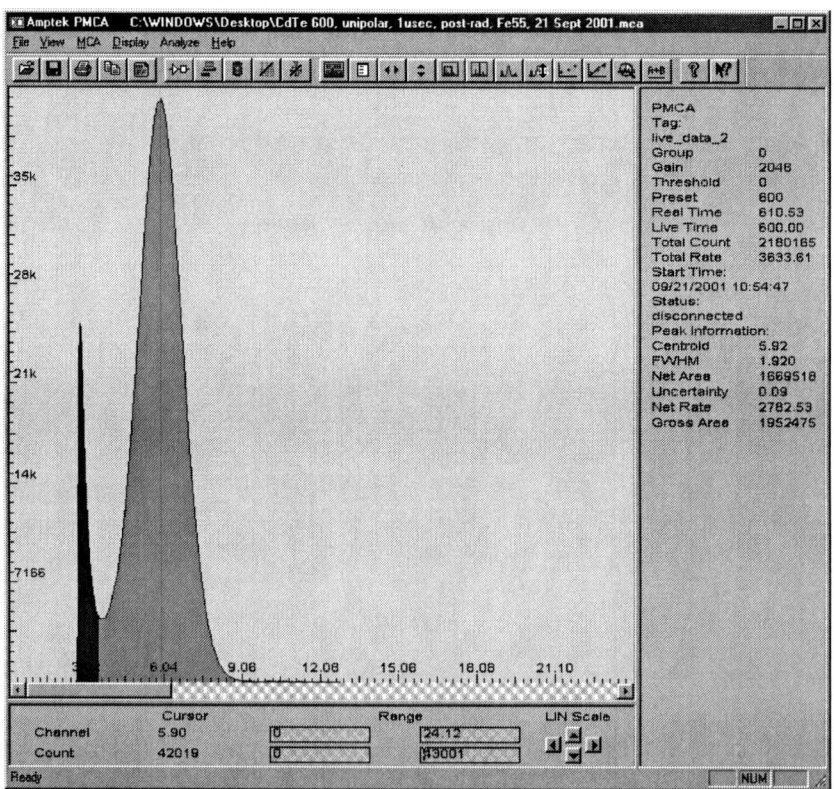

FIGURE 3. ^{55}Fe spectrum at room temperature after irradiation to 10^{11}p/cm^2 and post annealing.

The energy resolution at 59.6keV is 1.5keV FWHM. We may conclude that the annealed detector at relatively low temperature of 65^0C, partially recovered from the radiation damage caused by a fluence of $10^{11}p/cm^2$.

Prior to the irradiation to protons, the stability of the detector at -30^0C was investigated during a period of 120h. As shown in Figure 5 no polarization effects were observed and the energy resolution did not change in time. Also the peak position was independent of bias above 100V both at room temperature and at -30^0C. We further investigated the stability of the detector at -30^0C after proton irradiation to $10^{11}p/cm^2$ and post annealing. The stability measurement was carried out for more than 3 hours. As seen in Figures 6a and 6b, we observed a polarization effect at 600V during the above time period. The 59.6keV line shifted by 5% to lower energies and its energy resolution degraded from 1.5keV to 2.7keV FWHM. We have also noticed that in order to depolarize the detector, that is, bringing the detector to its initial state prior of applying the bias, the high voltage had to be shut off for 15 minutes at -30^0C. After the irradiation with protons, we also observed a change in peak position as a function of the applied bias. We investigated the change with bias at room temperature, irradiating the detector with 6.4keV photons from a [57]Co source and varying the bias between 300V to 800V. A peak position increase of about 48% was observed. Prior to proton irradiation no shift in peak position was observed when varying the voltage between 300V and 800V. Comparing the absolute peak position for 6.4keV prior to proton irradiation and after irradiation and annealing, it was found that at 600V the peak shift amounts to 20%.

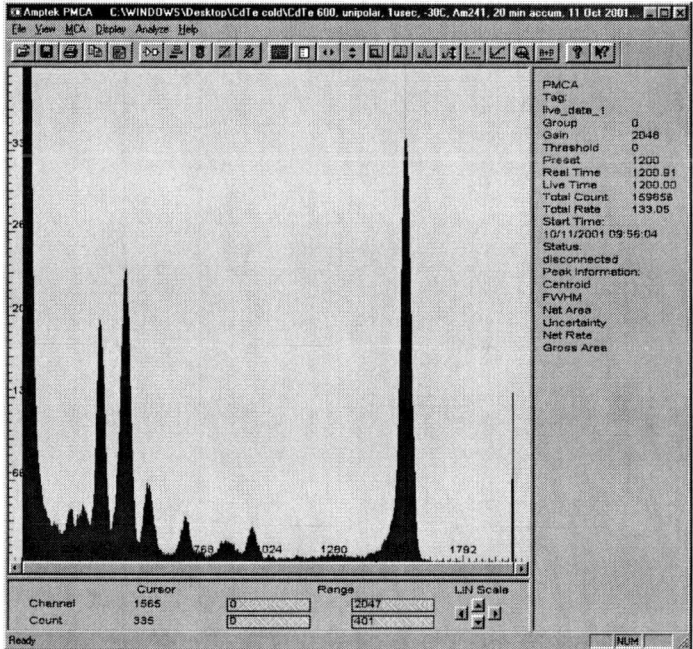

FIGURE 4. [241]Am spectrum at -30^0C after irradiation to $10^{11}p/cm^2$ and post annealing.

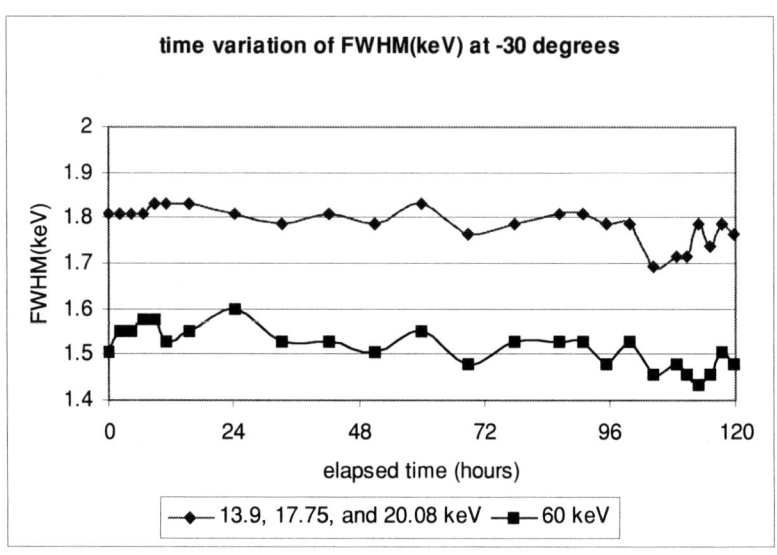

FIGURE 5. Time variation of the energy resolution (FWHM) of ^{241}Am low energy and high energy peaks prior of proton irradiation. Values displayed are at -30^0C.

We concluded that Schottky CdTe detectors are severely damaged by 200MeV proton fluence of 10^{11}/cm^2 and cannot be completely recovered by anneal at 65^{0}C for 40 hours. This should not be a surprise because a fluence of 10^{11}/cm^2 is very high and the silicon PIN detectors chosen for the MESSENGER mission did not fully recover after annealing either. MESSENGER will also carry a Ge gamma ray detector. Our experience with Ge detectors indicates that it would not survive such fluence either.

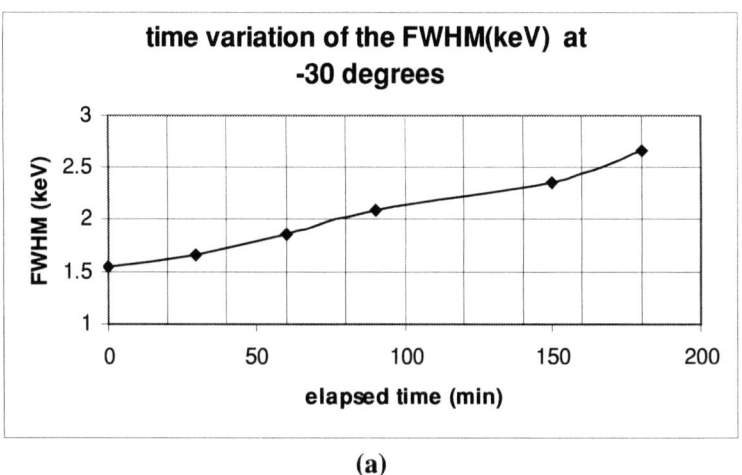

(a)

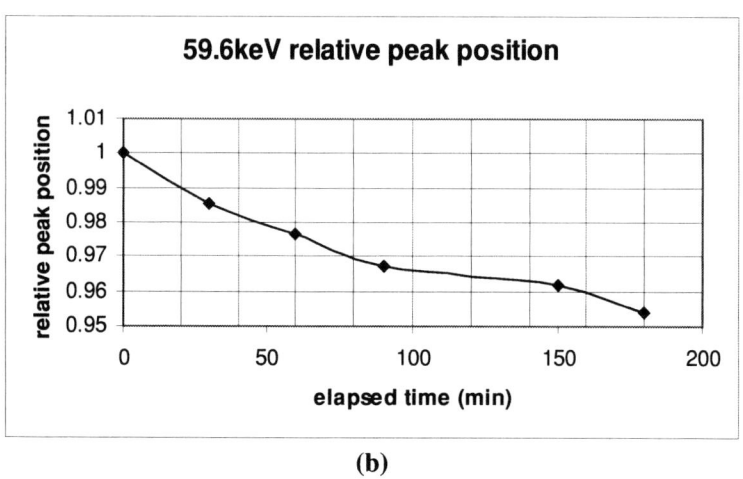

(b)

FIGURE 6. Time variation of CdTe detector at -30^0C after proton irradiation to 10^{11} p/cm^2 and post annealing. a) Energy resolution (FWHM) at 59.6 keV. b) Relative peak position.

After anneal, the Schottky CdTe detector partially recovered and showed good spectroscopic characteristics when operated at -30^0C. The high density of created traps increased the polarization rate even at -30^0C. An increase of the applied bias above 300V caused an increase in peak position and improved the charge collection efficiency and the peak to valley ratio. These are indications that the proton irradiation created a partially depleted detector.

The MESSENGER test conditions of 10^{11}p/cm^2 were worst case for that mission. Most foreseeable future missions should be at least an order of magnitude lower. Calculations based on an assumption of the deep ionized traps show that the Schottky detector irradiated by a fluence of 10^{10} p/cm^2 and followed by a thermal anneal can still function as a good x-ray spectrometer when operated at -30^0C.

Summary

The Schottky CdTe(Cl) detector based instrument system shows great potential for long duration space missions. The detector has good radiation damage tolerance. Even when the damaged can not be fully repaired with thermal annealing, it can perform well with proper system design considerations. Simple bias voltage power cycling, judiciously managed, mitigates the enhanced polarization problem caused by radiation damage. The CdTe crystal growth process has reached a development stage where approximately 4cm diameter by 5cm long single crystals are obtainable. This will enable the fabrication of large area detectors, either from tiling many small chips or using a contact segmented large monolith.

REFERENCES

1. Fichtel C. E. and Trombka J. I., *Gamma Ray Astrophysics, New Insight into the Universe, Second Edition,* NASA Reference Publication 1386, September, 1997, 367 pp.
2. Nittler L. R., et al., Meteorit. Planet. Sci. **36**, 1673-1696 (2001).

TERRESTRIAL APPLICATIONS

Subsurface Remote Sensing

Jeffrey S. Schweitzer[*] and Joel L. Groves[¶]

[*]Department of Physics, University of Connecticut, Storrs, CT 06269-3046
[¶]Schlumberger/EMR, Princeton Junction, NJ 08550

Abstract. Subsurface remote sensing measurements are widely used for oil and gas exploration, for oil and gas production monitoring, and for basic studies in the earth sciences. Radiation sensors, often including small accelerator sources, are used to obtain bulk properties of the surrounding strata as well as to provide detailed elemental analyses of the rocks and fluids in rock pores. Typically, instrument packages are lowered into a borehole at the end of a long cable, that may be as long as 10 km, and two-way data and instruction telemetry allows a single radiation instrument to operate in different modes and to send the data to a surface computer. Because these boreholes are often in remote locations throughout the world, the data are frequently transmitted by satellite to various locations around the world for almost real-time analysis and incorporation with other data. The complete system approach that permits rapid and reliable data acquisition, remote analysis and transmission to those making decisions is described.

INTRODUCTION

Subsurface remote sensing measurements are widely used for oil and gas exploration, for oil and gas production monitoring, and for basic studies in the earth sciences [1]. An enormous variety of measurements involving acoustic, electromagnetic and nuclear sensors are deployed for evaluation of the heterogeneous material surrounding the borehole. Nuclear radiation sensors, often including small accelerator-based neutron sources, are used to obtain bulk properties of the surrounding strata as well as to provide detailed elemental analyses of the rocks and fluids in rock pores [2]. These instruments operate in harsh environments where the temperature can reach 175 °C or hotter, the pressure can be 20,000 psi, and the surrounding atmosphere can be a corrosive gas containing up to 20% H_2S. Thus, these remote sensing instruments must be constructed such that they are immune to environmental hazards. The information output from a tool string in a borehole is transmitted to a surface data acquisition system via a cable or other downhole communication mode.

Today, all these borehole data acquisition systems are connected to a high-speed network that spans the globe. The most remote location can communicate with the all other locations. A petroleum engineer in New York and a geologist in Paris can review the data acquired at the well head in northwest China or offshore in Nigeria at the time of acquisition. Oil and gas wells are often located great distances from the

CP632, *Unattended Radiation Sensor Systems for Remote Applications,* edited by J. I. Trombka et al.
© 2002 American Institute of Physics 0-7354-0087-3/02/$19.00

places where experts exist to analyze the data and where the people making the decisions on future operations are located. Since waiting to make decisions on further operations can be very expensive (offshore oil platforms can cost $2000/hour), it is essential that the data be rapidly transferred and analyzed to permit decisions to be made rapidly on future operations. Thus, it is important to have a complete system in place to provide reliable instruments to obtain data, securely transmit the data to a location where experts can analyze the data, and provide equally rapid communication between the experts who perform the analysis and the people who will reach the decision on future operations.

SYSTEM APPROACH

The main features of the system design can be considered in five functions. Data are (1) acquired downhole by Smart Sensors, (2) transmitted to the surface through a digital communication line, (3) processed and stored by a Local Computer, (4) analyzed in depth at a regional center by a Regional Computer and (5) presented to the client in an easy-to-read format. The network involved in carrying out the above tasks will involve many different communication links including a wireline between the Smart Sensors and the Local Computer, a RF transmission link to a satellite transceiver, and a worldwide high speed digital network.

The first step to building a satisfactory system is to start with measurement hardware that has a high degree of reliability. Since instruments must operate at temperatures of 175 $^{\circ}$C or more, and the instrument must be placed inside a small diameter (1.7 to 3.0 inch) pressure housing, the minimum hardware that can perform the required measurements is the design goal. This can be realized by ensuring that as much processing of signals can be done in software. This minimizes the components that can fail. The design also attempts to provide a safeguard, in that the design goal is to operate successfully at higher temperatures than the instrument is exposed to. Where possible, hardware redundancy is included so a single failure does not stop all data acquisition.

These borehole measurement tools are operated over a long cable that provides power and two-way digital telemetry. The operation of the borehole instrument is controlled by a Local Computer at the surface; however, there are often microprocessors in the borehole instrument in order to minimize the data transmission during acquisition. The Local Computer equipment is completely redundant so a single failure does not interfere with data acquisition and telemetry. This Local Computer also provides a real-time analysis of the data to guide further acquisition.

The data and preprocessed results from the Local Computer are then sent by satellite and/or land network to a location where experts are available to analyze the data, frequently to a regional location where extensive data bases are maintained that include all the known information about the geological region and the instrument readings for a wide variety of possible environmental conditions. As soon as the analysis is complete, it can be transmitted to the people who must make the decision on the following operations. This can, for example, include further drilling, stopping drilling and setting casing to proceed to production, or shutting down operations and abandoning the well. This set of decisions is appropriate for an exploration scenario.

During production other decisions must be made but the process is the same. Such a system is illustrated in Fig. 1.

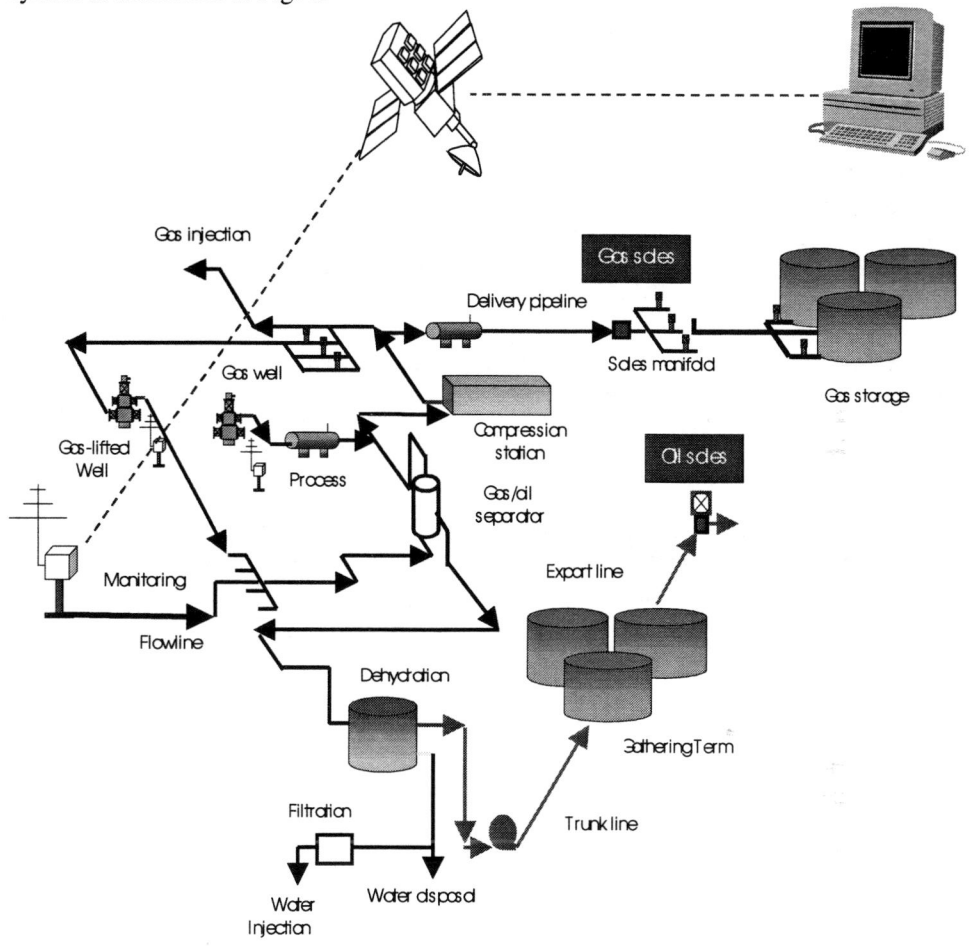

FIGURE 1. Representation of complete system from data monitoring to analysis and control.

A recent development is the installation of permanent sensors in a borehole that can obtain data over many years. These devices are installed so that they do not interfere with other operations in a borehole or in the surface equipment and are not damaged by other instruments that may be put in a borehole at various times. Such devices must have high reliability, must have access to a reliable source of power, and must have access to telemetry for transmitting the data to a surface location, where the same satellite systems described above can transmit them to locations anywhere in the world. Simple versions of such permanently installed sensors have already been tested in a few fields to allow control of subsurface valves to regulate the rate of fluid extraction from the reservoir.

155

The data obtained from subsurface measurements are extensive and the analysis can be difficult. An example of the analyzed results from a number of different instruments is shown in Fig. 2, for approximately 2000 feet of measurements. This output can be presented to the clients immediately through a satellite transmission. This allows the client to make rapid decisions concerning operations and thus, to eliminate costly delays from waiting for information.

One of the most important characteristics of such remote sensing systems is the total system reliability. Where possible redundancy must be included in the system design. Most of the communication networks have built in redundancy. A failure of one component or one node in the network will not shut down the system – an alternate communication path is found. Where redundancy is not possible, then super reliable systems must be built. For example, in the oil field, systems are now being installed on the sea floor for monitoring the separate flow rates of the oil, water and gas leaving the subsea wellhead. These subsea flowmeters are too expensive to build with duplicate hardware. In these cases, the engineering development must include a significant effort at establishing the reliability before the installation. A failure of one component likely will make the flowmeter useless.

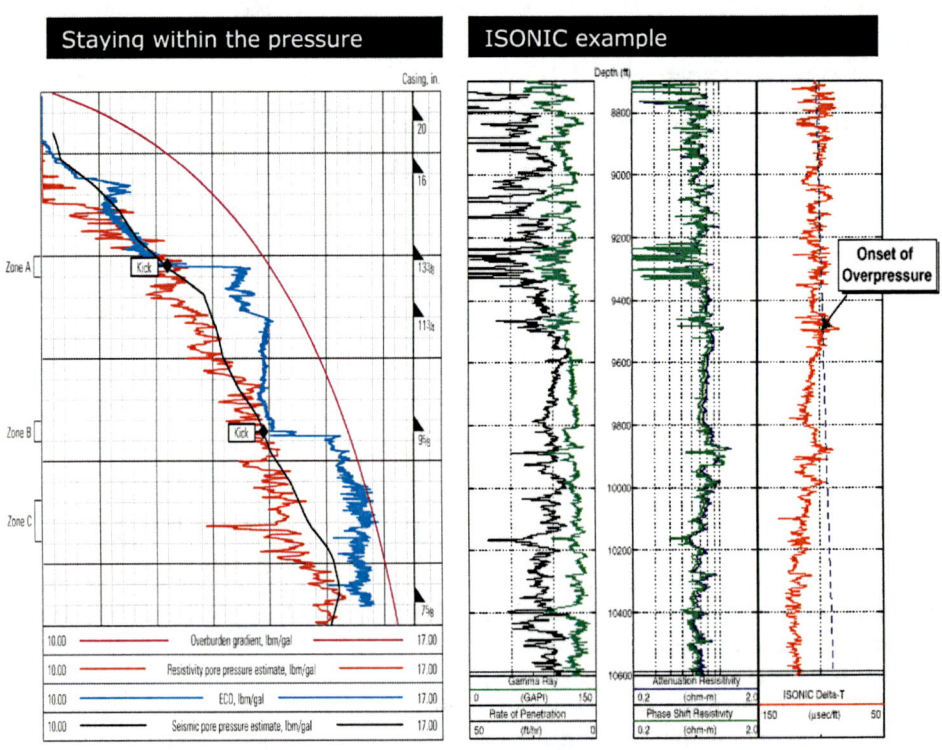

FIGURE 2. Example of about 2000 feet of data from instruments.

CONCLUSIONS

The technology exists today to build completely unattended remote sensing systems as has been developed for the oil and gas exploration and production industry. These systems permit data to be (1) acquired at remote locations, (2) transported by satellite from remote locations to regional or national centers, (3) analyzed and integrated with other measurements and databases by experts at the processing centers, and (5) presented to decision makers at other distant locations in a timely manner.

The development of such systems requires careful attention to all elements of the system, as failure in any part of the system makes the entire system non-functioning. Reliability must be ensured from initial design to final fabrication, and must be capable of verification. Self-calibrating designs are preferred. Remote sensing systems require a greater investment during development and engineering since repair of the deployed system is often not possible.

However, the long-term trends in instrumentation development are in the directions required for remote sensing operations. Highly reliable and redundant communication networks already cover the world. A mountain climber on the summit of Mt. Everest is able to talk to his wife in Australia. Smarter instruments, self-calibrating and self-repairing instruments, low power and very high reliability systems are in development.

REFERENCES

1. Clark, B., and Kleinberg, R., *Physics Today* **55** (4), 48-53 (2002).
2. Schweitzer, J. S., "Subsurface Nuclear Measurements for Geochemical Analysis" in *Remote Geochemical Analysis: Elemental and Mineralogical Composition*, edited by Carle M. Pieters and Peter A. J. Englert, Cambridge University Press, New York (1993).

A Ground Systems Template for Remote Sensing Systems

Timothy P. McClanahan[a] , Jacob I. Trombka[1], Samuel R. Floyd[1], Walter Truskowski[1] Richard D. Starr[2], Pamela E. Clark[2], Larry G. Evans[3]

[1]NASA, Goddard Space Flight Center, AstroChemistry Branch, Greenbelt MD 20771, USA
[2]The Catholic University of America, Department of Physics, Washington, DC 20064, USA
[3]Computer Sciences Corporation, Science Programs, Lanham, MD, 20706, USA

Abstract. Spaceborne remote sensing using gamma and X-ray spectrometers requires particular attention to the design and development of reliable systems. These systems must ensure the scientific requirements of the mission within the challenging technical constraints of operating instrumentation in space. The Near Earth Asteroid Rendezvous (NEAR) spacecraft included X-ray and gamma-ray spectrometers (XGRS), whose mission was to map the elemental chemistry of the 433 Eros asteroid. A remote sensing system template, similar to a blackboard systems approach used in artificial intelligence, was identified in which the spacecraft, instrument, and ground system was designed and developed to monitor and adapt to evolving mission requirements in a complicated operational setting. Systems were developed for ground tracking of instrument calibration, instrument health, data quality, orbital geometry, solar flux as well as models of the asteroid's surface characteristics, requiring an intensive human effort. In the future, missions such as the Autonomous Nano-Technology Swarm (ANTS) program will have to rely heavily on automation to collectively encounter and sample asteroids in the outer asteroid belt. Using similar instrumentation, ANTS will require information similar to data collected by the NEAR X-ray/Gamma-Ray Spectrometer (XGRS) ground system for science and operations management. The NEAR XGRS systems will be studied to identify the equivalent subsystems that may be automated for ANTS. The effort will also investigate the possibility of applying blackboard style approaches to automated decision making required for ANTS.

1. INTRODUCTION

On February 14, 2000 the Near Earth Asteroid Rendezvous (NEAR) spacecraft was inserted into orbit around the asteroid 433 Eros (EROS). A year long encounter ensued, and brought the first close-in measurements of an asteroid. The mission culminated with NEAR's successful landing and sampling of the surface by the Gamma-ray spectrometer. Operating at times farther than 2 AU from Earth, NEAR required hardware, software and personnel that were able to remotely monitor, identify and quickly adapt to changing spacecraft conditions and mission

CP632, *Unattended Radiation Sensor Systems for Remote Applications*, edited by J. I. Trombka et al.
© 2002 American Institute of Physics 0-7354-0087-3/02/$19.00

requirements. Several times the spacecraft and instrument subsystems were reconfigured to add new, and modify existing functionality, illustrating the robust nature of the NEAR spacecraft, the instruments and its support systems.

The initial design and development of the X-ray/Gamma-Ray (XGRS) ground system was prompted in many ways by the lack of pre-encounter knowledge of important factors required to interpret the XGRS signals. To obtain these measurements XGRS ground processing was based on a ground-system architecture metaphorically similar to a blackboard architecture. Blackboard systems, utilized in a number of artificial intelligence applications where there exist large or complex problem domains, are used to partition a problem space into independent research processes known as knowledge sources. As the solutions to the sub-problems evolve they are posted to a globally accessible blackboard or data repository for integration into an overall problem solution. A control module or scheduler is then used to review the blackboard and integrate the appropriate combinations of partial solutions to use in a particular task. As these solutions mature, they tend to become static with fewer revisions as resource requirements and performance characteristics become known. An added benefit to this type of process is the realization of step-wise refinements in the accuracy and resolution of information generated from the system. Where applicable, these systems can then be migrated from human intensive efforts to automated processes.

In the case of NEAR XGRS, only rough estimates of factors such as EROS's shape and mass, solar, instrument and other characteristics existed prior to encounter. As each factor was identified, they were assigned to specific XGRS team members who acted as the knowledge sources that researched and generated solutions for each problem. As solutions to the were generated, the algorithms and feature sets were designed and built into the XGRS ground processing system, which served as the team's common repository for science and operational planning activities. Much of this research encompassed the analysis of datasets from other instrumentation, resulting in a cooperative data fusion process in which knowledge solutions were developed, and stepwise refinements made to the accuracy and resolution of the XGRS database during the course of the mission. This proved an effective strategy for the XGRS science and planning activities, but was however a human intensive effort in the research and development of the XGRS ground system. It may then be viewed as a human based blackboard system.

Presently the ideas and results of the NEAR XGRS mission are being used to facilitate the next generation of NASA's exploration of space in the Autonomous Nano-Technology Swarm (ANTS) project. In this mission spatially distributed sensor arrays perhaps comprising >1000 independent nano-satellites will be launched to encounter the main asteroid belt and categorize asteroids greater than 1 km in diameter.

The mission, currently in the concept phase, will utilize the significant advances that have taken place in the fields of artificial intelligence and robotics to automate the process of cooperative encountering of target asteroids. Using sensors, similar to the types used in NEAR, groups of satellites, analogous to a swarm of ants, will target and encounter asteroids. Communications with the earth will necessarily be limited and mission planning activities will take place within the collective intelligence of the

ANT swarm. Individual ants (satellites) will maintain differing operational and remote sensing roles, with intelligence to collaborate and cooperate in the remote sensing process. The set of instruments planned for use in the ANTS mission have similar roles, objectives, and features, similar to that of the NEAR instrument payload. An initial step in the ANTS mission will be to study the instantiation of knowledge sources from NEAR XGRS, such as science and operations planning, and to transition from the human based blackboard paradigm to more automated approaches, which may be realized within an ANT swarm. To realize this transition we seek to identify the resources, algorithms, and feature sets used in the NEAR XGRS ground systems architecture to begin planning the automation of these processes in a truly automated blackboard architecture for the ANTS mission.

2. Blackboard Systems Architecture

The blackboard architecture was originally developed to deal with the complexities of automated speech understanding: a very large search space, errors, incomplete data, and incomplete problem solving knowledge. These are characteristics that require a problem solving approach that will support incremental and adaptable development of different solutions from possibly disparate knowledge sources. *Figure 1.* illustrates the main concepts behind the blackboard architecture, metaphorically similar to a group of experts collaborating around a blackboard to solve a particular problem. In this scenario the blackboard represents facts, assumptions and deductions made by the system. Experts view the blackboard and try to contribute information leading to a solution. A facilitator reviews the set of solutions and integrates the most appropriate solutions for a given task and the latest results are posted in the database. User interfaces are used for human intervention and tracking of the system.

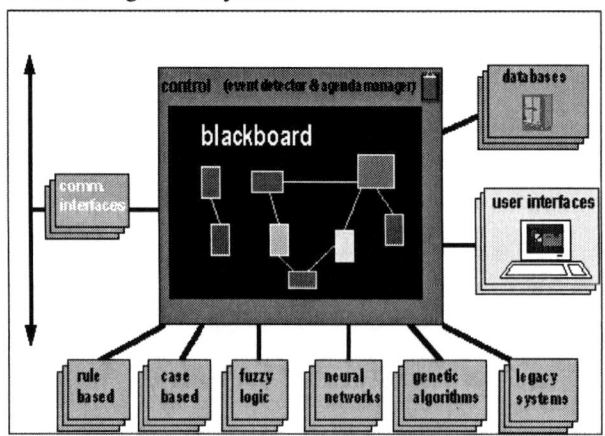

Figure 1. Overview of a Blackboard system in which expert systems are used to solve complicated problems. Expert systems are used to define algorithms and features that find correlation between disparate information sources and enhance the solution and perception of the system.

The blackboard architecture features 3 major components: (1) a hierarchically organized memory (the blackboard) which saves the solutions generated by the different knowledge sources. (2) a collection of knowledge sources that generate solutions using expert systems and moves them to the blackboard as they become available. (3) a separate control module which integrates the most appropriate solutions for a given task. The problem is partitioned among the various knowledge sources (KS) where each (KS) is an expert within their problem domain. They identify the required resources needed for solving

160

their problem, the algorithm defining the solution and the features that are generated. The solutions are then posted to the blackboard where a control module is used to select the most appropriate total solution to the overall problem.

3. NEAR XGRS Ground Systems and Modeling Issues

The NEAR XGRS ground processing system was designed for flexible feedback from various channels of communication from knowledge sources to the centralized XGRS ground processing system at the University of Arizona, Lunar and Planetary Laboratory (UA-LPL). Each knowledge source provided information that was used to update or indirectly generate critical parameters contained within the logical definition of an XGRS integral record. The ground processing system fed a Sybase relational database (UA-RDBMS), which was maintained on-line and was partitioned into XRS and GRS ground data processing systems. In each XRS or GRS integral record is encapsulated a critical feature set identified by the NEAR XGRS team that meets all of the operational and science requirements of the mission. In *Figure 2.0* the logical record structure is defined within the UA-RDBMS with Mission Event Time (MET) or 'Time' acting as the primary key for database organization and query access. SPECTRA, SCI, ENG partitions of the database comprise portions of the logical record that are directly supplied by the XGRS instrument system, representing the spectra collected by the instrument at each integral time, the instrument scientific housekeeping, and instrument engineering housekeeping. DER_ENG, SPATIAL and SOLAR partitions represent information required from various knowledge sources which include derived engineering and calibration information, navigation and pointing information, shape modeling, plus solar modeling information.

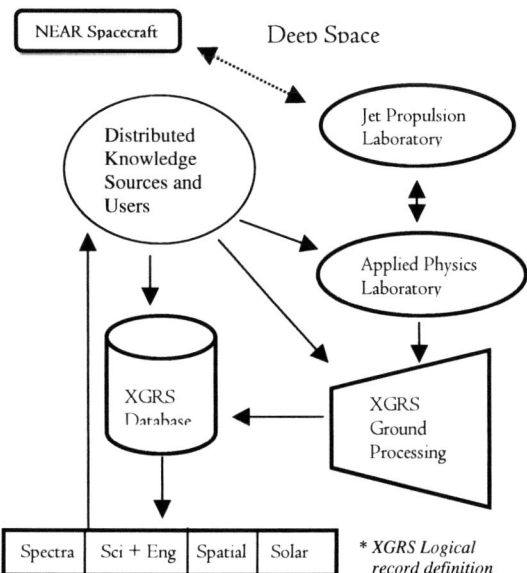

Key elements include the explicit use of feedback cycles between the NEAR spacecraft, designated NEAR knowledge sources, and the UA-RDBMS that provide the information to the user community where new ideas and refinements are hypothesized, synthesized, validated and then implemented as part of the XGRS ground processing enterprise. Old data is then reprocessed with the updated solution set realizing stepwise improvements to the XGRS data products.

Figure 2. Diagram illustrating the flow of information from the NEAR spacecraft through the ground processing system to knowledge sources who heuristically modify the ground processing system and plan future science activities.

Critical knowledge sources for XGRS ground processing include the EROS Shape, Calibration, Navigation/ Pointing, and Time drift models that were directly derived from collaborative analysis of the NEAR Multi-spectral imager, NEAR Laser rangefinder and NEAR radio-astronomy observations during the mission. The models were periodically generated as a set of NASA Ancillary Information Format (NAIF), SPICE kernels that were produced during the mission by the NEAR mission operations personnel residing at the NEAR Science Data Center (SDC) at APL and the Jet Propulsion Laboratory (JPL). The knowledge sources provided science and operations support for generating commands for future science and mission operations. These command scripts were verified at the APL SDC and routed through the DSN back to NEAR. Other knowledge sources provided support to ground processing systems by generating new and upgraded solutions that were incorporated in the XGRS ground processing system.

Of significant note is the highly irregular, non-spherical shape of the EROS asteroid, which prior to encounter was loosely approximated using a tri-axial ellipsoid of 33x13x13 km dimension. As the mission progressed however, the shape model incrementally improved in both resolution and accuracy, facilitating more accurate XGRS surface simulations. The model was used in conjunction with navigation, pointing and timing information to simulate the location and orientation of the instrument and sun with respect to the asteroid surface at the time of integration. In that simulation, the MET time for a particular spectral integration was used to access NAIF SPICE kernels, which provide a numerical transformation that generates the position and orientation of the spacecraft and sun about the asteroid in body fixed coordinates. The simulation then projects the XRS instrument field of view, a 2.5° half angle from the spacecraft position along the path of the instrument boresite vector to the modeled surface. The subtended surface, called a footprint, is obtained as the set of contiguous model plates that discretely approximate the surface. The ground processing system generates key surface measurements such as incidence angle between the sun and surface and emission angle between the spacecraft and surface, which are used in the science analysis process. During optimal sampling configurations the number of simulations exceeded 2000 integrations per day. A similar surface processing system was also generated for the GRS subject to a larger field of view, 30° half angle, and

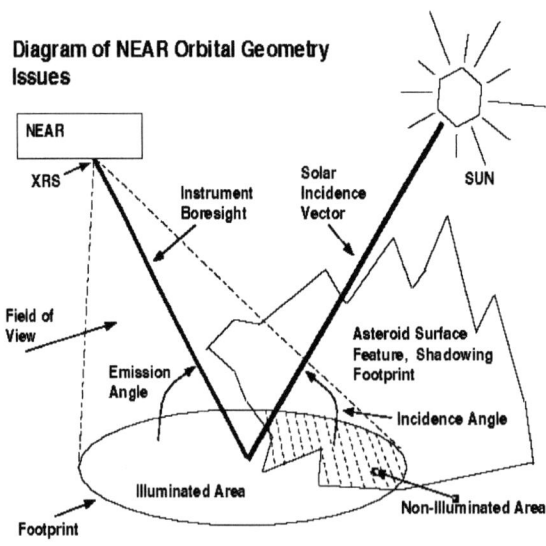

Figure 3. Diagram of NEAR orbital geometry factors Derived from XGRS ground processing simulation

isotropic assumptions of incidence angle from cosmic ray induced activity. *Figure 3* illustrates the most common orbital issues that were required of the NEAR XGRS ground processing system.

Other knowledge sources that were used in the XGRS ground processing system include: (1) the GRS ground calibration system, which tracked the detector energy calibration through the course of the encounter and was used for automated gain correction. This allowed the capability to automatically co-add calibrated spectra served from the UA-RDBMS. (2) Modeling of NEAR XRS gas proportional counter solar monitor spectra to infer solar temperature from spectra. (3) Identification and flagging of bad spectra which were either corrupted or otherwise deemed unacceptable.

4. Autonomous Nano-Technology Satellites (ANTS)

A significant constraint to the ANTS mission is the lack of communications bandwidth available to command, control and downlink data from large constellations of satellites. This constraint mandates the use of intelligent spacecraft systems that can operate autonomously. Once these traffic types are offloaded the available bandwidth can then be distributed to other telemetry modes. This realizes dual benefits of reducing the human intensive ground processing costs and facilitates the goal of deploying of larger constellations of autonomous spacecraft.

In ANTS, the proposed communications paradigm is that individual ANTS will have omni directional antennae with high and low bandwidth capabilities. High bandwidth communications will be used for spatially local, ANT to ANT operations. Low bandwidth capabilities will be dedicated to collective swarm processes. The ANTS system will also have to communicate both the science data collected as well as its operational status back to the swarm for review and analysis.

The initial ANTS design is for a cooperative system in which ants (satellites) maintain a hierarchical social structure among individual ant workers with specialized responsibilities and collective mission goals. As each encounter ensues the swarm will be designed to adapt to the conditions that evolve by using a set of virtual blackboards that will employ multiple intelligent software agents that heuristically review both the status of the encounter and the swarm's operational requirements. This information can then be used as part of the autonomous analysis process that will be required to provide knowledge and decision making capacity. The control module or scheduler for any given task will be required to make decisions on the appropriate partial solutions to use while using incomplete and possibly erroneous information as a database for such decisions. The results of this analysis are then used to extrapolate the information into commands to be scheduled for future swarm operations. A significant navigational challenge will be the automated planning, commanding and encounter of asteroids, where ANTS will target asteroids, ascertain their trajectories, plan intersecting trajectories and plan and execute the encounter process. This will entail many different mission scenarios and priorities. Any system related task may then require the use of many features that will have variable importance to the decision making process. Such a system will probably be required to collectively learn through the effects of correlating the data that has been collected with decisions that were made in the past.

The NEAR XGRS ground system implementation is currently being studied to identify the knowledge sources, algorithms and feature sets that were implemented to support science and operations management. These systems are being used to provide an important initial step in automating similar processes for the ANTS mission.

- Navigation, Pointing and Shape Definition
- Solar Temperature
- Instrument Calibration and Data Quality
- Command and Data Handling

These studies will require in depth study of the knowledge sources, algorithms and feature sets that were used for NEAR XGRS. The studies will employ the use of analysis techniques that will include the use of artificial intelligence, trend and statistical analysis of the database to automatically identify and weight datasets based on the occurrence and distributions of feature sets within the data. These terms may then be used as part of the higher level processes to generate automated approaches to mission and science planning in ANTS.

References

1. Carver, N.C. et al, The Evolution of Blackboard Control Architectures, *CMPSCI Technical Report 92-71*, October 1992.
2. Cheng, A. F., Near Earth Asteroid Rendezvous: Mission Overview, *The Near Earth Asteroid Rendezvous Mission*, C. T. Russell, Space Science Reviews **82**:3-29, 1997.
3. Curtis, S.A. et al., ANTS (Autonomous Nano-Technology Swarm): An Artificial Intelligence Approach to Asteroid Belt Resource Exploration, *Proceedings of the 51st International Astronautical Congress*, October 2000.
4. Evans L.G et al, Calibration of the NEAR Gamma-ray Spectrometer, *Icarus*, **148**: 95-117, (2000).
5. Evans L.G et al, Elemental Composition from Gamma-ray spectroscopy of the NEAR-Shoemaker Landing Site on 433Eros, *Meteoritics and Planetary Science*, **36**:1639-1660, (2001).
6. Floyd, S. F., et al, Radiation effects on the proportional counter X-ray detectors on board the NEAR spacecraft. *Nuclear Instruments and Methods in Physics Research*, pp, 577-581, A 422, 1999.
7. Goldsten, J. O., et al., The X-ray/Gamma-ray Spectrometer on the Near Earth Asteroid Rendezvous Mission, *The Near Earth Asteroid Rendezvous Mission*, C. T. Russell, Space Science Reviews, 82:3-29, 1997 8. Heeres, K. J., et al., The NEAR Science Data Center, *The Near Earth Asteroid Rendezvous Mission*, C. T. Russell, Space Science Reviews, **82**:3-29, 1997.
9. McClanahan, T. P., et al., Data Management and analysis techniques used in the NEAR X-ray and Gamma-ray Spectrometer Systems, *Nuclear Instruments and Methods in Physics Research*, pp. 482-485, A 422, 1999.
10. McClanahan, T.P. and Mikheeva I. et al, Data Processing for the Near Earth Asteroid Rendezvous (NEAR), X-ray and Gamma-ray Spectrometer (XGRS) ground system, *Hard X-ray, Gamma-ray and Neutron Detector Physics*, (1999).
11. Nittler L.R. et al., X-ray Flourescence Measurements of the Surface Elemental Composition of Asteroid 433Eros, *Meteoritics and Planetary Science*, **36**:1673-1695 (2001).
12. Starr R. et al, Instrument Calibrations and Data Analysis Procedures for the NEAR X-ray Spectrometer, *Icarus*, **147**:498-519, (2000).
13. Trombka, J. I., et al., The NEAR-Shoemaker X-ray /Gamma-ray Spectrometer Experiment: Overview and Lessons Learned, *Meteoritics and Planetary Science*, **36**:1605-16 (2001).
14. Trombka, J. I., et al., Compositional Mapping with the NEAR X-ray / Gamma-ray spectrometer, *Journal of Geophysical Research*, vol. 102, No. E10, pages 23,729-23,750, October 25, 1997

ENVIRONMENTAL APPLICATIONS

Algal Genetic Systems as Sensors for Radiation Detection

Tanya Kuritz

Chemical Sciences Division, Oak Ridge National Laboratory, Oak Ridge, TN 37831, USA

Abstract. Numerous attempts were made to create biosensors for the detection of a variety of signals. However, no biosensor for radiation detection has been developed that would be suitable for field applications. We have generated a collection of potentially radiation-responsive genetic elements from cyanobacteria for their use in mix-and-match application-driven configurations of biosensors. Use of different biosensor system elements is discussed.

INTRODUCTION

In the past two decades, numerous attempts were made to create biosensors for the detection of a variety of signals. Successes in the detection of individual chemical substances or groups of chemical compounds, as well as temperature were reported, and some of the sensors have been successfully applied or are currently being tested for applications.

Biological sensors of chemical substances utilized two major approaches. The first approach built on the natural, individual sensitivity of microorganisms (predominantly algae) to a variety of chemicals that affected their photosynthetic functions. In these systems, change in chlorophyll or other pigment fluorescence served as a reporter. The algal fluorescence-based systems report physiological stress in the organism [1] and therefore are non-specific and provide identical responses to warfare agents [2], pesticides [1] or chemical pollutants [3]. A recent work that attempted application of fluorescence-based sensors to study space radiation provided no specific details on the effect of different doses on fluorescence [4]. Applicability of these systems is limited to the environments, natural backgrounds of which can be constantly monitored, and for which historic record exists to enable validation of the nonspecific signal.

Another approach that enabled specificity of biosensors to certain agents, or classes of chemicals, involved the use of genetically-modified microorganisms with transcriptional reporters. The general design and operation principles of these systems are illustrated in Figure 1. A chemical causes cell damage and the cell responds by synthesis of a protein that can either degrade the chemical, or detoxify it. The protein is translated from mRNA, therefore, biosynthesis of mRNA is the first step in the response. Replacement of the damage-

CP632, Unattended Radiation Sensor Systems for Remote Applications, edited by J. I. Trombka et al.
© 2002 American Institute of Physics 0-7354-0087-3/02/$19.00

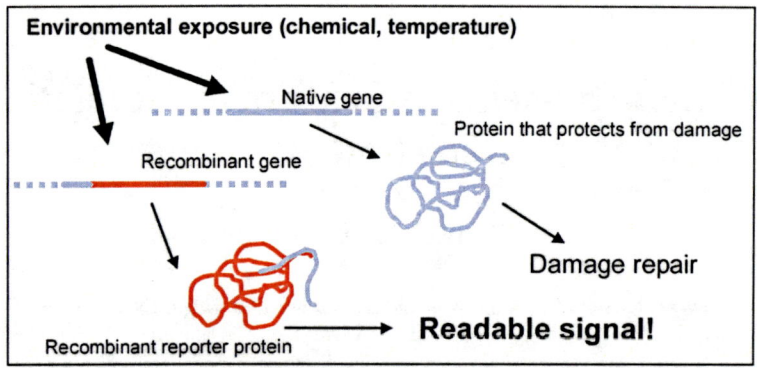

FIGURE 1. Principle of operation of transcription-based biosensor.

response protein with a protein that can deliver an optically-recordable signal (reporter protein) allows monitoring of the cell response and deduces the presence of the chemical in the environment. Although these biosensors are more specific to environmental impact, they depend on a variety of factors, including host organism viability.

Current molecular biology has developed a number of standardized tools and procedures enabling the combinatorial design of biological systems. We propose to create a set of genetic tools (responding elements, reporter genes, and hosts) that would allow a mix-and-match design of biological sensors of radiation capable of sensing and reporting exposure to the wide range of energies.

MATERIALS AND METHODS

Complete genomic sequences of *Synechocystis* PCC6803 and *Anabaena* PCC7120 were obtained from Kazusa Institute website [5]. All other sequences were obtained from the National Center for Biotechnology Information (NCBI) site [6]. Analysis of genetic sequences was carried out using web-based tools on the NCBI site [6].

Strains of cyanobacteria *Anabaena* PCC7120, *Synechococcus* PCC803 and *Synechocystis* were grown in BG11 medium with constant illumination (2000 lux) and agitation as described earlier [7].

For DNA amplification, 500 µl of stationary-phase cultures were pelleted by centrifugation at 14 000 rpm for 30 sec, and 1 µl of each pellet was added to the standard PCR cocktail (Sigma Chemical Co., St. Lois, MO).

PCR amplification was carried out as follows: 94 °C (5 min), 35 cycles at 94 °C (1 min), 50 °C (2 min), 72 °C (2 min) followed by 5 min final extension at 72 °C. The Taq polymerase and a standard PCR buffer were from Sigma Chemical Company. Nucleotides were from Life Technologies. PCR products were resolved by electrophoresis in 1.5 % agarose in TBE and visualized by staining with ethidium bromide (both from Fisher). The amplified fragments, with or without purification, were subcloned into a standard vector pCR2.1 (Invitrogen, Carlsbad, CA).

DNA was sequenced using the ABI Prism Dye Terminator Cycle Sequencing Ready Reaction Kit on an ABI 373 DNA sequencer (both of PE-Applied Biosystems). DNA sequences were analyzed by BLAST [8].

RESULTS

Table 1 lists reported cyanobacterial SOS-induced genetic systems that are upregulated in response to UV-B irradiation (*psbA* family of genes), by strong environmental stresses (*nblR*) and by DNA damage.

TABLE1. Presumptive Radiation-Responsive Genes of Cyanobacteria

Organism	Stress-responsive gene	Function	Nature of stress	Proof of function	Source
Synechococcus sp. PCC7942	*nblR*	Phycobilisome degradation	Intensive light, starvation	Confirmed	[9]
Synechocystis sp. PCC6803	*psbA*	Replacement of a PSII protein	UV-B irradiation	Confirmed	[10-13]
Anabaena sp. PCC7120	*recA*	DNA repair, recombination	DNA damage	Prediction	[14, 15]
Anabaena sp. PCC7120	*UV-endonuclease*	DNA repair	DNA damage	Prediction	[14, 15]
Anabaena sp. PCC7120	*DNA damage repair*	DNA repair	DNA damage	Prediction	[14, 15]
Synechocystis sp. PCC6803	*recA*	DNA repair, recombination	DNA damage	Prediction	[16]
Synechocystis sp. PCC6803	*relB*	DNA repair	DNA damage	Prediction	[16]

Using DNA sequence data from the GenBank, we designed biased primer pairs for the amplification of the fragments of cyanobacterial genes involved in nonspecific stress response, DNA and protein repair and response to far UV range.   PCR amplification from these allowed us to generate a library consisting of the following gene fragments:

psbA of *Synechocystis* sp. PCC6803 (700 bp; primers: PSBA DIR 5'-agacaacgactctccaaca-3' PSBA REV5'-gtaaccgtagttctggga-3')

nblR of *Synechococcus* sp. PCC7942 (primers: NBLR DIR 5'-ttcgcgtagg acataaccga-3'; NBLR REV 5'-caaatcgatcttcagaagc-3')

recA of Anabaena sp PCC7120 (primers: 7120 recA dir 5'-Attgagcgsagcttcgg-3'; 7120 recA rev 5'-accgtaggtmacaccaa-3')

UV dependent endonuclease of *Anabaena* sp. PCC7120 (primers: 7120 uvend dir 5'-gacaatgacgcgcaca-3' 7120 uvend rev 5'-tcggtgctgtaagcgta-3')

DNA damage repair protein of *Anabaena* sp. PCC7120 (primers: 7120 dnadam dir 5'-tcaagagagcgagaacca-3'; 7120 dnadam rev 5'-acttcaccgactaactg-3')

169

All amplified gene fragments were subcloned into standard vectors and deposited into the laboratory collection. The identity of the subcloned fragments was validated by sequencing. These fragments will be used for the construction of radiation responsive sensors.

DISCUSSION AND CONCLUSIONS

Application of biosensors in open systems, under field conditions, has been impacted by several general limitations. The important limitation is due to the variability in the individual tolerance of organisms to radiation or other environmental stresses. Therefore, in order to be able to provide reliable sensing over a range of energies, we should consider using arrays of sensing genetic systems and of host organisms with different individual sensitivities to radiation and environmental factors.

Rational development of radiation sensors was confined to the use of a limited number of genetic systems proven to be sensitive to radiation. Limitations of the radiation sensory ranges are in part due to the lack of knowledge of the microbial physiological response to radiation: both of the range and types of radiation that induce response in organisms and of the genetic systems involved in such a response.

Initial attempts to develop a transcription reporter based system for radiation detection used common heterothrophic organisms as hosts, such *Escherichia coli*, which provided linear responses to gamma-ray doses of 1 to 50 Gy [17]. The majority of the reported radiation-sensing systems employed *lux* as a reporter gene fused to different radiation/oxidative stress-sensitive genetic elements that were involved in SOS response in *E. coli*, including *recA, grpG, katG, uvrA, alkA,* and *cda* [17 - 19]. These systems have contributed to our understanding of requirements to biosensor design, however, they were not usable in the field due to:

- Low viability of *E. coli* under field conditions;
- Nonspecific nature of the SOS-response (it can be induced by factors other than radiation);
- Dependence of the intensity of the *lux* system on partial pressure of oxygen in the system and on other environmental parameters, which leads to reporter (not sensor!) signal modulation by a variety of environmental factors;
- Conditional pathogenicity of *E.coli*.

Host organism survivability presented one of the problems in biosensor design. Survival of commonly used hosts *Escherichia coli* and *Pseudomonas* sp. is impaired by a number of factors, including drought and UV radiation. These limitations call for the use of organism(s), as individual hosts, or arrays of hosts, which will tolerate wide ranges of environmental impacts due to variations in individual tolerance of organisms to radiation (Figure 2). Microalgal isolates from desert environments are more tolerant to radiation and desiccation than *E. coli* and can evolve linear responses to gamma-radiation doses between 2 and 15 kGy [20], which is comparable with the resistance of *Deinococcus radiodurans* [21]. In addition to DNA damage-repair genetic systems, algae possess systems that are known to respond to radiation damage only.

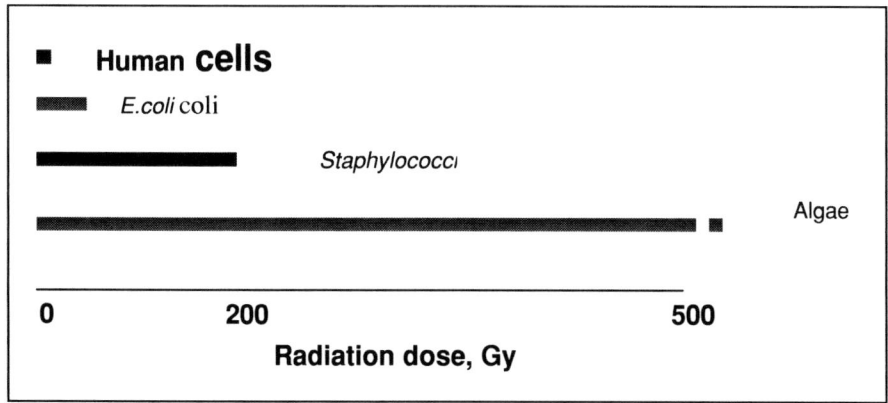

FIGURE 2. Viability of different organisms exposed to ionizing radiation

Similar to other organisms, cyanobacteria possess genetic systems for repair of DNA and proteins damaged by a variety of oxidative stresses. Annotation of complete genomic sequences of *Synechocystis* sp. PCC6714 and *Anabaena* sp. PCC7120 identified, by homology, such genes as *recA, uvrA, B, kat* and other sequences potentially involved in oxidative stress response [5, 6, 14-16].

Cyanobacteria are photosynthetic microorganisms, and the genetic systems involved in photosynthesis account for over 2/3 of cyanobacterial genome [22]. Photosynthetic apparatus is a complex system, which provides cyanobacteria with energy, with photoautotrophic cyanobacteria being dependent solely on photosynthesis as the energy source. Conversion of light energy into chemical is carried out by multiprotein complexes called photosystems, which, and respond to a wide range of radiation. Radiation-induced damage of photosystem II calls for the need to restore its functions. The reported radiation-response system involves SOS-synthesis of multiple copies of the photosystem II protein PsbA (synthesized from *psbAI* gene under normal conditions) from two radiation-responsive genes *psbAII* and *psbAIII* [10 - 13]. This response is induced by UV-B but not by UV-A or visible light. Differential biosynthesis of pigments in cyanobacteria in response to the exposure to different energy ranges in the UV spectrum [23] also points to the potential existence of specific sensory systems to shorter wavelengths radiation. We see the need for testing responses of the *psbA* system and gene encoding protective pigments' biosynthesis to other energy ranges, including γ-rays and x-rays.

Compatibility of biologically-generated signals with existing remote interrogation systems is another problem, which should be addressed by co-designing of "bioware" and hardware elements of the sensor system. Biosensor systems of the future will be built of mix-and-match elements that will be interacted as systems comprising arrays of different host organisms, sensing genetic systems, reporting genetic systems, and interrogating and signal-processing hardware. And collections of proven radiation-responsive genetic elements will provide critical elements for different biosensor configurations.

ACKNOWLEDGMENTS

This work was performed under DOE NA-22 funding. The author is grateful to Vivian Baylor, David Spears and Brian Davison for support and helpful discussions. Oak Ridge National Laboratory is managed for the US Department of Energy by UT-Battelle, LLC under the contract DE-AC05-00OR22725.

REFERENCES

1. Van Rensen, J.J., and Curwiel, V.B. *Indian J. Biochem. Biophys.* **37**, 377-382 (2000).
2. Sanders, S.A., Rodriguez, M.R., and Greenbaum, E. *Biosen. Bioelectron.* **16**, 439-446 (2001).
3. Yatsenko, V. *J. Chromatogr. A.* **722**, 233-243 (1996).
4. Angelini, G., Ragni, P., Esposito, D. Giardi, P. Pompili, M.L., Moscardelli, R., and Giardi, M.T. *Phys. Med.* **17,** Suppl 1, 267-268 (2001).
5. Kazusa DNA Research Institute, Chiba, Japan. Site: www.kazusa.or.jp
6. National Center for Biotechnology Information website www.ncbi.nlm.nih.gov.
7. Kuritz, T., and Wolk, C.P. *Appl Environ Microbiol* **61**, 134-138 (1995).
8. Altschul, S. F. , Madden, T.L., Schaffer, A.A., Zhang, J., Zhang, Z., Miller, W., and Lipman, D.J. *Nucl.Ac.Res.* **25**, 3389-3402 (1997).
9. Schwarz, R., and Grossman A.R. *Proc Natl Acad Sci USA* **95**, 11008-11013 (1998).
10. Vass, I., Kirilovsky, D., Perewoska, I., Mate, Z., Nagy, F., and Etienne, A.L. *Eur J Biochem.* **267**, 2640-2648 (2000).
11. Mate, Z., Sass, L., Szekeres, M., Vass, I., and Nagy, F. *J. Biol. Chem.* **273**, 17439-17444 (1998).
12. Schaefer, M.R., and Golden, S.S. *J. Bacteriol.* **171**, 3973-3981 (1989).
13. Campbell, D. , Eriksson, MJ., Oquist, G., Gistafsson, P., and Clarke, A.K.. *Proc Natl Acad Sci USA* **95**, 364-369 (1998).
14. Kaneko, T., Nakamura, Y. Wolk, C.P., Kuritz, T. Sasamoto, S., Watanabe, A., Iriguchi, M., Ishikawa, A., Kawashima, K., Kimura, T., Kishida, Y., Kohara, M., Matsumoto, M., Matsuno, A., Muraki, A., Nakazaki, N., Shimpo, S., Sugimoto, M., Takazawa, M., Yamada, M., Yasuda, M., and Tabata, S. *DNA Res.* **8**, 205-213 (2001).
15. Kaneko, T., Nakamura, Y. Wolk, C.P., Kuritz, T. Sasamoto, S., Watanabe, A., Iriguchi, M., Ishikawa, A., Kawashima, K., Kimura, T., Kishida, Y., Kohara, M., Matsumoto, M., Matsuno, A., Muraki, A., Nakazaki, N., Shimpo, S., Sugimoto, M., Takazawa, M., Yamada, M., Yasuda, M., and Tabata, S. *DNA Res.* **8**, 227-253 (2001).
16. Kaneko, T., Sato, S., Kotani, H., Tanaka, A., Asamizu, E., Nakamura, Y., Miyajima, N., Hirosawa, M., Sugiura, M., Sasamoto, S., Kimura, T., Hosouchi, T., Matsuno, A., Muraki, A., Nakazaki, N., Naruo, K., Okumura, S., Shimpo, S., Takeuchi, C., Wada, T., Watanabe, A., Yamada, M., Yasuda, M. and Tabata, S. *DNA Res.* **3**, 109-136 (1996).
17. Min, J., Lee, C.W., Moon, S.-H., LaRossa, R.A., and Gu, M.B. *Radiat. Environ Biophys.* **39**, 41-45 (2000).
18. Vollmer, A.C., Yatsenko, V. *J. Chromatogr. A.* **722**, 233-243 (1996).Belkin, S., Smulski, D.R., Van Dyk, T.K., and LaRossa, R.A.. *Appl Environ Microbiol* **63**, 2566-2571 (1997).
19. Ptistyn, L.O., Horneck, G., Komova, O., Kozubek, S., Krasavin, E.A., Bonev, M., and Rettenberg, P. *Appl. Environ. Microbiol.* **63**, 4377-4384 (1997).
20. Billi, D., Friedman, E.I., Hoffer, K.G., Caiola, M.G., and Ocampo-Friedman, R. *Appl. Environ. Microbiol.* **66,** 1489-1492 (2000).
21. Mattimore, V., Udupa, K.S., Berne, G.A, and Battista, J.R. *J. Bacteriol.* **177**, 5232-5237(1996).
22. Kuritz, T., A. Ernst, T.A. Black, and Wolk, C.P. *Mol. Microbiol.* **8,** 101-110 (1993).
23. Ehling-Schulz, M., Bilger, W., and Scherer, S. *J. Bacteriol.* **179**, 1940-1945, (1997).

Resonance Ionization Mass Spectrometry System for Measurement of Environmental Samples

L. Pibida, C. A. McMahon[+], W. Nörtershäuser[→] and B. A. Bushaw[*]

National Institute of Standards and Technology, Gaithersburg, MD 20899-8462, USA
[*] Pacific Northwest National Laboratory, Richland, WA 99362, USA
[+] Present: Radiological Protection Institute of Ireland, Dublin 14, Ireland
[→] Present: Gesellschaft für Schwerionenforschung, D-64291 Darmstadt, Germany

Abstract. A resonance ionization mass spectrometry (RIMS) system has been developed at the National Institute of Standards and Technology (NIST) for sensitive and selective determination of radio-cesium in the environment. The overall efficiency was determined to be 4×10^{-7} with a combined (laser and mass spectrometer) selectivity of 10^8 for both ^{135}Cs and ^{137}Cs with respect to ^{133}Cs. RIMS isotopic ratio measurements of $^{135}Cs / {}^{137}Cs$ were performed on a nuclear fuel burn-up sample and compared to measurements on a similar system at Pacific Northwest National Laboratory (PNNL) and to conventional thermal ionization mass spectrometry (TIMS). Results of preliminary RIMS investigations on a freshwater lake sediment sample are also discussed.

INTRODUCTION

A resonance ionization mass spectrometry (RIMS) system at the National Institute of Standards and Technology (NIST) has been implemented for detection of radionuclides in the environment. The determination of the isotopic ratio of ^{135}Cs and ^{137}Cs and their concentrations in low-level natural matrices are important for evaluating migration, age, and source conditions of radioactive contamination in the environment. Cesium-135 and ^{137}Cs are high-yield fission products from ^{235}U, ^{233}U and ^{239}Pu [1] with half-lives of $t_{1/2} = 2.3 \times 10^6$ years and $t_{1/2} = 30.07$ years respectively. Due to the difference in the half-lives, the isotopic ratio can be used as a clock for dating radioactive waste. This ratio can also yield information on the neutron flux conditions during production of the radio-cesium [5]. Cesium-137 can be detected by the associated 661 keV gamma-ray emission of the ^{137m}Ba daughter, while for ^{135}Cs the pure beta decay (210 keV) and low specific activity make radioactive decay counting impractical for low-level measurements. Mass spectrometry techniques are an alternative method for simultaneous detection of both radioisotopes. Thermal ionization mass spectrometry (TIMS) measurements using a triple-sector mass

CP632, *Unattended Radiation Sensor Systems for Remote Applications*, edited by J. I. Trombka et al.
© 2002 American Institute of Physics 0-7354-0087-3/02/$19.00

spectrometer [2] have demonstrated abundance sensitivity as low as 10^{-11}, which is sufficient to address stable ^{133}Cs interference, but careful chemical separation is required to remove possible isobaric interferences. TIMS with a double-sector instrument has been used for the determination of ^{135}Cs / ^{137}Cs ratios in low-level environmental samples [3]; although careful chemical separations were used to remove barium interferences, the measurements were ultimately limited by other molecular isobars. Inductively coupled plasma mass spectrometry (ICP-MS) also has been applied to determine ^{134}Cs / ^{135}Cs / ^{137}Cs ratios in high activity nuclear materials [4]. For environmental samples extremely high-quality chemical separations would be needed to remove the Ba isobars since there is no thermal selectivity in the ICP ion source. Neutron activation analysis has also been used to measure ^{135}Cs / ^{137}Cs ratios in nuclear materials [5], but interference from ^{134}Cs, the activation product of ^{133}Cs, make these measurements very difficult in samples containing large excesses of stable Cs. RIMS avoids isobaric interferences by using tunable lasers to selectively ionize the element of interest [7-9]. In the RIMS system discussed here, high-resolution continuous-wave (cw) lasers significantly improve the sensitivity and selectivity with respect to pulsed laser systems [8, 9]. Previous work on Sr [10-12], Kr [13], and Ca [14-16] illustrate the potential of this technique.

For high-resolution RIMS, the overall isotopic selectivity is obtained from both the mass spectrometer and optical isotopic selectivity provided by the laser excitation process. The overall detection efficiency, which is an important parameter for ultra-trace measurements, depends on a number factors including atomization efficiency of the source, transport of atoms to the laser ionization region, laser induced ionization efficiency, and mass spectrometer transmission. In this work, we have investigated chemical procedures to optimize the production of neutral Cs atoms and have used a single-resonance excitation scheme combined with optical pumping techniques to improve optical isotopic selectivity [6, 17]. This method yields an optical selectivity greater than 10^2 for ^{135}Cs and ^{137}Cs against ^{133}Cs. In combination with a 90° magnetic sector mass spectrometer an overall selectivity of $\approx 1 \times 10^8$ was observed. A separate high-power argon ion laser is used for ionization of the excited state atoms to improve ionization efficiency. The NIST system was compared with a similar RIMS system at Pacific Northwest National Laboratory (PNNL) [6]. The PNNL system combines a quadrupole mass spectrometer with a similar laser system and the experimentally observed overall selectivity was $\approx 1 \times 10^9$ and the overall detection efficiency was $\approx 2 \times 10^{-6}$. For comparison, both RIMS systems were used to measure the ^{135}Cs / ^{137}Cs isotopic ratio in a standard nuclear burn-up sample and the results showed good agreement and are consistent with conventional TIMS measurements.

EXPERIMENTAL

The experimental system consist of a graphite crucible for sample loading and atomization, a laser system for excitation and ionization of the isotopes and a 90° magnetic sector mass spectrometer for mass separation and detection of the ions. The atomization crucible is a graphite tube of 2 mm inside diameter and 20 mm depth. The exit orifice of the crucible is ≈ 14 mm away from the laser interaction region. The atomic beam emitted from the crucible is collimated to ≈ 10° full angle of divergence by a set of apertures and is irradiated in perpendicular geometry with the laser beams passing across the small dimension of the apertures. Single-resonance excitation of the $6s\,^2S_{1/2}\,(F = 4) \rightarrow 6p\,^2P_{3/2}\,(F' = 3,4,5)$ transition of Cs was performed with an external cavity diode laser (EOSI model 2010[1]) operating near 852 nm and delivering a power of 7 mW. Direct ionization of the excited atoms was accomplished with ≈ 5 W from the 488.0 nm line of an argon ion laser. The ions were accelerated to 8 kV and focused by the source lens system before entering a 90° magnetic sector analyzer. A schematic diagram of the system is shown in Fig. 1. In the initial phase of a measurement, at relatively low crucible temperature and atomization rate, the line center of the ^{133}Cs resonance was located with the help of a wavemeter (Burleigh WA1000). The diode laser frequency was then actively stabilized by fringe-offset-locking [18, 19] in a 300 MHz free-spectral-range confocal interferometer (Coherent 216-C), relative to a stabilized single-mode He:Ne laser (Melles Griot 05-STP-910). This computer-controlled stabilization system limits frequency drifts to < 1 MHz/hour and allows the diode laser to be shifted and locked at an arbitrary frequency or to be scanned in an accurately calibrated manner. The PNNL system is similar and has been described previously in [6, 10].

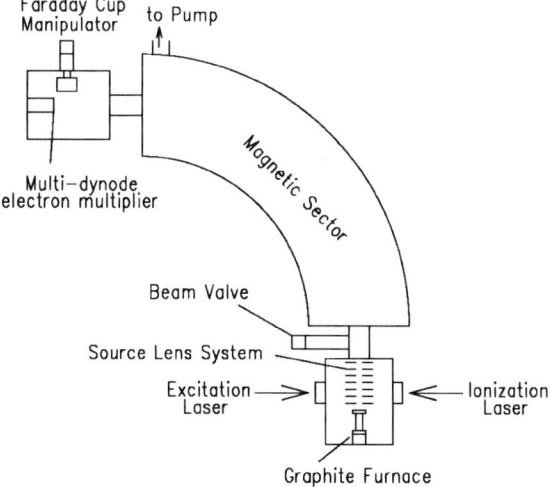

Figure 1. Schematic diagram of the NIST RIMS system.

For TIMS measurements at NIST, the graphite crucible and the source lens system were replaced by a system that uses rhenium filaments for thermal ionization [20].

---

[1] Mention of commercial products does not imply recommendation nor endorsement by the National Institute of Standards and Technology, nor does it imply that the products identified are necessarily the best available for the purpose.

This system has an overall efficiency for Cs detection of 7×10^{-3}. However, the isotopic selectivity for ^{135}Cs and ^{137}Cs relative to ^{133}Cs is $\approx 10^6$, limited by a constant background of scattered ^{133}Cs ions. This background level is typical for a single-sector mass separator, which generally cannot provide the extremely high abundance sensitivity as obtained in [2, 3]. Thus, TIMS isotopic ratio measurements were used to verify the RIMS measurements only for samples where ^{133}Cs was present at excesses $< 10^6$.

A NIST ^{137}Cs standard reference material, SRM 4233B-1, prepared from a nuclear burn-up sample was used for test isotopic ratio measurements made with the PNNL and NIST systems. This sample has a measured ^{135}Cs / ^{137}Cs ratio of 1.59717(46) determined by TIMS with a high-level sample at a reference date of August 22, 1977, and is known to have an approximate 10^2 excess of stable ^{133}Cs. For the RIMS measurements at NIST a reference solution was prepared from SRM 4233B-1 by dilution in 0.1 mol/L NH$_4$OH with a total activity of ≈ 1600 Bq/mL, without addition of extra stable Cs. Sample loads were 10 µL and corresponded to activity per load of approximately 16 Bq ($\approx 2\times10^{10}$ atoms ^{137}Cs). TIMS measurements used sample sizes of ≈ 2 µL, prepared from the same stock solution, and loaded directly onto the rhenium ionization filament. For the measurements at PNNL, the same standard, but with lower activity of ≈ 70 Bq/mL, was used and stable ^{133}Cs was added using a solution made from reagent grade CsNO$_3$ dissolved in deionized water. Measurement samples were then prepared by dilution of the stock solutions with 1 mol/L NH$_4$OH, as basic solutions were found to give a factor of 10 higher atomization yield than acidic nitrate solutions.

The NIST freshwater lake sediment SRM 4354 has a certified ^{137}Cs content of 8.104×10^7 atoms/g at February 14, 1986 12:00 PM EST. Samples for analysis were prepared using a simple chemical separation for cesium. The sediment was dried at 80°C for 24 hours and divided into 6 aliquots of 2 g. Two aliquots were spiked with ^{134}Cs (9.19×10^2 Bq and 7.04×10^2 Bq), two with ^{133}Cs (3.49×10^{17} atoms in both), and two were not spiked. Each aliquot was digested in 200 mL warm *aqua regia* for 24 hours and then centrifuged for 30 min at 3000 rpm. Two different procedures were used to isolate the cesium out from the supernatant and each procedure was applied to one aliquot of each type:

(a) 60 mg AMP(ammonium molybdophosphate) ion-exchange crystals were added at a pH ≈ 1.5. The samples were shaken vigorously for 5 min, then centrifuged and the liquid portion discarded. The AMP, containing the cesium, was air-dried and dissolved in 250 µL concentrated NH$_4$OH.

(b) the solutions were evaporated and 20 mL of concentrated HNO$_3$ was added to convert the cesium would be nitrate form. The solution was evaporated again and the residue dissolved in 120 mL of 8M HNO$_3$ at ≈ 50°C. The resulting solution was then passed through an EICHROM$^©$ TRU column to separate iron from the sample. Finally, the AMP was added and procedure (a) was carried out to separate the cesium.

The overall chemical yield for the sample preparation was measured for the [134]Cs spiked aliquots using a high-purity germanium (HPGe) gamma-ray detector system. The recovery was found to be 91% for procedure (a) and 82% for procedure (b).

RESULTS AND DISCUSSIONS

Analytical measurements were performed by repetitively setting the excitation laser to the $6s\,^2S_{1/2}$ ($F = 4$) $\rightarrow 6p\,^2P_{3/2}$ ($F' = 5$) for all three isotopes. This transition is preferred because it is the strongest component and it is the only transition that represents a closed two-level system. The hyperfine structure (HFS) in the $6s\,^2S_{1/2} \rightarrow 6p\,^2P_{3/2}$ resonance transition of [133]Cs is well-known [21] and the HFS and isotope shifts (IS) have been measured for radioisotopes over the range [118]Cs to [145]Cs [22]. The [133]Cs, [135]Cs, and ^{137}Cs isotopes all have nuclear spin $I = 7/2$ so they exhibit the same general level structure. The frequency shifts between the hyperfine components for the isotopes of interest were calculated using the data from [21, 22], and are given in Table 1. The diode laser beam diameter was increased to approximately 4 mm to allow pre-pumping of the atomic beam before it was exposed to the ionization laser. This allows optical pumping, which is advantageous because the [133]Cs ($F = 4 \rightarrow F' = 4$) transition frequency is almost degenerate with the desired [135]Cs ($F = 4 \rightarrow F' = 5$) component [6]. Thus, when the resonance laser is tuned for [135]Cs ($F = 4 \rightarrow F' = 5$), the [133]Cs signal is reduced via pumping into the $F = 3$ ground state component. The [137]Cs ($F = 4 \rightarrow F' = 5$) component does not coincide with an optically pumped transition of [133]Cs, however, it is far enough in the wing that the response is similar to that at the [135]Cs position. For both the NIST and PNNL systems, the [133]Cs response is reduced by 2 to 3 orders of magnitude when the resonance laser is tuned to either the [135]Cs or [137]Cs ($F = 4 \rightarrow F' = 5$) positions.

Isotopic ratio measurements were performed on samples prepared from the NIST ^{137}Cs SRM 4233B-1. For these measurements, the sample was first heated to just above the atomization threshold; then the laser frequency, the overlap of the resonance and ionization lasers, ion optics, and the mass spectrometer were tuned and optimized on the [133]Cs signal. The atomization crucible temperature was then slowly increased under computer control to evolve the sample while the mass spectrometer and diode laser frequency were switched every 10 seconds between positions for [135]Cs, ^{137}Cs, and a background measurement at mass 139 (with the laser still tuned for ^{137}Cs). The choice of 139 as a background position is somewhat arbitrary, as the background appears to be independent of the mass setting. These measurements were performed several times to test the accuracy and reproducibility of the determinations. Figure 2 shows the results, where open circles represent the individual RIMS measurements performed at PNNL and the filled circles represent the RIMS measurements at NIST. The error bars on the individual points are 1σ uncertainties derived purely from counting statistics. Additionally, TIMS measurements were performed with 5 different sample loads and 3 to 4 different ion count integration periods performed for each load. The TIMS average is represented by the solid triangle and the error bar is the standard deviation of the values obtained from each signal integration period. From the

TIMS result, a decay-corrected average ^{135}Cs / ^{137}Cs ratio of 2.79(6) was obtained. This is in excellent agreement with the original high-level high-precision TIMS measurement performed in 1977, yielding an expected value for the ^{135}Cs / ^{137}Cs ratio of 2.7946 (23), which is represented by the dashed horizontal line in Figure 2. The average value of the ^{135}Cs / ^{137}Cs isotopic ratios determined by RIMS at PNNL is 2.82 (5), while the average value from the NIST RIMS measurements is 2.83 (13). The two sets of RIMS measurements agree well with one another and with the TIMS measurements, lying well within the stated uncertainties. From the value of the reference ratio measured in 1977 and the RIMS ratios, one calculates an elapsed time of 24.6(8) years, which is in excellent agreement with the actual elapsed time of 24.3 years for the current measurements. If we use an initial value for the Cs ratio produced from ^{235}U of 1.0565 given in [1], an effective sample age of 42.6 (8) years is calculated.

Table 1: Frequency shifts for HFS components in the $6s\ ^2S_{1/2} \rightarrow 6p\ ^2P_{3/2}$ transition for ^{133}Cs, ^{135}Cs and ^{137}Cs. Only components originating in $F = 4$ of the ground state are shown and shift values are given relative to the $F = 4 \rightarrow F' = 5$ transition of ^{133}Cs.

			Shift (MHz)	
Isotope	F'(upper level)	Calculated[a]	PNNL (Ref. 7)	NIST
133	3	-452.25	-452.5(4)	-451.7 (10)
133	4	-251.00	-251.0(3)	-252.4 (12)
133	5	0	0	0
135	3	-733.12		
135	4	-520.18		
135	5	-250.03	-249.4(13)	
137	3	-1020.81		
137	4	-802.53		
137	5	-524.74	-522.1(35)	

[a] Calculated from [21] and [22].

Preliminary RIMS and TIMS measurements were performed on the freshwater lake sediment SRM 4354 samples. It was generally found that sample atomization occurred at higher temperatures than for the cleaner nuclear burn-up samples. However, removing the iron with procedure (b) lowered the atomization temperatures and reduced the number of background ions produced at high temperatures in the RIMS measurements, as compared to the samples prepared with procedure (a). Examples of mass spectra obtained by scanning the magnetic field for TIMS and RIMS measurements are shown in Fig. 3 (a) and (b), respectively. The SRM 4354 has a certified value only for ^{137}Cs, but not for stable ^{133}Cs. Thus the ^{133}Cs-spiked sample

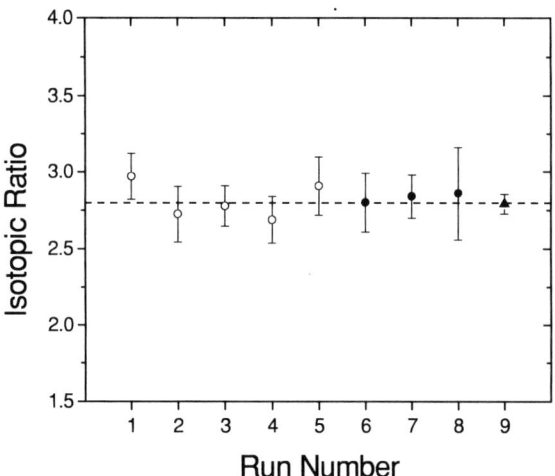

Figure 2. Results of repeated ^{135}Cs / ^{137}Cs isotope ratio measurements for the NIST 4233B-1 standard reference material. Open and filled circles are individual values from RIMS measurements at PNNL and NIST, respectively; uncertainty estimates are from counting statistics. The solid triangle is the average value obtained from repeated TIMS measurements with an uncertainty given by the standard deviation. The dotted line represents the expected value calculated from the 1977 reference ratio and the radioisotope half-lives. All values are decay corrected to a reference date of November 29, 2001.

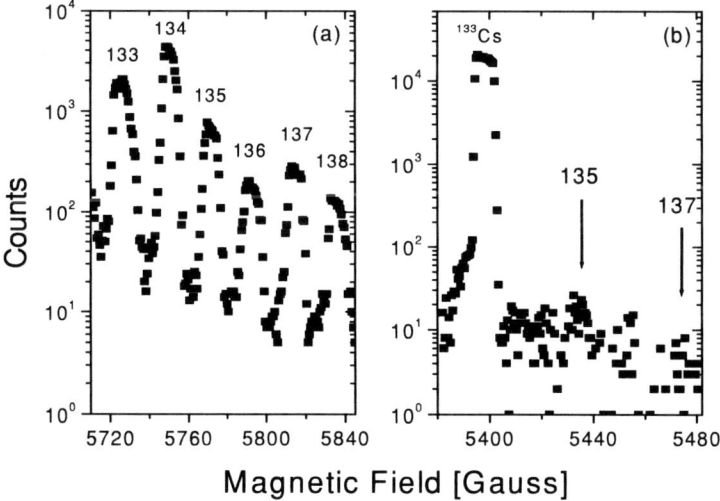

Figure 3. Magnetic field scan (a) TIMS measurements using a sample spiked with ^{134}Cs with the iron removed and (b) RIMS measurement on a non-spiked sample with the iron removed. The laser was tuned to the resonance of ^{133}Cs. The positions for ^{135}Cs and ^{137}Cs are indicated, but with the laser tuned for ^{133}Cs the signals at these positions are not attributed to the radioisotopes, but rather indicate the level of background that is to be expected during their measurement. The difference in magnetic fields required to transmit the same masses it due to differences in geometry and electric potentials for the RIMS and TIMS sources.

was used to determine an overall RIMS detection efficiency of $\approx 9 \times 10^{-8}$ by integrating the ion counts as the crucible temperature was increased until vaporization of the sample was complete. Using this efficiency for similar measurements on the unspiked samples yields a [133]Cs content of $\approx 4 \times 10^{14}$ atoms/g. In the TIMS measurements, lack of elemental selectivity and wide ranging interferences at all masses (see Fig. 3a) made it impossible to derive reliable information on cesium content in the sample. Since [133]Cs is present in the sample at a $\sim 10^5$ excess above the [134]Cs spike and $\sim 10^7$ above the ^{137}Cs, it is clear that the observed signals must be attributed to interfering isobars rather than cesium. In contrast, the elemental selectivity and isobaric suppression obtained with the RIMS approach are apparent in Fig. 3b. However, some modest improvements in overall efficiency are still needed to address low-level environmental measurements of [135]Cs / [137]Cs. In this regard, modifications to the existing system and changes in the sample preparation are being made to address measurement of [135]Cs / [137]Cs ratios for the lake sediment using reasonably small sample sizes (on the order of a few grams).

CONCLUSIONS

The capability of the NIST RIMS system has been investigated for ultra-trace analysis of radio-cesium, which requires ultra-high isotopic and isobaric selectivity. Measurements of [135]Cs / [137]Cs ratios in a nuclear burn-up standard reference material were found to be in excellent agreement for RIMS systems at both NIST and PNNL and for conventional TIMS. From these isotope ratio measurements, we were able to accurately reproduce the elapsed time from the present measurements to the high-level calibration measurements that were performed in 1977, as well as to calculate the age of the sample when produced from [235]U. The NIST system has only recently added the RIMS capability to existing TIMS instrumentation and results given in this work are preliminary. Initial measurements on a freshwater lake sediment reference material were performed and demonstrated the capability for reducing interferences when dealing with complex sample matrices; however, further optimization of overall efficiency and sample preparation procedures are needed for performing [135]Cs / [137]Cs isotope ratio measurements on real environmental samples.

ACKNOWLEDGMENTS

We would like to thank Dr. Larry Lucas for preparation of the stock standard solutions used in this work. B. A. Bushaw, W. Nörtershäuser, and the RIMS measurements performed at PNNL were supported by the U.S. Department of Energy, Office of Science under contract DE-AC06-76RLO 1830. The portion of this work performed at NIST was supported by the Office of Standard Reference Materials.

REFERENCES

1.   F. W. Walker, J. R. Parrington, F. Feiner, *Nuclides and Isotopes*. GE Nuclear Energy 1989.

2.	J. J. Stoffel(s), D. R. Ells, L. A. Bond, P. A. Freedman, B. N. Tattersall, C. R. Lagergren, "A triple-sector mass spectrometer with high transmission efficiency and 10^{-11} isotope-abundance sensitivity", *Int. J. Mass Spectrom. Ion Proc.* **132**, pp. 217-224, 1994.

3.	T. Lee, T. -L. Ku, H.-L Lu, J. -C. Chen, "First detection of fallout Cs-135 and potential applications of $^{137}Cs/^{135}Cs$ ratios", *Geochim. Cosmochim. Acta* **57**, pp. 3493-3497, 1993.

4.	J. M. B. Moreno, M. Betti, G. Nicolaou, "Determination of caesium and its isotopic composition in nuclear samples using isotope dilution-ion chromatography-inductively coupled plasma mass spectrometry", *J. Anal. At. Spectrom.* **14**, pp. 875-879, 1999.

5.	J. -H. Chao, C. -L. Tseng, "Determination of ^{135}Cs by neutron activation analysis", *Nuclear Instruments and Methods in Physics Research A* **372**, pp. 275-279, 1996.

6.	L. Pibida, W. Nörtershäuser, J. M. R. Hutchinson and B. A. Bushaw, "Evaluation of resonance ionization mass spectrometry for the determination of $^{135}Cs/^{137}Cs$ isotope ratios in low-level samples", *Radiochim. Acta* **89**, pp. 161-168, 2001.

7.	G. S. Hurst, M. G. Payne, S. D. Kramer, J. P. Young, "Resonance ionization spectroscopy and one-atom detection", *Rev. Mod. Phys.* **51**, pp. 767-819, 1979.

8.	B. A. Bushaw, "High-resolution laser-induced ionization spectroscopy", *Prog. Analyt. Spectrosc.* **12**, pp. 247-276, 1989.

9.	M. G. Payne, L. Deng, N. Thonnard, "Applications of resonance ionization mass spectrometry", *Rev. Sci. Instrum.* **65**, pp. 2433-2459, 1994.

10.	B. A. Bushaw, B. D. Cannon, "Diode laser based resonance ionization mass spectrometric measurement of strontium-90", *Spectrochim. Acta B* **52**, pp. 1839-1854, 1997.

11.	B. A. Bushaw, B. D. Cannon, "Diode-Laser-Based RIMS Measurements of Strontium-90", *AIP Conf. Proc.* **454**, pp. 177-182, 1998.

12.	B. A. Bushaw, W. Nörtershäuser, "Resonance Ionization Spectroscopy of Strontium Isotopes via $5s^2 \, {}^1S_0 \rightarrow 5s5p \, {}^1P_1 \rightarrow 5s5d \, {}^1D_2 \rightarrow 5s11f \, {}^1F_3 \rightarrow Sr^+$", *Spectrochim. Acta. B,* **55**, pp. 1679-1692, 2000.

13.	G. R. Janik, B. A. Bushaw, B. D. Cannon, "Resonant isotopic depletion spectroscopy", *Opt. Lett.* **15**, pp. 266-268, 1989.

14.	P. Müller, B. A. Bushaw, K. Blaum, W. Nörtershäuser, N. Trautmann, K. Wendt, "Ultratrace Determination of the Long-lived Isotope ^{41}Ca by Narrowband CW-RIMS", *AIP Conf. Proc.* **454**, pp. 73-78, 1998.

15.	P. Müller, K. Blaum, B. A. Bushaw, S. Diel, Ch. Geppert, A. Nähler, W. Nörtershäuser, N. Trautman, N., and K. Wendt, "Trace Detection of ^{41}Ca in Nuclear Reactor Concrete by Diode-laser-based Resonance Ionization Mass Spectrometry", *Radiochim. Acta,* **88**, pp. 487-493, 2000.

16.	B. A. Bushaw, W. Nörtershäuser, P. Müller, K. Wendt, "Diode-laser-based resonance ionization mass spectrometry of the long-lived radionuclide ^{41}Ca with abundance sensitivity $< 10^{-12}$", *J. Radioanal. Nucl. Chem.,* **247**, pp. 351-356, 2001.

17.	B. D. Cannon, T. J. Whitaker, G. K. Gerke, B. A. Bushaw, "Anomalous linewidths and peak-height ratios in ^{137}Ba hyperfine lines", *Appl. Phys. B* **47**, pp. 201-206, 1988.

18.	B. A. Bushaw, B. D. Cannon, G. K. Gerke, T. J. Whitaker, "Laser-enhanced electron impact ionization spectroscopy", *Opt. Lett.* **11**, pp. 422-424, 1986.

19.	W. Z. Zhao, J. E. Simsarian, L.A. Orozco, G. D. Sprouse, "A computer-based digital feedback control of frequency drift of multiple lasers", *Rev. Sci. Instrum.* **69**, pp. 3737-3740, 1998.

20.	W. R. Shields, *NBS Technical Note 546*, Analytical Mass Spectrometry Section, Analytical Chemistry Division, Institute of Materials Research, National Bureau of Standards, Washington, DC, Nov. 1970.

21.	C. E. Tanner, C. Wieman, "Precision measurement of the hyperfine structure of the $^{133}Cs \, 6P_{3/2}$ state", *Phys. Rev. A* **38**, pp. 1616-1617, 1988.

22.	C. Thibault, F. Touchard, S. Büttgenbach, R. Klapisch, M. De Saint Simon, H. T. Duong, P. Jacquinot, P. Juncar, S. Liberman, P. Pillet, J. Pinard, J. L. Vialle, A. Pesnelle, G. Huber, "Hyperfine structure and isotope shift of the D_2 line of $^{118\text{-}145}Cs$ and some of their isomers", *Nucl. Phys. A* **367**, pp. 1-12, 1981.

An Intelligent Radiation Detector System For Remote Monitoring

Norman Latner, Norman Chiu and Colin G. Sanderson

Environmental Measurements Laboratory
U. S. Department of Energy
201 Varick Street, 5th Floor
New York, New York 10014-4811

Abstract. A unique real-time gamma radiation detector and spectroscopic analyzer, specifically designed for a "Homeland Security Radiological Network", has been developed by the Environmental Measurements Laboratory (EML). The Intelligent Radiation Detector's (IRD) sensitivity and rapid sampling cycle assure up-to-the minute radiological data, which will indicate fast changes in atmospheric radioactivity. In addition, an immediate alert will occur within seconds to signal rapid changes in activity or levels elevated beyond a preset. This feature is particularly valuable to detect radioactivity from moving vehicles. The IRD also supplies spectral data, which allows the associated network computer to identify the specific radionuclides detected and to distinguish between natural and manmade radioactivity. To minimize cost and maximize rapid availability, the IRD uses readily available "off the shelf" components combined with an inexpensive, unique detector housing made of PVC plastic pipe. Reliability with no required maintenance is inherent in the IRD, which operates automatically and unattended on a "24/7" basis. A prototype unit installed on EML's roof has been in continuous operation since November 27, 2001.

INTRODUCTION

While the need for a radiation detection system for Homeland Security is obvious, the method of detection is far less obvious. However, by defining the ideal system, we can eliminate many competing techniques and close in on the best choice.

The ideal system should be sensitive, robust and reliable, and yet relatively inexpensive and easy to fabricate. It should operate automatically and unattended, require no maintenance, and should supply complete weather data. The ideal system should respond to a rapidly rising radiation level within seconds and give complete results in 15 minutes or less. These results should include a spectrum to allow identification of the detected nuclides. Upon sensing elevated radiation, it should alert nearby personnel with audible and visual alarms, as well as communicating with a network. All data should be transmitted to the network and also be backed up on the system's hard drive.

The Intelligent Radiation Detector (IRD) is the result of our efforts to incorporate all these features into a realizable instrument (see Figure 1).

CP632, *Unattended Radiation Sensor Systems for Remote Applications,* edited by J. I. Trombka et al.
2002 American Institute of Physics 0-7354-0087-3

FIGURE 1. The IRD installed on EML's roof in downtown Manhattan, NY.

Description

The IRD, a unique gamma radiation detector and spectroscopic analyzer, specifically geared to the needs of a Homeland Security network, has been developed by the Environmental Measurements Laboratory (EML).

The IRD takes real-time measurements by using a continuously repeating 15-minute measurement sequence. The resulting gamma-ray spectrum, along with a single "activity number" that indicates the intensity of the radiation measured, is then transmitted to a network. If the "activity number" appears elevated, the network is alerted and performs a computer analysis of the spectra, which can then distinguish between natural and manmade radioactivity, as well as identify specific isotopes. The network can also signal the IRD to take shorter, more frequent measurements. Such a response would enable the rapid detection and tracking of a radioactive plume or cloud.

In addition, <u>at any time</u> during the normal measurement sequence, an <u>immediate alert</u> will occur within seconds to signal rapid changes in activity, or levels elevated beyond a 2 second preset limit. Thus, warnings of elevated radioactivity do not have to wait until a 15-minute sequence has been completed. When warnings do occur, both audible and visual alarms will alert nearby personnel. These same alerts are issued to the network. This mode is particularly useful for units installed at entrances to bridges and tunnels, at border checkpoints or near tollbooths to detect the passing of a vehicle carrying radioactive material.

In contrast with most conventional systems that simply show total gamma radiation activity, the IRD also supplies spectral data [1], which allows the specific radionuclides detected to be identified (see Figure 2). This important information can be extremely useful in minimizing radiological health effects and planning remediation efforts. For example, by identifying a radioactive release as isotopic iodine, protection against damaging thyroid uptake is possible by ingesting potassium iodide pills.

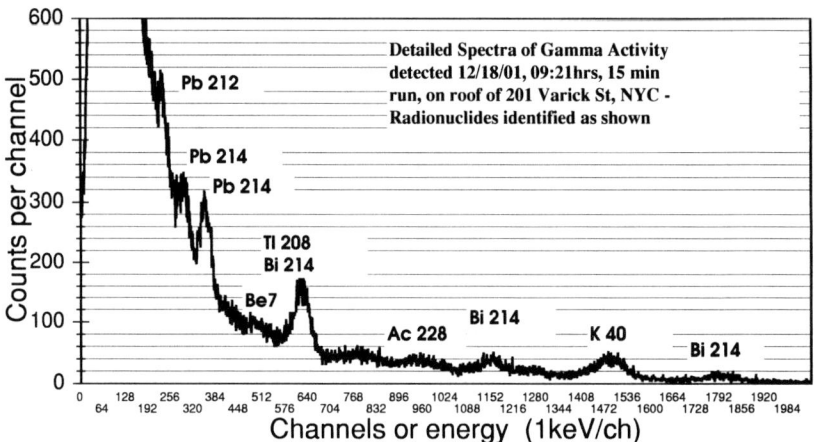

FIGURE 2. Gamma spectral data detected 12/18/01, 15 minute run (25 keV to 2035 keV).

Automatic and unattended operation is an important feature of the IRD. Once set up, this highly reliable unit needs no further attention, and in addition to transmitting all results to the network, the computer used with the IRD stores backup readings on its hard drive, which can hold some 30 years' worth of data.

Full meteorological capabilities are incorporated into the system as an aid in plotting the speed and direction of any radioactive clouds detected, as well as correlating normal activity peaks with precipitation (see Figure 3). Measured parameters include wind speed and direction, barometric pressure, rainfall, temperature, and relative humidity.

In the event of a domestic nuclear event, vast quantities of real time radiation data will be required for emergency response and crisis management. Radiation levels before the event (T-1), at the time of the event (T0), and immediately after the event (T++) will be required. A dense national network of IRDs can supply continuous real-time reporting of dose exposure related to a nuclear event, and will provide the required data. The network offers wide geographic coverage and because of multiple alarms, verifies any "above normal" radiation conditions. The network also acts as the communication link that will transmit data to a Central Station, which serves as a data

FIGURE 3. Weather station for meteorological data.

collection facility. The Central Station would coordinate the data, and perform spectral unfolding and isotopic identification. The spectral unfolding program can also account for detector temperature effects by shifting and stretching the spectra to find the best fit using naturally occurring peaks. In addition, by using both the meteorological and nuclear data from the sites, the Central Station could accurately map the movement of any radioactive release in real time.

To minimize unit cost and maximize rapid availability, the IRD uses "off the shelf" components and a rugged, weather resistant, yet inexpensive detector housing made of readily available PVC plastic pipe. The system uses any low-end computer, a Davis "Vantage Pro" weather station, and a relatively inexpensive multichannel analyzer, the Radiation Safety Associates – Universal Radiation Spectrum Analyzer, URSA (Figure 4) [2]. While the URSA has had a number of new and useful features added to optimize the IRD concept, they are now included as standard features in all units.

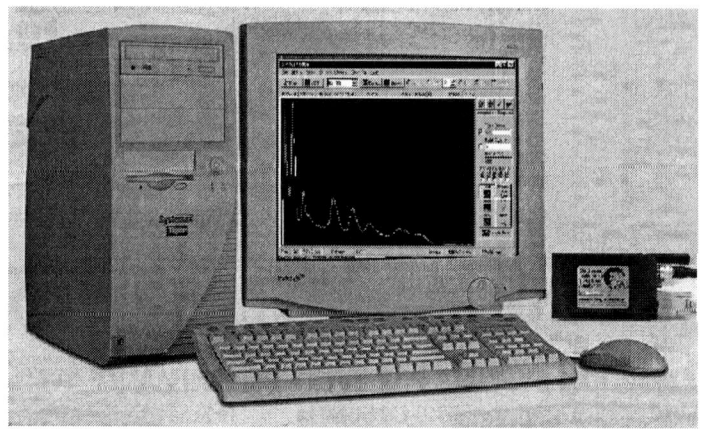

FIGURE 4. IRD computer system with URSA multichannel analyzer.

MATERIALS AND METHODS

Detector and Housing

The detector used in the IRD is a commercially available 3 by 3 inch inline, sodium iodide/ photomultiplier combination. This detector size was determined to be the best choice to maximize sensitivity, cost and physical configuration [3]. It is used in conjunction with a base/preamplifier that allows a long cable run between the detector and the URSA. At present, 50 feet of cable is being used.

The choice of sodium iodide (NaI) as the detector, rather than a competing technology, was in keeping with the goals of rapid response, reliability, efficiency, spectral information, economy, and zero maintenance [4]. Integrating systems, such as aerosol collection followed by counting, thermoluminescent dosimeters or film badges, were quickly eliminated for being far too slow to supply the results needed. Ionization chambers, Geiger counters and most other gas filled detectors cannot match the density, and thus the efficiency of NaI detectors, nor do they supply spectra. Germanium detectors, while capable of high resolution, are 35 to 50 times more costly than NaI detectors and add great complexity and maintenance effort to keep these detectors cooled. More exotic detectors, such as cadmium zinc telluride are expensive and only available in small sizes, and thus low efficiencies.

The detector and preamp are contained within a 24-inch high by 4-inch diameter weatherproof housing, constructed of readily available PVC Schedule 40 plastic pipe and associated fittings (Figure 5). The bottom of the housing consists of a cast iron flange, which lends great stability to the unit. Internally, two 3 to 2 inch adapters are cemented together and glued inside the 4-inch pipe to form a solid support for the detector/preamp. The cables going to the URSA are led out of a small slot near the base. The housing has proven itself to be thoroughly weather resistant, and should be able to sit in 6 inches of water without affecting operation.

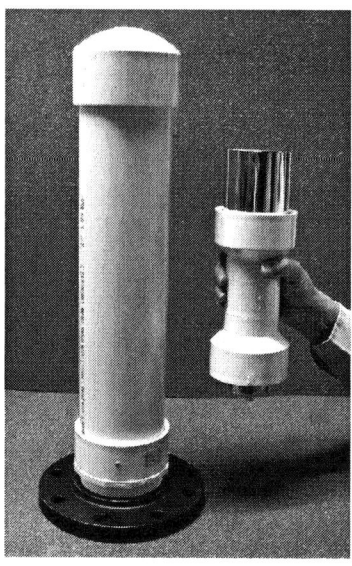

FIGURE 5. Sodium iodide inline detector and preamplifier shown outside of housing.

Multichannel Analyzer

The multichannel analyzer used is the URSA, a compact, relatively inexpensive, stand-alone unit that operates with most computer operating systems. It communicates with the computer via the serial port and offers very user friendly and intuitive software. The URSA supplies detector bias and preamplifier power. The normal operating mode consists of a continuously repeating 15-minute spectrum acquisition, followed by a network transmission of the results, as well as a local "save" on the computer hard drive. The transmitted and saved data include both the detailed spectra and a single activity number that is the sum of all the counts in the spectrum from 25 keV to 2035 keV. This sum, the integrated gamma activity, is plotted against time and offers a quick and easy way to assess rising levels of activity (Figure 6). The spectrum associated with any activity number can be easily retrieved so that specific isotopes may be identified. The URSA also offers up to 12 rapid alarm settings to cover all or part of the spectrum. Each alarm can be set to trigger above a preset value or a rate-of-rise occurring in a selectable time period of 1 second or more.

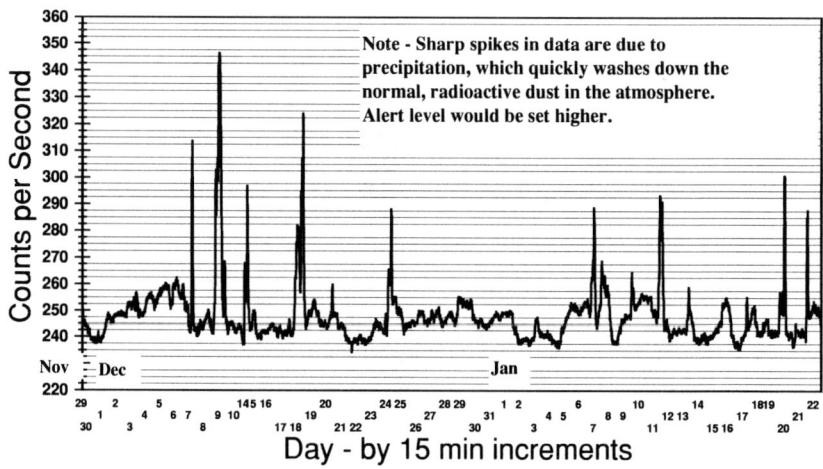

FIGURE 6. Integrated gamma activity (25 to 2035 keV) – roof data from November 2001 – January 2002.

RESULTS AND CONCLUSIONS

More than 16,000, 15-minute gamma-ray spectra have been recorded by the IRD since it was installed on EML's roof at 201 Varick Street, in lower Manhattan, NY, on November 27, 2001. The unit has been subjected to temperature variations from 20°F to 80°F with only minor gain shifts. The spectral unfolding program easily compensates for these small shifts.

Typical results during dry periods showed no isotopic activity above background. During periods of precipitation, either snow or rain, data analysis showed measurable amounts of naturally occurring bismuth-214 and lead-214 (from radon-222), as would be expected.

The IRD ideally meets the needs of a Homeland Security Radiological Network. It is inexpensive and completely maintenance free. Its only moving part is the computer hard drive, which stores the detector data. Since each IRD has its own computer, network operation, and communication with a central data center can be very flexible. Assembled with off-the-shelf components, deployments in large numbers will not be delayed because of production problems.

Data availability is extremely rapid due to the systems 15-minute measurement cycle and 2-second alert analysis protocol. Because the system measures and transmits a gamma-ray spectrum, it is possible to determine the composition of both natural and manmade radiation.

REFERENCES

1. Sanderson, C. G., Latner, N., Larsen, R. J., *Nucl. Instrum. Methods Res. A* **339**, 271-277 (1994).

2. Steinmeyer, P. R., "Universal Radiation Spectrum Analyzer Operation Manual," Radiation Safety Associates, Hebron, CT (2001).

3. Rozsa, C. M., "Energy Calculations for Selected Scintillation Products," Bicron Inorganic Scintillation Products, SGIC, Inc., FP1096R398, Newbury, Ohio (1996).

4. Knoll, G. F., *Radiation Detection and Measurement*," John Wiley & Sons, New York, 1989, ISBN 0-471-81504-7.

Low Cost Autonomous Field-Deployable Environment Sensors

Robert L. Kremens, Andrew J. Gallagher, Adolph Seema

Center for Imaging Science, Rochester Institute of Technology, 54 Lomb Memorial Drive, Rochester NY, 14623

Abstract. An Autonomous Environmental Sensor (AES) is a miniature electronic package combining position location capability (using the Global Positioning System (GPS)), communications (packet or voice-synthesized radio), and environmental detection capability (thermal, gas, radiation, optical emissions) into a small, inexpensive, deployable package. AESs can now be made with commercial off-the-shelf components. The AES package can be deployed at a study site by airdrop or by workers on the ground, and operates as a data logger (recording data locally) or as a sentry (transmitting data real-time). Using current low-power electronics technology, an AES can operate for a number of weeks using a simple dry battery pack, and can be designed to have a transmitting range of several kilometers with current low power radio communication technology. A receiver to capture the data stream from the AES can be made as light, inexpensive and portable as the AES itself. In addition, inexpensive portable repeaters can be used to extend the range of the AES and to coordinate many probes into an autonomous network.

We will discuss the design goals and engineering restrictions of an AES, and show a design in a particular application as a wildland fire sentry.

INTRODUCTION AND CONCEPT

Studies in ecology, biology, geology and other fields often require collection of data in widely dispersed remote locations. The recent attack on the United States by extremist groups has created new requirements for remote and sometimes clandestine monitoring of radiation, airborne chemicals and microbes. Current data collection methods are cumbersome, expensive and manpower intensive and do not usually provide measurements over an extended period of time. For example, monitoring the surface temperature of a body of water to study its hydrodynamic and biologic properties is commonly done by field biologists, most often by crews collecting data manually from boats. A system that could automatically measure and process a number of physical parameters in a remote setting would thus be embraced by workers in a large number of fields. A tremendous advantage is obtained if such a system could be built inexpensively and adapted for a variety of applications without extensive redevelopment. We call such a low cost environmental re-configurable data collection system an autonomous environmental sensor (AES).

Recently, advances in positioning capability (using the global positioning system), electronics design, component packaging, networking and radio communication technology have made it possible to construct the AES in lightweight, compact inexpensive packages. The AES has the ability to record and analyze data and report the data and the position of the device by radio to a collection station. These devices

CP632, *Unattended Radiation Sensor Systems for Remote Applications,* edited by J. I. Trombka et al.
© 2002 American Institute of Physics 0-7354-0087-3/02/$19.00

can be deployed using a variety of means, and could collect data continuously, periodically or on command. The collected data can be reported immediately, stored and forwarded at a known time, or stored and retrieved later if the AES is recovered. Technology is available for recording a wide variety of physical parameters using inexpensive, compact sensors. Some of the environmental parameters and the corresponding sensors suitable for monitoring by AES are shown in Table 1.

An AES can be constructed using commercially available off-the-shelf components and software, speeding development time and minimizing the cost of the devices. Many of the devices and software developed over the last ten years for the cellular telephone and Internet computer infrastructure can be applied directly to the AES concept. The savings in development time and cost are substantial allowing simple AES systems to be deployed for as little as $100 each. (US$2002)

AN EXAMPLE APPLICATION: WILDLAND FIRE SENTRY

One of the major problems in combating wildland fires is monitoring the time history of the fire [1]. Understanding the size, location, and progression of the fire front is critical to optimal allocation of fire fighting resources and maintaining safety of the fire crew. Investigation of major wildland fire accidents involving loss of life indicates that the crews became imperiled because of insufficient or untimely information about the location and progression of the fire [2].

An AES can be used as a field deployed fire alarm that has the ability to report its location and whether a fire is in the vicinity. A fire can be detected by one or more inexpensive sensors in the AES that detect smoke, carbon monoxide, methyl chloride or temperature. These devices may also be equipped to record and transmit other data affecting fire spread like humidity and wind speed. The data gathered by the AES can be recorded locally to get a post-fire time history and is also transmitted by radio to individual firefighters equipped with appropriate receivers or to a central control receiver.

At present, once firefighters are on the ground near the fire site, they are effectively blind to the activity of the fire. Spotter planes and other aircraft may periodically over

Table 1. Environmental measurements suitable for monitoring with inexpensive sensor systems.

Environmental Parameter	Sensor	Cost (US$2002)
Temperature	Thermistor and signal conditioning (+/- 0.2K)	3 - 40
Humidity	Capacitive humidity sensor	10
Solar Flux	Cosine-corrected photodiode and signal conditioning	30
Radioactivity (α,β,γ)	Large area PIN photodiode and signal conditioning	50
Wind Speed	Thermistor cooling-rate monitoring	30
Water Turbidity, Particulate Smoke	IR 90° scattering cell	5
Gas Detection	Catalytic or electrochemical sensor and signal conditioning	5 - 30

191

fly the area and report the movement and location of the fire to the incident commander, but often even this rudimentary data is lacking. In its simplest use model, AESs provide direct real time voice data to firefighters on the ground. The time history of the fire can be kept manually by the incident commander by recording the position and time of AES fire alarms on paper maps.

Much effort has been expended in modeling the movement of fires in wildland settings [3,4] but these models are only as good as the detailed weather, terrain and fuel load information. Lacking precise information of the fire site, these complex fire models can predict fire behavior for short time periods, but must then be 'tuned' with actual data to obtain long-term accuracy. These fire models are similar to modern weather simulations that are similarly adjusted periodically with weather data to provide long-term modeling.

Using AESs, and armed with handheld computers running these fire models, firefighters will have accurate real-time data for model 'tuning', and may be able to more accurately predict the behavior based on past fire movement even when only very imprecise weather, fuel and terrain information is initially available. The ability to predict the movement of the fire is a powerful advantage to fire logistics and firefighter safety.

The use of satellites to obtain fire data for model tuning is feasible, but complications are imposed by limited satellite spatial resolution, complicated ground link equipment, and short satellite loiter time (for low Earth orbit satellites) over the target area. Real time data can be obtained using unmanned or remotely controlled unmanned flying vehicles (UFVs) flying over the fire site, but this solution is both

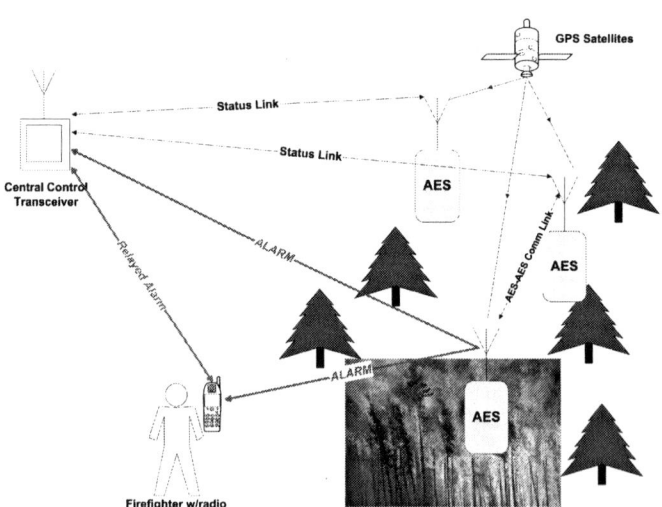

FIGURE 1. Deployment and communication between AESs and base units, other AESs and firefighters in a fully networked system.

complex and difficult to support in the field, and would require additional worker training to operate and maintain the UFV. A small number of AESs that are located in the forest could provide this data at low cost and with little additional effort in training or support. We present both advanced and simple AES communication concepts and show the initial design of a prototype.

OPERATIONAL CONSIDERATIONS

In use, AESs can be dropped from an aircraft or unmanned airborne vehicle (UAV) or placed manually by crews in a study area. The mechanical package of the AES can be designed for any of a number of applications, including urban environmental monitoring, wildland monitoring, or as drifters on bodies of water. The devices periodically report their position and status to each other, to a central receiver, or to radio receiving equipment carried by individuals.

After they are deposited in the fire area, AESs will find their location (via their internal GPS receiver) and report their initial position and fire alarm status via a radio link.

Networked Operation

One option for communication is a digital link with a network protocol. In this application, the radio link allows AES-to-AES as well as AES-to-base unit communication. A diagram of the communication links between the various units is shown in Figure 1. The AESs will periodically report their status to each other and a central control transceiver unit. Upon detection of a fire, the reporting AES or AESs will transmit an alarm to other AESs in the area and to the central transceiver. Crews in the area can be alerted either directly from the reporting AES, or through alarm messages that are relayed from the control transceiver. The control transceiver has the capability of overlaying geographical information system (GIS) maps with the location and alarm state of the AES, and presenting this data to the incident commander or other personnel at the fire site. This system represents a relatively complex configuration of the AES system.

Non-Networked Operation (Point-to-Point)

There are many simpler modes of operation of the AES system, one of which is depicted in Figure 2. In this mode, the AESs operate independently of each other, without a central control transceiver, instead reporting synthesized voice messages to firefighters on the ground. This message might contain the ID number of the AES, its GPS position and the alarm state of the device. Since the AESs have highly accurate synchronized clocks via their GPS receivers, each can be programmed to transmit at a slightly different time in order to avoid AES message collisions and interference, even when operating on a single radio frequency. The firefighter, equipped with nothing more than the present VHF/UHF FM radio transceiver ('handi-talki'), and with no

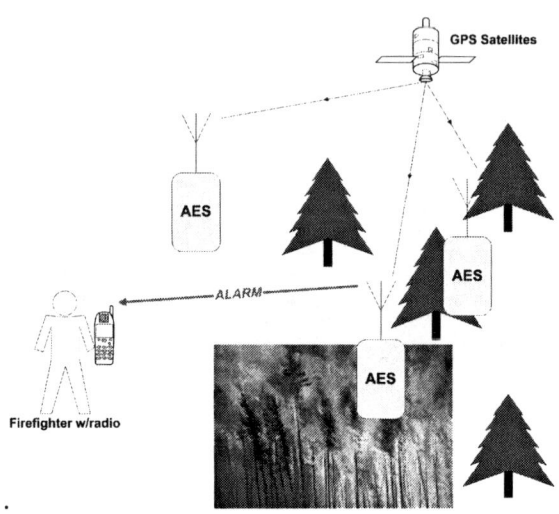

FIGURE 2. Simple point-to-point AES operation. The AES units report via synthesized voice messages directly to radio receivers at the fire site.

additional infrastructure, would be able to receive voice AES status on one of the normal radio communication channels. This mode of operation is most suitable when a few (~10) AESs are used on geographically small fires.

A block diagram of the AES is shown in Figure 3. A microprocessor coordinates inputs from the global positioning system receiver and the fire sensors, and generates the communication and modulation stream for the radio transceiver system. This communication stream can be digital, can employ packet or radio-teletype encoding technology or can be a synthesized voice, as discussed above. Several fire sensors may be used in an AES. These sensors can be smoke detectors (photoelectric or ionization), gas detectors (combustion precursor gases, carbon monoxide, etc.), thermal (temperature), passive microwave or optical radiation detectors. The AES internally measures the strength of signals from the sensors and makes decisions as to whether or not to issue an alarm. The use of more than one inexpensive detector can greatly reduce the probability of false alarms while not significantly increasing the cost. We are currently planning a test program to evaluate and optimize several fire sensor configurations during controlled wildland fires. The AES will be programmed to observe and report sensor input periodically and to 'sleep' in the interim to conserve battery energy.

PROTOTYPE DESIGN

We have constructed a prototype AES that conforms to the basic design discussed above. Key prerequisites of the design are cost effectiveness, durability, low power consumption and adequate transmitting range. Any design must be sensitive to the multiple design constraints of low cost, ruggedness, low power consumption, and

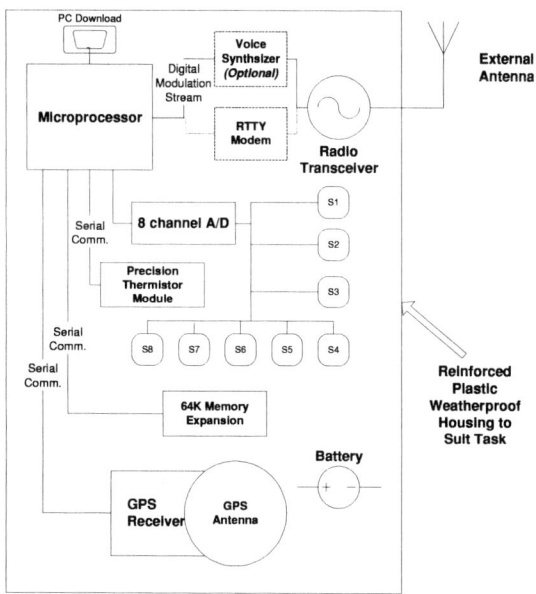

FIGURE 3. Block diagram of the AES, showing the logical and signal interconnections between the fire sensors, global positioning system receiver, radio transceiver and power system.

adequate transmitting range. The prototype uses commercially available components and systems where possible. The prototype device is currently transmit-only and uses a simple voice messaging system to report the unit ID number, position and time (as determined by the GPS) and alarm status or sensor level for up to eight sensors. The alarm information is transmitted at a predetermined time that is unique for each unit to avoid data collisions. The units use VHF (145 MHz) frequency modulated (FM) radio transceivers employing audio frequency shift keying (AFSK) to transmit digital information at 9600 bits per second. The entire transmission lasts under a second, conserving battery power. A radio receiver-demodulator attached by a serial link to a laptop computer receives the data stream and displays the messages. All of the components, including the radio transmitter, were obtained commercially. A photograph of the circuit boards of the prototype is shown in Figure 4, and a detailed block diagram is shown in Figure 5.

In order to speed development, a Parallax, Inc. Basic Stamp 2 microcomputer module (BS2) was used for the central processing unit. The BS2 has 16 digital input/output lines that can be programmed individually to perform a number of functions. This microcomputer module has been optimized for control applications and has several programmable 'sleep' modes that reduce power consumption to a very low level during idle periods (e.g. when no alarm or status information is being transmitted). The BS2 has a programming port and comes with development software using the powerful PBASIC programming language. Programs are developed on a PC-compatible computer and downloaded to the BS2, where they reside on an electrically

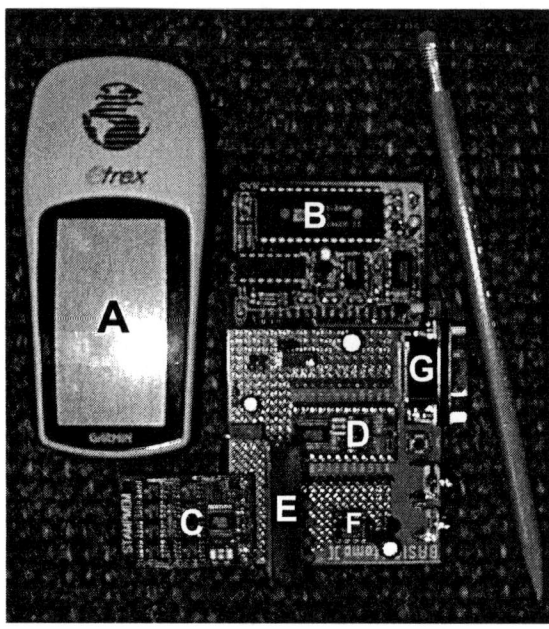

FIGURE 4. Photograph of the prototype AES. A - Packaged GPS unit; B - Voice synthesizer board; C - 64 kbyte non-volatile memory expansion unit; D - Parallax BS2 microcomputer module; E - Precision thermistor interface module; F - LTC1298 2 channel 12 bit analog to digital converter; G - Programming/ data downloading RS232 port.

erasable programmable memory (EEPROM). The combination of powerful input/output based programming language and self-contained development system makes the BS2 very easy to use for rapid development of simple applications. Detailed information about this processor can be found at the company's web site [5].

The prototype AES has two 12-bit analog input channels to accept input from fire sensors. A Linear Technology, Inc. LTC1298 ADC communicates with the microprocessor through a 3-wire synchronous serial link. We are currently evaluating several sensor types for suitability in this application. We have successfully detected test fires using commercial ionization chamber smoke and carbon monoxide detector modules. In addition to the analog inputs, several switch inputs provide access to test routines, enable the device after deployment, and halt the device for storage. Four status LEDs provide visual indication of unit ID number, self-test results and AES state information (alarm, transmitting, idle, etc.). An audio alert is also included to provide local indication of an alarm condition.

Asynchronous serial communication in RS-232 format is used to communicate with the transmitter and GPS unit. A commercial Garmin Model 12 GPS unit sends ASCII information indicating time, latitude and longitude over one serial link to the microprocessor. The messages transmitted by the GPS unit conform to the National Marine Electronics Association NMEA-0183 standard. The actual AES will use one of the readily available unpackaged GPS 'decks' that do not have displays or keyboard input and are smaller and consume less power. Another RS-232 link is used to communicate ASCII information to the radio modulator/transmitter. One of two modulators may be used, depending on the demands of the application. One modulator employs an AMD AM7910 audio frequency shift keying modulator/demodulator that allows transmission of an ASCII digital data stream. The other modulator we have tested is a voice synthesizer manufactured by Quradravox, Inc. Serial commands to this synthesizer allow direct voice communication. Both of these modulators drive a commercial VHF-FM transceiver unit which, in our application, operates at an output power of 5 watts in the 2-meter (145 MHz) amateur radio portion of the VHF spectrum. With a moderate antenna, this transmitter has a worst case range of more than 5 miles over water and about 5 miles on land, depending on the terrain.

A flow diagram of the software written for the prototype AES is shown in Figure 6. The powerful control-oriented PBASIC language simplified programming and reduced the initial effort to just a few days. There are other branch points in the software flow (not shown in Figure 6) that provide test functions (such as sensor and battery test) and readiness verification. Another Basic Stamp BS2 has been programmed as an input test set. This device produces simulated signals for the input switches and two analog voltages to represent sensor outputs, and allows us to test the AES without directly stimulating the sensors.

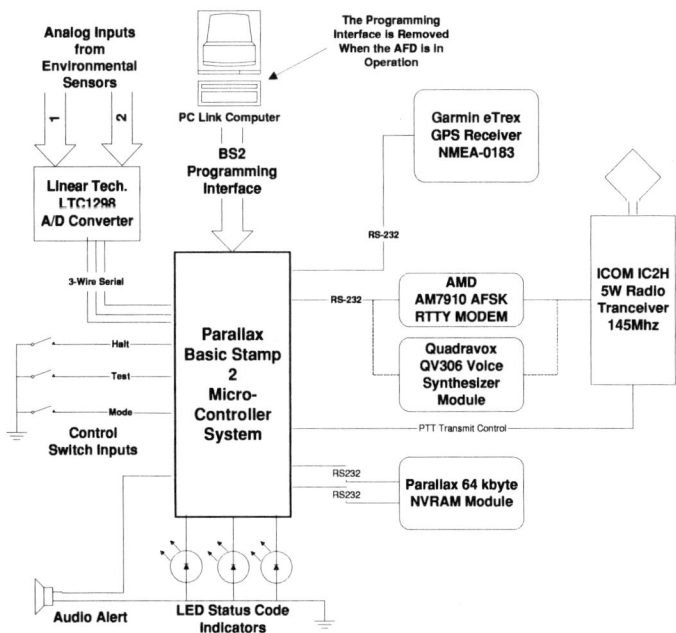

FIGURE 5. Detailed block diagram of the prototype AES. Fourteen of the sixteen available input/output pins of the BS2 are used in this application.

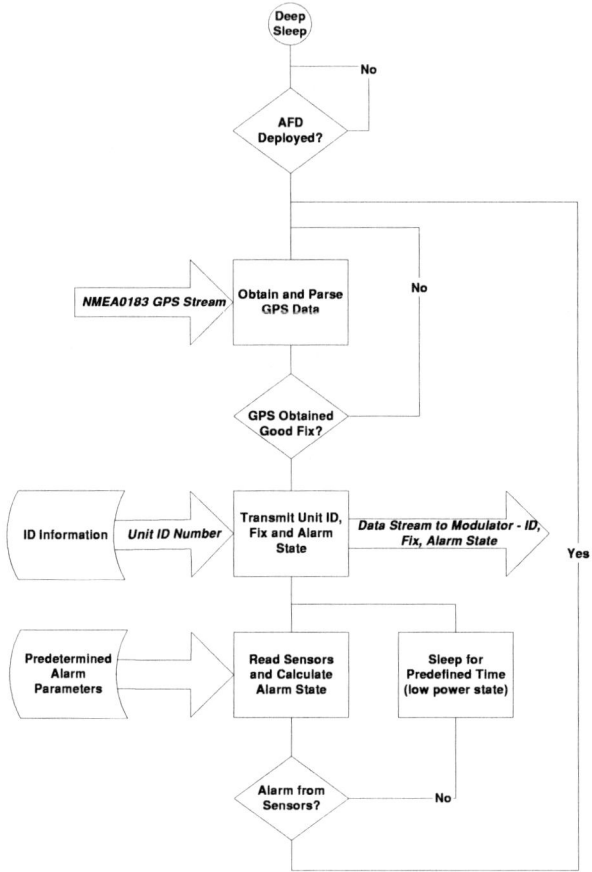

FIGURE 6. Software flow diagram for the AES prototype. The software makes use of the powerful instructions in the Parallax, Inc. PBASIC programming language.

CONCLUSION AND FUTURE DEVELOPMENT

We have demonstrated the concept and electronics for an autonomous environmental sensor, with a particular application as a wildland fire sentry. By maintaining the simplicity of both the physical hardware and operational principles, the device has been prototyped rapidly. We are currently evaluating 1 and 2-sensor wildland fire detector packages using small test fires, and will further test a completed mechanical and electrical package during controlled wildland burns planned with the United States Forest Service for the Spring of 2002.

ACKNOWLEDGEMENTS

The National Aviation and Space Administration supported this work under Grant NAG5-10051. Its financial support has been greatly appreciated.

REFERENCES

1. Chandler, C., Cheney, P., Thomas, P., Traubaud, L., Williams, D. *Fire in Forestry*, John Wiley and Sons, New York, 1983, Vol. II.
2. Rothermel, R.C., *Mann Gulch Fire: A Race That Couldn't Be Won*, General Technical Report, INT-299, United States Department of Agriculture, Forest Service, Intermountain Research Station, Ogden, Utah, 1993.
3. Andrews, P.L., *BEHAVE: Fire Behavior Prediction and Fuel Modeling System - BURN Subsystem, Part 1*, General Technical Report, INT-194, United States Department of Agriculture, Forest Service, Intermountain Forest and Range Experimental Station, Ogden, Utah, 1986
4. Finney, M.A., FARSITE: A Fire Area Simulator for Fire Managers, in the *Proceedings of The Biswell Symposium*, Walnut Creek, California, 1994
5. Parallax Corporation http://parallaxinc.com (2001)

A Comparative Study of Proposed Human Cytogenetic Fingerprints for Radiation LET

Joe N. Lucas[1] and Wen Deng[2]

[1]University of California, P.O. Box 2573, Dublin, CA 94568
[2]Department of Anatomy, Faculty of Medicine, The University of Hong Kong, Hong Kong

Abstract. A stable, easily measurable bioassay for past exposure to densely ionizing radiation would be of significant value to estimate environmental radiation risk. However, bioassays at present can only determine radiation doses if the radiation type or linear energy transfer (LET) is known. Our objectives are to identify the most effective cytogenetic "fingerprint" that strongly correlates with radiation LET, and is independent of dose. We *in vitro* irradiated human lymphocytes with 3.0 Gy ^{60}Co γ-rays, 0.9 Gy ^{3}H β-rays and 0.2 Gy 2.7 Mev neutrons, and conducted a detailed chromosome aberrations analyzed by combined fluorescence *in situ* hybridization with pan-telomere staining and specific whole chromosome painting. Among the 6 proposed radiation cytogenetic fingerprints, the ratio of total simple translocations to insertions (I ratio), showed the largest difference between low-LET ^{60}Co γ-ray and high-LET neutron radiation. The ratios of complete exchanges to incomplete rejoinings (S(I) ratio) and dicentrics to interstitial deletions (H ratio), showed a similar significant difference between low- and high-LET radiation. Other ratios measured showed no significant difference. We conclude that Pan-telomere staining with specific whole chromosome painting allows simultaneous and objective detection of complete or incomplete chromosome exchanges and interstitial or terminal deletions in human peripheral lymphocytes. Developing a distinctive clastogenic fingerprint should facilitate better detection and estimates of high-LET radiation exposure, as well as establishing a causal connection between early exposure to densely ionizing radiation in the environment and late development of cancer.

INTRODUCTION

Bioassays for ionizing radiation play critical roles in radiation exposure detection in nuclear radiation accidents, retrospective health studies of nuclear workers or veterans involved in early development of nuclear weapons, the nuclear power industry, and dose reconstruction as well as health risk assessment of astronauts and atomic bomb survivors. Unique compared to physical assays, bioassays utilize human tissues and cells as dosimeters for radiation exposure. However, bioassays at present can only determine radiation doses if the radiation type is known. If the quality of radiation, especially low linear energy transfer (LET) or high-LET, is unknown, the present bioassay methods cannot be used. However, cytogenetic measurements based on the following hypothesis will solve the problem.

Our overall hypothesis is that cytogenetic fingerprints can distinguish low-LET from high-LET radiation based on the synergy of the differential spatial distribution of ionizing events produced by low-LET and high-LET radiation and the proximity effect, which is facilitated by the topographic structure of closely folded interphase chromosomes.

Six cytogenetic "fingerprints" have been proposed recently based on this hypothesis. The first two ratios are based on the speculated mechanism that the higher proportion of difficult-to-repair

CP632, *Unattended Radiation Sensor Systems for Remote Applications*, edited by J. I. Trombka et al.
© 2002 American Institute of Physics 0-7354-0087-3/02/$19.00

double strand breaks (DSB) induced by high-LET radiation leads to a higher fraction of incomplete chromosome translocations, and small chromosome elements produced by high-LET radiation account for an increase in hidden complete translocations; thus S(I) ratio, complete translocations to incomplete translocations, and S(II) ratio, apparent complete translocations to hidden complete translocations, are expected to be fingerprints for distinguishing low-LET from high-LET radiation. The second two ratios are based on the mechanism that the small sized, densely clustered broken chromosome elements produced by high-LET radiation are more likely to undergo exchanges, in contrast to sparsely distributed broken chromosome elements; thus G ratio, acentric interstitial deletions to centric rings, and H ratio, dicentrics to acentric interstitial deletions, are expected to be another two fingerprints for radiation LET. The fifth ratio is based on the speculated mechanism that multiple breaks on different interphase chromosomes after high-LET radiation produce more chromosomal insertions, in contrast to aberrations derived from single breaks on different chromosomes; thus I ratio, complete translocations to insertions, is expected to be a chromosomal fingerprint for distinguishing low-LET from high-LET radiation. The sixth ratio, F ratio, is centric rings to dicentrics.

We recently developed a new technique of combined fluorescence in situ hybridization (FISH) for human lymphocytes with telomeric peptide nucleic acids (PNA) and whole chromosome-specific DNA probes, which allows clear staining of 100% of telomeres together with specific whole chromosome painting [1]. This FISH technique provides a powerful tool to make accurate measurements of all proposed cytogenetic fingerprint candidates for low- and high-LET radiation in the same cells. All intact human chromosomes contain telomeres, the repetitive DNA sequences at both terminal ends, maintaining chromosome structural integrity. Visualizing color patterns with whole-chromosome painting combined with telomere staining allows chromosome aberrations to be analyzed objectively, as illustrated in Fig. 1. The telomere length of human chromosomes is about 5–15 kb [2], much shorter than the minimum detectable size for whole-chromosome painting (11–14 Mbp) [3]. Thus, the combined pan-telomere detection and whole chromosome painting significantly improves the accuracy in detecting interstitial and terminal deletions, incomplete and hidden complete exchanges. This new technique enabled us to carry out a comparative study of the 6 proposed potential cytogenetic fingerprints and to identify the most effective fingerprint for distinguishing low-LET from high-LET radiation.

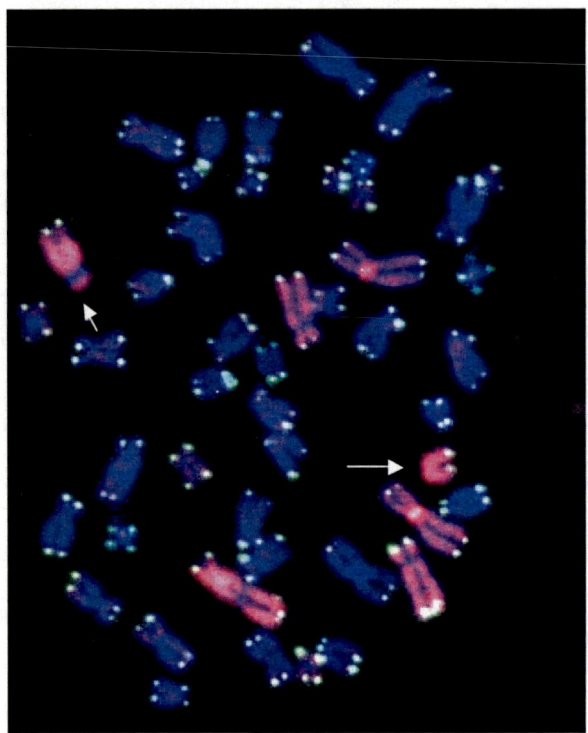

FIGURE 1. An example of a terminal deletion on chromosome 1 q arm (arrows): telomeric signals are present only at the terminal ends of both broken chromosome elements and no extra telomeric signal is observed at interstitial ends. A total of 92 chromosome telomeric signals are present in this human lymphocyte metaphase. The non-telomeric incomplete ends are rounded (arrows). Each of the other chromosomal ends shows separated chromatid ends with telomeric signals.

RESULTS

We detected of all telomeres (green signals) and whole chromosomes 1, 2 and 4 (pink signals) in normal human lymphocytes using our FISH method with pan-telomeric PNA probes and whole chromosome-specific DNA probes. All unpainted chromosomes were counter-stained blue with DAPI. Centromeres were distinct as bright white-blue signals viewed through a DAPI filter. Consistent with the previous observation for tritium β-ray-induced chromosome aberrations [1], our further analyses for ^{60}Co γ-ray and neutron-induced 714 chromosome aberrations showed no evidence of telomere addition or regeneration in peripheral lymphocytes after 52 h cell culture. These results show that complete or incomplete chromosome exchanges, interstitial deletions and terminal deletions can be analyzed accurately from telomere signals and whole chromosome painting color patterns (illustrated in Fig. 1).

The measured results for the proposed cytogenetic fingerprints for radiation LET are presented in Table 1. The ratios of total translocations to insertions (I ratios) were obtained from two kinds of FISH data, multi-color whole chromosome painting (see

"Materials and methods" section), and single color painting plus telomere staining, for γ-ray- and neutron-irradiated samples. The relative ratios of total translocations to insertions measured by multi-color whole chromosome painting or combined telomere detection plus whole chromosome painting did not differ significantly, thus the I ratios from pooled data obtained by the two FISH methods are presented in Table 1.

From Table 1, of the 6 proposed potential cytogenetic fingerprints for radiation LET, the I ratios showed the largest difference between low- and high-LET radiation. The I ratio for γ-rays was about three times as high as that for neutrons ($p \leq 0.01$). The H ratios (dicentrics to interstitial deletions) and S(I) ratios (complete exchanges to incomplete rejoinings) for ^{60}Co γ-rays were about two times as high as that for neutrons. The differences for both ratios are significant ($p \leq 0.05$). The G ratios (interstitial deletions to centric rings) showed a trend of LET related difference, although the difference is not significant by the criterion of $p \leq 0.05$. This is because centric rings occurred at relatively low frequencies, thus the statistical errors for centric rings were relatively high. The F ratios (dicentrics to centric rings) showed no significant difference between low LET radiation (^{60}Co γ-rays) and high LET radiation (2.7 Mev neutrons). This is also true for S(II) ratios (apparent complete exchanges to hidden complete exchanges). None of the ratios showed a significant LET related difference between ^{60}Co γ-rays and ^{3}H β-rays, at the statistical power of our data.

TABLE 1. Ratios Of Chromosome Aberrations Induced By ^{60}Co γ-Rays, ^{3}H β-Rays And 2.7 Mev Neutrons In Human Lymphocytes

Radiation	F Dic/CR	G Int. del./CR	H Dic/Int. del.	S(I) Ec/Inc. rej	S(II) AEc/HEc	I t/ins.
3.0 Gy ^{60}Co γ-rays	5.3 ±1.1	2.0 ± 0.5	2.6 ± 0.5 *	11.3 ± 2.2*	2.7 ± 0.3	23.1 ± 5.5**
0.9 Gy ^{3}H β-rays	5.8 ± 1.0	2.4 ± 0.5	2.4 ± 0.3	10.7 ± 1.6	2.5 ± 0.2	33.1 ± 11.3
0.2 Gy neutrons	5.2 ± 1.5	3.8 ± 1.1	1.4 ± 0.2 *	6.3 ± 1.3 *	2.5 ± 0.4	8.0 ± 1.1**

* $p \leq 0.05$; ** $p \leq 0.01$

Ec = Atc + Htc + ACdic + HCdic; Inc. rej = ti + Inc. dic. + Ter. del.; AEc = Atc + ACdic; HEc= Htc + HCdic; t = Atc + Htc + ti. The I ratios were obtained from multi-color whole chromosome painting as well as single color painting plus telomere staining, for gamma-ray- and neutron-irradiated samples. By multi-color whole chromosome painting, 305 translocations, 34 insertions were scored in 9833 cells for neutron irradiation, and 265 translocations, 12 insertions were scored in 1606 cells for gamma-ray irradiation. The I ratios measured by multi-color whole chromosome painting or combined telomere detection plus whole chromosome painting did not differ significantly, thus the I ratios in Table 1 were calculated from pooled data obtained by the two FISH methods.

DISCUSSION

To develop an effective cytogenetic "fingerprint" that strongly correlates with radiation LET, and is independent of dose, would be most useful in radiation detection, protection and health management; as well as identifying individuals suspected of smuggling or working with nuclear materials. We conducted a comparative study of the 6 proposed potential cytogenetic fingerprints in the same cells, and identified the most effective fingerprint for distinguishing low-LET from high-LET radiation.

A big concern in using cytogenetic fingerprints to distinguish low-LET from high-LET radiation is that, as low-LET radiation dose increases, multiple tracks of this low-LET radiation passing through a cell nuclei will produce more multiple chromosome breaks and complex aberrations. An immediate question raised by the concern is: can that look similar to high-LET radiation? To answer this question, we performed acute 3.0 Gy ^{60}Co γ-ray irradiation and compared the results with the cytogenetic ratios obtained for low dose neutron irradiation. The acute 3.0 Gy ^{60}Co γ-ray radiation was chosen because it is representative of a high dose and dose rate that is potentially lethal to humans [4]. Even at this high dose and dose rate of γ-rays, it still produced a much lower relative fraction of insertions compared with low dose high-LET radiation. Among the 6 proposed potential radiation cytogenetic fingerprints, the ratio of total translocations to insertions, I ratio, showed the largest difference between low- and high-LET radiation (Table 1). The difference between I ratios for low-LET and high-LET radiation is consistent with the results calculated from Grigorova *et al.* [5] data for Chinese hamster splenocytes after 1.0 Mev neutron and 200 kvp X-ray irradiation (they presented the fraction of insertions relative to translocations). Further support of the LET related difference for I ratio can be found from the report by Griffin *et al.* [6] on chromosome aberrations in human fibroblasts after α particle and X-ray irradiation, and from the data in human lymphocytes after 1 GeV/u Fe ion and γ-ray irradiation by Wu *et al.* [7], although they did not present I ratios.

High-LET Radiation Is More Effective In Producing Incomplete Rejoinings Than Low-LET Radiation

To compare the percentages of incomplete exchanges for low- and high-LET radiation, recently Wu *et al.* [7, 8] conducted FISH in human lymphocytes using telomeric and whole chromosome-specific DNA probes. However, in their studies only "potential incomplete exchanges" were measured, and "true incomplete exchanges" were estimated. They estimated that the percentage of "true incomplete exchanges" was about 3% for γ-rays and 4-8% for 1 GeV/u Fe ions. In their calculations, insertions were counted as complete exchange events. The percentage difference for incomplete exchanges between low- and high-LET radiation was not as remarkable as our directly measured results for percentages of simple incomplete to total simple exchanges (5% for γ-rays and 10% for 2.7 Mev neutrons, insertions were not included). This discrepancy is understandable considering the extra errors introduced by correction for percentage of telomere ends displaying telomere signals in the estimation of true incompletes from potential incompletes by Wu *et al.* [7, 8].

In this comparative study of candidate cytogenetic radiation fingerprints, we used incomplete rejoinings which included incomplete exchanges plus terminal deletions to calculate S(I) ratios. Our data showed a significant difference for S(I) ratios between low- and high-LET radiation. It is generally believed that DSB induced by ionizing radiation are the most important lesions leading to the formation of chromosome aberrations and cell lethality. Several studies have shown that rejoining kinetics of DSB induced by high-LET radiation were much slower than that of low-LET radiation, which may be due to the complexity of DNA damage that can occur in a nucleosome-sized structure (in a range of a few nanometers), including one or more DSB as well as associated single strand breaks, damaged bases and DNA-protein and DNA-DNA cross links induced by high-LET radiation [9, 10, 11]. *We postulate that the high fraction of slowly rejoined or un-rejoined DSB after high-LET radiation may result in more un-repaired chromosome elements before DNA replication in cell culture, compared with low-LET radiation.* The un-repaired, broken sister-chromatids formed after DNA replication are very close to each other, thus they finally fuse together, not allowing further chromosome exchange- or rejoining-repair. This may result in the formation of incomplete exchanges and terminal deletions, which were measured in the radiation cytogenetic fingerprint S(I) in human metaphase lymphocytes.

High-LET Radiation Produces A Higher Relative Fraction Of Interstitial Deletions Than Low-LET Radiation

In agreement with Boei *et al.* [12], and based on the low percentage of incomplete rejoinings after both low- and high-LET radiation as well the disk-shape of most of interstitial deletions (conventionally termed double minutes), we reason that the two ends of most of the interstitial deletions must had rejoined, forming acentric rings. Recently, Sachs *et al.* [13] proposed that G ratio may be a more reliable fingerprint for radiation LET than F ratio, based on their rapid motion model. Similarly, Brenner (in Nakamura *et al.* [14]) suggested that H ratio, could be another stronger candidate fingerprint for radiation LET. The reason is that interstitial deletions or acentric rings are derived from more closely spaced chromosome lesions within one arm of a chromosome than centric rings from opposite arms of a chromosome and would be less sensitive to large-scale chromosome motion, if any occurs. Experimental data reported by Pandita *et al.* [15] using conventional staining showed a significant difference between H ratios for ^{137}Cs γ-ray and neutrons in human fibroblasts. Bauchinger and Schmid [16] re-analyzed their previous data obtained by conventional staining technique and found that both G and H ratios increased approximately linearly with the logarithm of LET. However, it could be argued that, with conventional staining, the true interstitial deletions and terminal deletions could not be discriminated objectively. The results from our present study with pan-telomere detection showed that high-LET radiation produced a higher relative fraction of interstitial deletions than low-LET radiation, which is in line with the speculations of Brenner (in Nakamura *et al.* [14]) and Sachs *et al.* [13. However, the data obtained by Wu *et al.* [7, 8] did not show a higher fraction of interstitial deletions for 1 GeV/u Fe ions than for γ-rays. The reason for the discrepancy between our data and those of Wu *et al.* [7, 8] is not clear.

CONCLUSIONS

In the observed 1426 chromosome aberrations after 52 h culture induced by ^{60}Co γ-rays, ^{3}H β-rays and 2.7 Mev neutrons, evidence for chromatid fusion but not telomere addition was observed. The combination of pan-telomere staining plus specific whole chromosome painting allows simultaneous and objective detection of complete or incomplete chromosome exchanges, interstitial or terminal deletions in human peripheral lymphocytes. Our comparative study on the 6 proposed potential radiation cytogenetic fingerprints allowed us to rank the proposed fingerprint by their effectiveness in distinguishing low-LET from high-LET radiation in human metaphase lymphocytes. We demonstrated that the ratios of dicentrics to centric rings (F ratio) and apparent complete exchanges to hidden complete exchanges (S(II) ratio) showed no difference between low- and high-LET radiation. The ratios of centric rings to interstitial deletion (G ratio) showed a trend of LET related difference, but the difference was not significant at the statistical power of our data. The ratios of complete exchanges to incomplete rejoinings (S(I) ratio), dicentrics to interstitial deletions (H ratio) showed a similar significant difference between low- and high-LET radiation. Of the 6 proposed radiation cytogenetic fingerprints, the ratio of total translocations to insertions (I ratio), showed the largest difference between low-LET ^{60}Co γ-ray and high-LET neutron radiation. However, at the statistical power of our data, none of those ratios showed a significant difference between ^{60}Co γ-rays and ^{3}H β-rays (an intermediate LET radiation). This may indicate some limit of LET detection by using even the most effective radiation cytogenetic fingerprint, the I ratio. This important issue, distinguishing intermediate–LET from high or low–LET radiation, needs further work. In this report we have presented the proper tools to conduct such studies.

REFERENCES

1. Deng, W., and Lucas, J. N., *International Journal of Radiation Biology* **75**, 1107-1112 (1999).
2. Harley, C., Futcher, A. B. and Greider, C. W., *Nature* **345**, 458-460 (1990).
3. Kodama, Y., Nakano, M., Ohtaki, K., Delongghamp, R., Awa, A. and Nakamura, N., *International Journal of Radiation Biology* **71**, 35-39 (1997).
4. ICRP, Recommendations of the International Commission on Radiological Protection, Pergamon Press, New York, p. 93-193 (1991, 1990).
5. Grigorova, M., Brand, R., Xiao, Y. and Natarajan, A. T., *International Journal of Radiation Biology* **74**, 297-314 (1998).
6. Griffin, C. S., Marsden, S. J., Stevens, D. L., Simpson, P. and Savage, J. R. K., *International Journal of Radiation Biology* **67**, 431-439 (1995).
7. Wu, H., K. George, and Young, T. G., *International Journal of Radiation Biology* **75**, 593-599 (1999).
8. Wu, H., K. George and Young, T. G., *International Journal of Radiation Biology* **73**, 521-527 (1998).
9. Goodhead D. T., Thacker, J. and Cox, R., *International Journal of Radiation Biology* **63**, 543-556 (1993).
10. Jenner, T. J., deLara, C. M., O'Neill, P. and Stevens, D. L., *International Journal of Radiation Biology* **64**, 265-273 (1993).
11. Stenerlow, B., Blomquist, E., Grusell, E., Hartman, T. and Carlsson, J., *International Journal of Radiation Biology* **70**, 413-420 (1996).
12. Boei, J. J. W. A., Vermeulen, S., Fomina, J., and Natarajan, A. T., *International Journal of Radiation Biology* **73**, 599-603 (1998).
13. Sachs, R. K., Brenner, D. J., Chen, A. M., Hahnfeldt, P. and Hlatky, L. R., *Radiation Research* **148**, 330-340 (1997).
14. Nakamura, N., Tucker, J. D., Bauchinger, M., Littlefield, L. G., Lloyd, D. C., Preston, R. J., Sasaki, M. S., Awa, A. A., and Wolff, S., *Radiation Research* **150**, 492-494 (1998).
15. Pandita, T. K. and Geard, C. R., *Radiation Research* **145**, 730-739 (1996).
16. Bauchinger, M. and Schmid, E., *International Journal Radiation Biology* **74**, 17-25 (1998).

DETECTORS

Hand-Held Gamma-Ray Imaging Sensors Using Room-Temperature 3-Dimensional Position-Sensitive Semiconductor Spectrometers

Zhong He, Carolyn Lehner, Feng Zhang, David K. Wehe, Glenn F. Knoll
James Berry, and Yanfeng Du*

Nuclear Engineering and Radiological Sciences Department
The University of Michigan
Ann Arbor, Michigan 48109-2104

* Currently at GE Global Research Center
1 Research Circle – KWB617
Niskayuna, NY 12309

Abstract. This paper demonstrates the capability of compact gamma-ray imaging devices using 3-dimensional position sensitive CdZnTe semiconductor gamma-ray spectrometers, developed at the University of Michigan. A prototype imager was constructed and tested using two 1 cm cube 3-dimensional position sensitive CdZnTe detectors. Energy resolutions of 1.5% FWHM for single pixel events at 662 keV gamma-ray energy were obtained on both detectors, and an angular resolution of about 5° FWHM was demonstrated. The capabilities of proposed devices, which can cover a wider energy range up to 2.6 MeV, are discussed.

INTRODUCTION

The events of September 11, 2001, have increased national priorities for the development and deployment of instruments that could enhance capabilities to detect and monitor various nuclear materials. Advanced gamma ray sensors can play a role in this effort, particularly if they provide spectroscopic and imaging capabilities. For many applications, operation at room temperature and the avoidance of a requirement to cool the device would also be a significant advantage for hand-held portable instruments.

Isotopes associated with plutonium and uranium emit gamma rays in the energy range from 186 keV up to 2.6 MeV. Therefore, the measurement of characteristic gamma-ray lines can determine the presence of nuclear materials. Since the intensity of gamma rays from a point source is inversely proportional to the square of the distance to the source, hand-held gamma-ray sensors that enable inspection close to the source will increase the sensitivity of the measurement. In addition, if the instrument can provide the incident direction of gamma rays, gamma-ray spectroscopy can be performed for localized objects, thus further improving the sensitivity. In this paper, we investigate the performance of compact gamma-ray imaging sensors made using 3-dimensional position-sensitive CdZnTe gamma-ray spectrometers. These

CP632, *Unattended Radiation Sensor Systems for Remote Applications,* edited by J. I. Trombka et al.
© 2002 American Institute of Physics 0-7354-0087-3/02/$19.00

instruments allow room temperature operation, close-range inspection due to the compactness of the devices, and have imaging capability so that a source can be identified both by the spectroscopic measurement and by its location.

For systems in which small size and portability is important, the imaging method based on Compton scattering becomes a leading choice for gamma rays in the energy range between a few hundred keV to a few MeV. This technique does not require a collimator, which makes the device light-weight; covers a wide angular field of view, making inspection more efficient, and does not require compromise between angular resolution and device sensitivity. This paper demonstrates the capability of compact gamma-ray imaging devices using 3-dimensional position sensitive CdZnTe semiconductor gamma-ray spectrometers, which are being developed at the University of Michigan.

SYSTEM DESCRIPTION

The principle of a Compton imaging device can be described as follows: If the energy depositions and 3-dimensional coordinates of each gamma-ray interaction can be recorded, the original gamma-ray energy and the incident angle φ with respect to the axis defined by the first two interaction positions of the incident gamma ray can be obtained based on the Compton scattering formula:

$$\cos\varphi = 1 - \frac{m_e c^2}{E_{Scattering}} + \frac{m_e c^2}{E_{Incident}}$$

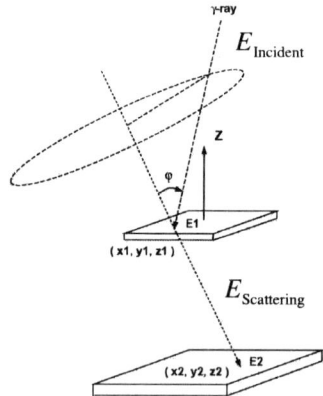

FIGURE 1. Illustration of a Compton-scattering imager.

In our first realization of the technique using CdZnTe, we employed two separate detectors as indicated in Figure 1. Two 1×1×1 cm cubic CdZnTe crystals were located in close proximity, and each provided 3-D spectroscopic readout. Pixellated anodes on each detector were connected to VA1 ASIC electronics through wire bonds [1]. The operational mode was to select single events in each detector, for which the first is a Compton scatter and the second is a photoelectric absorption. The second generation detectors were fabricated during 2000 to 2001 using 1.5×1.5×1 cm CdZnTe crystals, and the connection between pixel anodes and the inputs of ASIC are realized using plate-through-via technique. This anode-ASIC connection is much more rugged than the wire bonding used on first generation detectors. In a 3-D spectroscopic detector of sufficient size, a single detector can replace the function of the two individual detectors shown in Figure 1. Multiple interactions of each entering gamma ray photon are now recorded, and the sequence of these interactions reconstructed to provide the same kind of information. We are currently working on a large-volume CdZnTe detector that will begin to offer attractive detection efficiencies in this mode. It is based on a crystal grown by Yinnel Tech Inc. that was 2×2×1.5 cm in dimension when received. After etching and polishing at eV Products facilities, the final dimension will be ~ 1.8×1.8×1.3 cm.

ENERGY RESOLUTION

The overall energy resolution of a detector system is determined by the energy resolution of each individual detector. Energy resolutions of about 1.5% were demonstrated on the first generation 1 cm^3 detectors [1], and similar energy resolution is achieved on second generation detectors having 2 cm^3 detection volume. An example is shown in Figure 2.

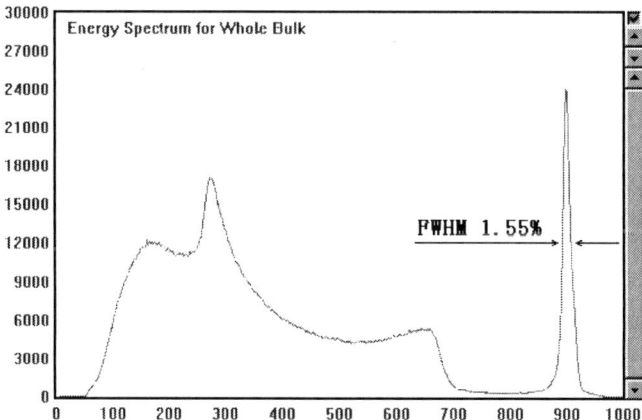

FIGURE 2. An energy spectrum of ^{137}Cs for single-pixel events obtained from a second generation 3-dimensional position sensitive CdZnTe detector. The spectrum was obtained from all 108 working pixel anodes, with a cathode bias voltage of −1000 V. The detector dimensions are 1.5×1.5×1 cm.

ANGULAR RESOLUTION

The angular resolution of a Compton scattering detector system is mainly limited by three factors: the detector energy resolution, position resolution (geometrical contribution) and the Doppler broadening effect [2]. Figure 3 shows the first prototype imaging system using two 1 cm^3 3-dimensional position sensitive CdZnTe spectrometers separated by a gap of 4 cm. The angular uncertainty due to each factor discussed above was investigated, and is shown in Figure 4 at a gamma-ray energy of 1 MeV. Experimental measurements showed an angular resolution of about 5 degrees FWHM at 662 keV gamma-ray energy, which is somewhat worse than 3.5 degrees predicted by Monte-Carlo simulations. It should be pointed out that the angular resolution improves at higher energies because of better energy resolution and the reduced effect of Doppler broadening on angular uncertainty.

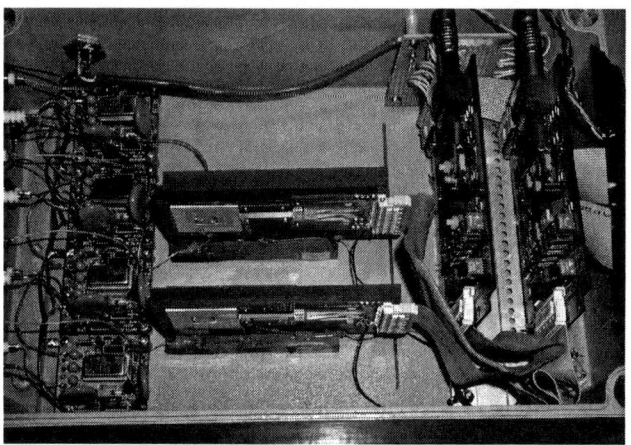

FIGURE 3. The top view of the first prototype Compton-scatter imaging system using two 1 cm cube 3-dimensional position-sensitive CdZnTe detectors.

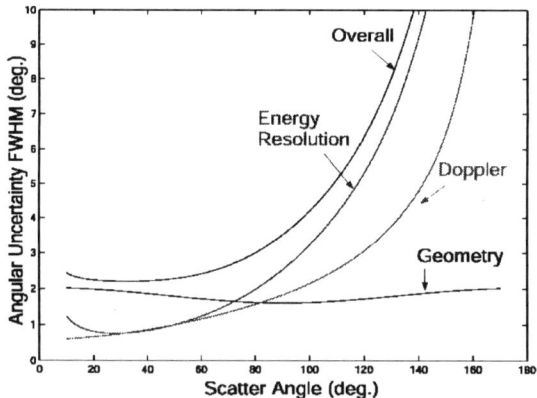

FIGURE 4. The angular resolution as a function of scattering angle for 1 MeV gamma rays predicted by Monte-Carlo simulations. Consistent results were obtained from experimental tests. The contributions to angular uncertainty from energy resolution, position resolution and the Doppler effect are shown individually.

When the new 1.8×1.8×1.3 cm CdZnTe detector is operated as an independent imager, the angular resolution will be dominated by geometrical uncertainty (position resolution). By assuming a position resolution of about 1.5 mm and an average separation of 6 mm (about half of the detector thickness) between two gamma-ray

212

scattering locations, an angular resolution of about $\Delta\theta \approx 24°$ FWHM should be expected (compared to 1 mm position resolution and 5 cm average separation between scatterings in the first prototype system). When a second detector is added to the system, and is separated from the first detector by 4 cm, similar angular resolutions of 3-5° FWHM at 662 keV, and 2.5°-4° FWHM at 1 MeV should be achievable. The addition of a third detector should not change the angular resolution significantly, but would increase the detection efficiency at higher energies.

DETECTION EFFICIENCY

An important property of a detector system is its sensitivity, which is determined by detector area, photopeak efficiency and photopeak fraction. Although it is still difficult to obtain single crystal CdZnTe with volumes larger than a few cm^3, the photopeak efficiency and in particular the photopeak fraction can be significantly improved by making intelligent use of the 3-D position resolution. An example is shown in Figure 5 at incident gamma-ray energy of 2.6 MeV, which are of interest in arms control applications. The energy spectra from a CdZnTe detector system with dimensions of 2×2×4.5 cm, which could be made using three 2×2×1.5 cm detector modules similar to the one being constructed, were investigated using Monte-Carlo simulations. The energy spectrum when the detector is operated as a simple gamma-ray spectrometer is shown on the left of Figure 5. A photopeak efficiency of about 10% and a peak-to-total ratio (photopeak fraction) of about 16% were obtained. The single and double escape peaks are clearly visible from pair production events. In a real environment, characteristic gamma-ray lines are superimposed on a gamma-ray background from the Compton continuum and escape peaks of higher energy gamma rays. The suppression of this background can significantly increase the sensitivity of the detection system. By using a 3-dimensional position sensitive detector, the complex signature of radiation interactions with the detection medium can be recognized so that events of interest can be selected and others rejected to reduce the background noise. For example, some events in the single escape peak can be recognized if a 511 keV energy deposition is detected at one location (through a photoelectric interaction) in coincidence with another energy deposition from the electron-positron pair generated in a pair production event. The correct photopeak energy can be obtained by adding 1022 keV energy (of two 511 keV gammas) to the energy deposition of the electron-positron pair. If there are three or more energy depositions detected in the system, the energy of the incident gamma-ray can be correctly reconstructed if the sequence of gamma-ray interaction can be recognized based on the kinematics of Compton scattering [3]. In addition, the majority of the Compton continuum and double escape events can be rejected if only a single energy deposition within the detector is recorded in the energy range for which the probability of a photoelectric interaction is very low. By implementing the event selection and reconstruction processes discussed above, the photopeak efficiency can be increased to about 15% (or improved by a factor of ~50%). Most importantly, the photopeak fraction is significantly increased to about 59%. As can be seen in Figure 5, the Compton continuum is reduced significantly, especially below the energy of the double escape peak. It should be noted that further reduction of the continuum between the double escape energy and the photopeak is possible. Some of the 511 keV gamma rays emitted from the annihilation of the electron-positron pair can be recognized if they deposit all their energy through Compton scattering in the device. This technique can be very useful if multiple gamma-ray lines must be observed, such as gamma rays at 2.6 MeV, 1.001 MeV and 911 keV for the identification of highly enriched uranium. Because the background continuum from higher energy gamma rays will be significantly reduced, the signal to noise ratio will be improved and the effective volume of a 3-dimensional position sensitive detector system can be several times larger than that of a simple spectrometer having the same detection volume. So far, we have not discussed the improvement of the system sensitivity due to the imaging capability of the instrument. By observing gamma rays from localized directions, the natural background, which tends to come from all 4π solid angles, can be further reduced.

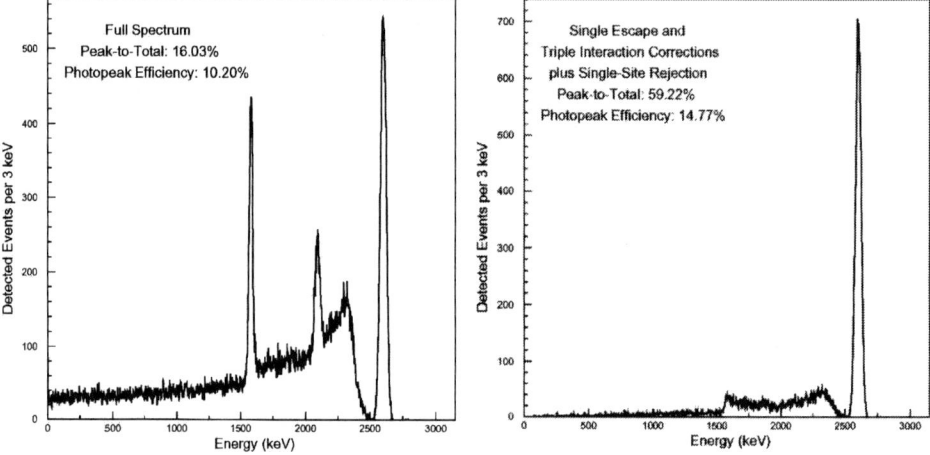

FIGURE 5. The comparison of energy spectra simulated from a 2×2×4.5 cm CdZnTe detector system, when the detector is operated as a simple spectrometer on the left, and as a 3-dimensional position-sensitive spectrometer on the right. 2.6 MeV Gamma rays normally incident on the cathode surface in Monte-Carlo smulations.

SUMMARY

The performance of a compact Compton imaging system using modular 3-dimensional position sensitive CdZnTe gamma-ray spectrometers has been investigated. Energy resolutions of about 1.5% FWHM at 662 keV gamma-ray energy have been demonstrated on prototype detectors, and angular resolutions of ~24° down to 3°-5° FWHM at 662 keV can be achieved depending on whether a single detector is employed, or 2 to 3 detector modules are used. A photopeak efficiency of ~15% and a photopeak fraction of ~59% can be achieved at 2.6 MeV, which should be of interest in nuclear non-proliferation applications.

ACKNOWLEDGMENTS

We want to acknowledge the support of DOE NEER program, grant number DE-FG07-98ID13645, and the support of DOE NN-20 office, grant number DE-FG03-98NV13357.

REFERENCES

1. He Z. et al., *Nuclear Instruments and Methods*, A 422 (1999) 173-178.
2. Du Y.F., "Development of a Prototype Compton Scattering Camera using 3-D Position Sensitive CZT Detectors," Ph.D Thesis, The University of Michigan, 2001.
3. Dogan, N., "Multiple Compton Scatter Camera for Gamma Ray Imaging," Ph.D Thesis, The University of Michigan, 1993.

New AIG Method of Growing Alkali Halide Crystals and Potential Application to CZT

A. Gleyzer* and E. Rhodes[†]

*PhotoPeak, Inc.,10180 Queens Way, Units 3 & 4, Chagrin Falls, OH 44023

[†]Space Department, Johns Hopkins University Applied Physics Laboratory,
11100 Johns Hopkins Road, Laurel, MD 20723-6099

Abstract. The new AIG (Advance Interface Growth) method has been successfully applied to alkali halide scintillation crystals at PhotoPeak, Inc. for the last four years. It produces single, stress-free crystals having a low level of defects and has resulted in increasing the yield of usable CsI(Tl) crystals to 75-85%. Essentially it is a low gradient method but has the capability to adapt the gradient to that needed by an individual crystal for the most successful growth. High quality crystals have been supplied to national laboratories and the nuclear medicine market. For example, a blank CsI(Tl) crystal 2" in diameter and 2" in length was produced having a measured energy resolution of 6.5% at 662 keV on a 2"-diameter PMT having a standard blue bialkali photocathode. This far exceeds the best resolution, 8.5-9.5%, obtained for CsI(Tl) crystals grown by the conventional Bridgman method. It is expected that this method can be successfully applied to grow high quality CZT crystals with substantially higher yield, 25-35%, than the presently existing 5-10%. The reasons for the expected improved yield of CZT crystals are that the phase diagram of CZT material has a narrow range of stability and CZT crystals should benefit from growth in a low gradient environment. Since the AIG method does not involve any moving parts, the temperature control and stability are much higher than for the conventional Bridgman method. The experience with CsI crystals indicates that imperfections like twinning, sparks, and multiplicities can be substantially reduced or even eliminated in CZT crystals. The expected higher yield and improved spectroscopic quality of CZT should allow many commercial applications to become a reality.

INTRODUCTION

The alkali halides scintillation crystals (NaI, CsI, KI, NaCl, etc.) have been known to the scientific and industrial communities for a long time. They are extensively used in high energy physics, nuclear medicine, space exploration, defense, security, and many other applications. The major characteristic of interest in these crystals is the ability to emit (scintillate) visible light as a result of interaction of a crystal with nuclear radiation. The amount of emitted light is proportional to the energy of a nuclear particle absorbed by the crystal, at the point of interaction. One of the most interesting features of scintillation, which sets scintillation crystals apart from another applications of crystals in industry, is that the light generated at a point in crystal could travel a long path before impinging on the photodetector. This fact requires that the whole crystal be of uniformly high quality and transparent to its scintillation light, in order to have good energy resolution.

It is well known that different manufactures of crystals have different level of quality, but it is not widely known that each manufacturer itself has quite large

CP632, *Unattended Radiation Sensor Systems for Remote Applications*, edited by J. I. Trombka et al.
© 2002 American Institute of Physics 0-7354-0087-3/02/$19.00

variation in quality of the crystals. For example, all major producers of alkali halide crystals give variation of afterglow in the range 0.5-5%, which points out that they do not have adequate control of this characteristic. The wide variation in quality of scintillation crystals points out that the technology of growing crystals is not very well established and the processes responsible for scintillation characteristics are not well understood. This paper addresses some of these problems.

DESCRIPTION OF THE METHOD

The majority of scintillation crystals are grown by two methods: Bridgman and Czochralski. These two methods have well known pros and cons, but it could be said with a good level of certainty that the Bridgman method (in the US) has been successfully used to produce large diameter (>20") crystals and that the Czochralski method (in the Ukraine) has been used to produce good crystals smaller (<15") in diameter but relatively tall (20-30"). Approximately five years ago one of the authors of this article (A. Gleyzer) has articulated the goal of growing consistently high quality alkali halide crystals with minimum variations.

A new type of furnace was designed and built to fulfill this goal. The furnace has two independent zones. A quartz ampoule with a charge in it is placed inside of an envelope, which provides good axial and radial gradients for the charge. The geometry of the envelope is variable to change the gradient from run to run. The furnace has no moving parts and the solidification is provided electronically by changing heat flow and field temperature inside the furnace. The temperature stability of the furnace is maintained within .03°C/day.

The macro-gradient of the furnace is changed with time, with the purpose of maintaining a constant gradient at the melt/solid interface. The solidification of crystal is provided by moving the melt/solid interface upward, and we call this method Advance Interface Growth (AIG). This method of growing crystals can be considered as a blind growth (no probe of position of interface). Using the described furnace and method, high quality single CsI(Na, Tl) crystals were manufactured. Fig. 1 shows a picture of a CsI(Tl) single crystal of diameter 4" and height 5.5". The cooling and annealing of the crystal is done in the furnace, and since the crystal is grown in the sealed ampoule, there is minimal interaction of the crystal with environment.

SCINTILLATION RESULTS FOR CsI (Tl, Na)

The AIG method was successfully developed and applied at PhotoPeak for the last four years. As of now the company has seven furnaces and routinely produces single crystals 4" in diameter and up to 6" long. The crystals grown by PhotoPeak have been used in nuclear medicine, space exploration, defense, and security. Fig. 2 shows the spectrum for 662 keV (Cs-137 source) taken with a CsI(Tl) crystal of diameter 4 cm and height 5 cm, using an 11-stage bialkali Hamamatsu PMT. The attained 6% resolution can be considered one of the best for this size of CsI(Tl) crystal. A resolution of 6.3-6.5% is routinely obtained with the AIG method for cylindrical crystals of size 2"x 2". These outstanding results were obtained without any special surface treatment or any special reflective materials. It should be pointed out that

217

FIGURE 1. CsI(Tl) single crystal of 4-inch diameter.

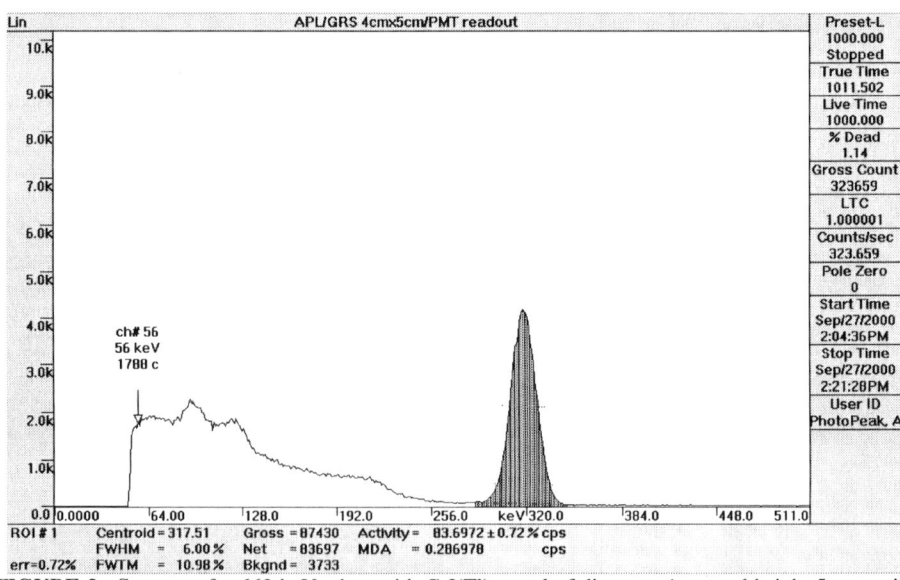

FIGURE 2. Spectrum for 662 keV taken with CsI(Tl) crytal of diameter 4 cm and height 5 cm, using an 11-stage bialkali Hamamatsu PMT. (Abscissa is uncalibrated.)

resolution on the level 6.2% is better than the resolution for the majority of NaI crystals of the same size, which is considered the standard in nuclear measurements.

CsI(Tl) has a scintillation maximum wavelength of 556 nm and so is a very good match to the response of a photodiode. A large (80 cm^3 volume) tapered crystal has measured 7.2% resolution at 662 keV on a 10x10 mm^2 Hamamatsu photodiode (private communication, Dr. M. Meier, Los Alamos). The crystals grown at PhotoPeak show consistent afterglow characteristics in the range of 0.4-0.8%, compared with 0.5-5% on the market. It is worth noting that CsI(Tl) grown by the AIG method has a light output of 70-75% of NaI(Tl). By the same token, CsI(Na) (λmax=420nm) grown by the AIG method had shown 125-135% of the light output of NaI(Tl), which is the highest light output reported in the literature to date.

CsI(Tl) crystals grown at PhotoPeak have shown a high transmission of 83-86% in the IR spectrum (1-10µm). It is interesting to note that pure CsI (produced by another vendor) used in optical applications has only 62-65% transmission in the same IR range. Excellent optical characteristics are mainly responsible for the very good scintillation characteristics.

WHY THE AIG METHOD WILL BE SUCCESSFUL FOR CZT

The AIG method has proved to be very successful for growing alkali halide crystals of high quality. In the course of development, it was found that there is an optimal range of gradients for manufacturing good CsI crystals. Too large a gradient produces large stresses and large dislocations (strata) in a crystal, which in turn could be light scattering centers. Small gradients could cause formation of small bubbles and micro-voids, which could impede scintillation as well. The AIG method allows one to vary the gradient experimentally and define the "optimum" range (for each given level of temperature stability and initial purity of powder) of gradient and heat flux. The AIG method has built-in flexibility to adapt the gradient to that needed by an individual crystal for the most successful growth and to provide good thermal conditions for growing a crystal as a whole. The conventional methods of growing crystals more or less control thermal conditions in the vicinity of interface melt-solid only, leaving the rest of volume without control. This approach is good for controlling local electronic characteristics of a crystal but does not provide optimal conditions for the crystal as a whole.

The analysis of the phase diagram of CZT shows a narrow range of temperature stability in equilibrium for the congruent melt-solid. This fact indicates that temperature stability and temperature range should be controlled with a high level of precision and accuracy in the process of the growth. There is no accurate data for the yield (% of usable material), although the price of material is pretty high (~$2,000-$3,000 per cc). Based on this price, it is reasonably to assume that the yield of CZT is in single digits. The major problems of growing CZT are multiplicities, twining, bubbles, and high density of imperfections. It is especially difficult to grow radiation spectroscopy grade crystals. We believe that the approach of the described AIG method, namely treating growth of a crystal as a whole, opens up new opportunities to grow CZT with much higher yield (25-35%).

Nanocrystal-based Scintillators for Radiation Detection

Sheng Dai,[1] Suree Saengkerdsub,[1] Hee-Jung Im,[1] Andrew C. Stephan,[2] and Shannon M. Mahurin[1]

[1]Chemical Sciences and [2]Nuclear Science and Technology Divisions,
Oak Ridge National Laboratory, Oak Ridge, TN 37831
Email: dais@ornl; Phone: 865-576-7307; Fax: 865-576-5235

Abstract. Several metal chalcogenides (e.g., ZnS, CdSe/ZnS) are known to be highly efficient scintillators. Concerns related to the use of these commercially available inorganic compounds (particle size: μm) include their low solubilities in organic and polymeric matrices. Their preparation in inorganic matrices, such as sol-gels, results in non-transparent gels, thus lowering their efficiency as scintillating devices. By reducing their particle sizes from commonly used micrometer- into nanometer-size regimes, their optical properties and solubilities in both polar and nonpolar solvents can be controlled. The combination of sol-gel technique and the chemistry of inorganic nanocrystalline quantum dots will be demonstrated as a powerful method in preparing scintillating devices.

INTRODUCTION

Uses of commercially available inorganic semiconductors (bulk particle size range: μm) as scintillators have been limited by their low solubilities in organic and polymeric matrices. Likewise, their preparation in inorganic matrices, such as sol-gel, results in optically opaque gels with the lowering of photoluminescence (PL) quantum yields.

During the past two decades, there have been extensive investigations on semiconductor nanocrystals or quantum dots (QDs). These nanocrystals are often composed of atoms from groups II-VI or III-V elements in the periodic table. When the sizes of these QDs become comparable to or smaller than the bulk exciton Bohr radius, unique optical and electronic properties occur [1–4]. These effects arising from the spatial confinement of electronic excitations to the physical dimensions of the nanocrystals are referred to as quantum confinement effects. One such effect is the quantization of the bulk valence and conduction bands which results in discrete atomiclike transitions that shift to higher energies as the size of the nanocrystal decreases. With the size-dependent optical properties of QDs (especially photoluminescence property), QDs with specific sizes can be made for specific detection wavelengths over the whole UV/Vis range.

In 1993, Bawendi and coworkers [5] were the first group to synthesize highly luminescent CdSe QDs by using high-temperature organometallic procedure. Later, the deposition of a surface-capping layer such as ZnS or CdS in the core/shell QD

CP632, *Unattended Radiation Sensor Systems for Remote Applications,* edited by J. I. Trombka et al.
© 2002 American Institute of Physics 0-7354-0087-3/02/$19.00

structure was also found to dramatically increase the quantum yields of CdSe nanocrystals up to 40 – 50 % at room temperature [6–8]. Other key advances in this area also include the synthesis of highly luminescent CdTe, CdSe, and CdS QDs in large quantities [9–10], better understanding of the QDs' surface chemistry and the preparation of water-soluble nanocrystals [11–13], enabling their applications in microelectronics, optoelectronics [1–4] such as light emitting devices [14–15], and in biological and medical applications such as multicolor fluorescent labels for ultrasensitive detection and imaging [11–13, 16–17].

We report here the application of highly luminescent II-VI semiconductor QDs (particle size range: ca 1 to 20 nm) as an entirely new class of neutron detectors. The QDs prepared were compatible to various inorganic and organic matrices. The quantum confinement effects enabled the QDs to be used with a wide variety of detectors. With the combination of advantages from nanochemistry and sol-gel chemistry, the QDs were embedded into sol-gel matrix and optically clear sol-gel scintillators were obtained.

Objectives of Nanoparticle Research

This research is expected to contribute to the development of new methodologies leading to significantly greater efficiencies of position-sensitive neutron detectors and will give rise to a new generation of scintillating materials. The specific objectives are: (1) to synthesize a new class of radiation scintillators using nanomaterials (ZnS, CdS, CdSe, ZnSe, and shelled nanoparticles); (2) to develop a scientific basis for a new methodology to greatly enhance fluorescent efficiencies induced by nuclear reactions through quantum confinement effect.

RESULTS AND DISCUSSION

Two types of QDs were synthesized, namely core/shell (CdSe)ZnS and ZnS QDs. The syntheses involve the use of long chain phosphine oxide or phosphinic acid coordinating agents. In the presence of these surfactant molecules, reverse micelle structures are formed and the nucleation and particle growth of semiconductors occur only in the centers of reverse micelles, resulting in the control of particle sizes. At elevated reaction temperatures (150 – 360 °C), highly crystalline QDs are formed.

The core/shell (CdSe)ZnS QDs with various particle sizes were synthesized by a two-step method modified from the literatures [6, 9]. The first step involves the size-controlled synthesis of CdSe QD cores and the second step involves the passivation of CdSe core surface defect sites by the ZnS QD shell. Upon the surface passivation of CdSe core by the ZnS shell, the QDs' photoluminescence was enhanced by three orders of magnitude and the reported quantum yields were as high as 50% [6]. To obtain water soluble QDs, the surface ligand exchange with a hydrophilic ligand, dithiothreitol (DTT), was performed according to the procedure reported by Thompson and coworkers [13]. Hydrophilic QDs were then added during the preparation of lithiated (^{6}LiOH) gels. The resulted nanocrystal-based scintillators in sol-gel matrices were transparent and highly luminescent (**FIGURE 1**).

FIGURE 1. Photoluminescence from (CdSe)ZnS QD sol-gel scintillators with various particle sizes of QDs.

A test scintillator sample was fabricated consisting of sol-gel doped with (CdSe)ZnS nanocrystals with a peak fluorescence emission wavelength around 590 nm. The sample was irradiated with alpha particles from a small Po-210 alpha source and the scintillation pulses were detected and recorded using a standard photomultiplier tube, amplifier, and multi-channel board setup. **FIGURE 2**, the resulting pulse height spectrum of the scintillation pulses, shows the distribution of the pulses across the different amplitude values. Comparison with the signal from background radiation showed that the Po-210 source produced pulses in the scintillator that were comparable in amplitude to those produced in existing inorganic scintillators.

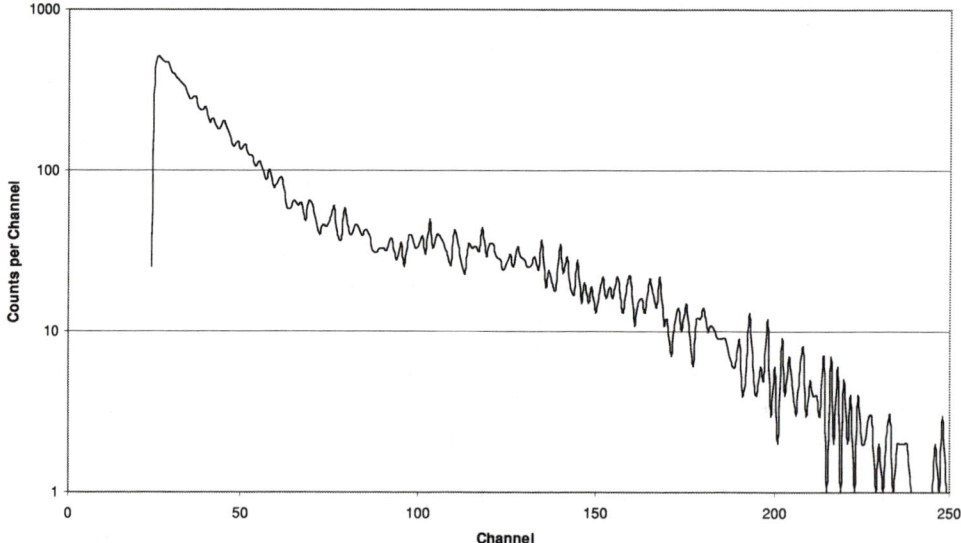

FIGURE 2. Pulse height spectrum of (CdSe)ZnS QD scintillator.

ZnS QDs were synthesized at an elevated temperature, giving rise to highly crystalline and highly luminescent QDs with a peak fluorescence emission wavelength at 380 nm (**FIGURE 3**, at the excitation wavelength of 337.1 nm using dye LASER). The photoluminescence band from the ZnS QDs presented here is much narrower

(FWHM 18 nm) than those reported from the ZnS nanocrystals prepared either at room temperature or by other methods (FWHM of at least 80 nm) [18–20]. Sol-gel based neutron sensors containing ^{6}Li and highly luminescent ZnS QDs with various dopants, including Ag^+, Mn^{2+}, and Ce^{3+}, are currently under an investigation in our group.

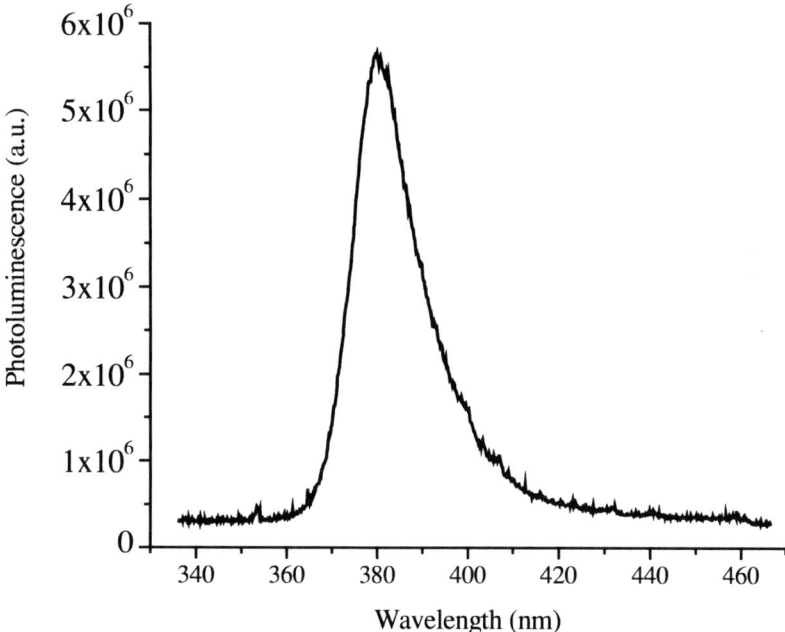

FIGURE 3. Photoluminescence spectrum of ZnS QD (excitation wavelength at 337.1 nm).

ACKNOWLEDGMENTS

The Oak Ridge National Laboratory is managed for the Department of Energy under contract No. DE-AC05-00OR22725 by UT-Battelle, LLC. Funding for this work is through the support of the Department of Energy NA-22. We also would like to thank Dr. D. Beach and Dr. D. Holcomb for helpful discussions.

REFERENCES

1. Henglein, A., *Chem. Rev.* **89**, 1861-1873 (1989).
2. Schmid, G., *Chem. Rev.* **92**, 1709-1727 (1992).
3. Alivisatos, A. P., *Science* **271**, 933-937 (1996).
4. Nirmal, M., and Brus, L. E., *Acc. Chem. Res.* **32**, 407-414 (1999).
5. Murray, C. B., Norris, D. J., and Bawendi, M. G., *J. Am. Chem. Soc.* **115**, 8706-8715 (1993).

6. Dabbousi, B. O., Rodriguez-Viejo, J., Mikulec, F. V., Heine, J. R., Mattoussi, H., Ober, R., Jensen, K. F., and Bawendi, M. G., *J. Phys. Chem. B* **101**, 9463-9475 (1997).
7. Peng, X., Schlamp, M. C., Kadavanich, A. V., and Alivisatos, A. P., *J. Am. Chem. Soc.* **119**, 7019-7029 (1997).
8. Hines, M. A., and Guyot-Sionnest, P. *J. Phys. Chem. B* **100**, 468-471 (1996).
9. Peng, Z. A., and Peng X., *J. Am. Chem. Soc.* **123**, 183-184 (2001).
10. Qu, L., Peng, Z. A., and Peng, X., *Nano. Lett.* **1**, 333-337 (2001).
11. Bruchez, M. Jr., Moronne, M., Gin, P., Weiss, S., and Alivisatos, A. P., *Science* **281**, 2013-2015 (1998).
12. Chan, W. C. W., and Nie, S., *Science* **281**, 2016-2018 (1998).
13. Pathak, S., Choi, S-K., Arnheim, N., and Thompson, M. E., *J. Am. Chem. Soc.* **123**, 4103-4104 (2001).
14. Schlamp, M. C., Peng, X., and Alivisatos, A. P., *J. Appl. Phys.* **82**, 5837-5842 (1997).
15. Lee, J., Sundar, V. C., Heine, J. R., Bawendi, M. G., and Jensen, K. F., *Adv. Mater.* **12**, 1102-1105 (2000).
16. Mattoussi, H., Mauro, J. M., Goldman, E. R., Anderson, G. P., Sundar, V. C., Mikulec, F. V., and Bawendi, M. G., *J. Am. Chem. Soc.* **122**, 12142-12150 (2000).
17. Mitchell, G. P., Mirkin, C. A., Letsinger, R. L., *J. Am. Chem. Soc.* **121**, 8122-8123 (1999).
18. Zhang, J. Z., *J. Phys. Chem. B.* **104**, 7239-7253 (2000).
19. Malik, M. A., O'Brien, P., and Revaprasadu, N., *J. Mater. Chem.* **11**, 2382-2386 (2001).
20. Tanaka, M., Sawai, S., Sengoku, M., Kato, M., Masumoto, Y., *J. Appl. Phys.* **87**, 8535-8540 (2000).

DATA NETWORK SYSTEMS

Infrastructure Needs to Support Unattended and Remote Detector Systems

Jacob Trombka, Timothy McClanahan, and Samuel Floyd

NASA/Goddard Space Flight Center, Laboratory for Extraterrestrial Physics, Code 691, Greenbelt, MD 20771

Abstract. The use of unattended and remote detection systems for use in telemedicine, teleforensics and applications to operations in homeland security and nuclear non proliferation programs will require the development of portable detector systems and public information network systems. With the availability of such networks, the deployment of relatively inexpensive sensor systems can be achieved.

INTRODUCTION

Unattended remote detection systems find important applications in a myriad of exploration and monitoring programs. These include, for example, planetary physics, solar physics, astrophysics, nuclear non-proliferation monitoring, Earth science exploration from orbit and on the surface, teleforensic investigation, dangerous material transportation monitoring, homeland security, highway integrity studies, mineral exploration, industrial non destructive testing. These application areas will require the development of remote sensor data acquisition, transmission, analysis and interpretation systems. Public information network systems need to be implemented to handle the data flow from remote sites to central data acquisition and interpretation centers. With the availability of such networks the deployment of relatively inexpensive sensor systems can be achieved.

REMOTE SENSOR DATA ACQUISITION, ANALYSIS INTERPRETATION, AND TRANSMISSION SYSTEM

In the following discussion, we will consider the various aspects of a remote sensor structure based on our experience in developing such systems for the Near Earth Asteroid Rendezvous (NEAR) X-ray and Gamma-ray Remote Spectroscopy (XGRS) system used for the exploration of the asteroid 433 Eros surface [1,2]. Figure 1 shows an outline of the analysis system developed for NEAR XGRS. There is a direct correlation with such a system to be used in a terrestrial remote sensing operation. The remote detection system would connect field operations with central operation and information systems. The detector systems at remote sites could then be calibrated in situ or remotely, health of the

CP632, *Unattended Radiation Sensor Systems for Remote Applications*, edited by J. I. Trombka et al.
2002 American Institute of Physics 0-7354-0087-3

operation system determined and repaired remotely, and data collected and transmitted to the central stations. The information can then be compared with data banks, interpreted and operational information returned to the field to help in operations[3]. Similar systems have been developed and are in operation in the military, space, and medical programs. More comprehensive systems are required for use in home security applications. Furthermore, affordable systems need to be developed so that remote and/or unattended sensing technology can be made more generally available.

In the following discussion, we will consider some general properties of various application areas in order to better define the overall infrastructure. A few operational and environmental constraints should be considered when developing the design specifications for remote sensing systems. These constraints include

- Long duration operation which may extend for many years
- Low power operation (e.g. using batteries and/or solar panels)
- Remote calibration, since the detectors may not be accessible or trained personnel available at the site
- Access to data transmission networks
- Very limited downlink bandwidth (e.g., high bandwidth needs in the transmitting of hyper spectral images)
- Designing for survival in harsh environments such as damaging radiation

On Board Data Systems

The output from the sensor in the detector system is sometimes analog. This sensor output could be, for example, from either an imager or spectrometer. The signal is conditioned and sometimes amplified before digitizing the signal. In some cases the output is directly digital. An in situ data processing unit (dpu) usually controls the operation of the detector system. Furthermore, the operation of the detector can be controlled, for example, by a central processing unit (cpu). If multiple detectors are used they may use separate or shared dpus and cpus. Many configurations are possible for multi instrument cases. The control of operations is accomplished using on board software systems. These tasks may include data accumulation and general timing modes; transient event detection; calibration protocols; tests for determining the health of the system; a certain limited amount of data analysis capability; data packaging, encryption and formatting for transmission; data compression capabilities if needed; inclusion of parity checks to verify the integrity of the transmitted data; and both upload and download capabilities. On board programming capability has been found to be very valuable in remote operations. During the NEAR mission, on board programmability was extremely important in the operation of the NEAR XGRS system. Two instances during the flight mission greatly enhanced the scientific capability of the system and also kept the x-ray spectrometer operational.

Two events will be described briefly to illustrate the importance of this capability. In the first case, a program upgrade was achieved after launch in order to upgrade the operation of the gamma-ray spectrometer to detect Gamma-Ray Bursts (GRBs). It was not possible to include this capability prior to launch because of strict time constraints required for mounting the XGRS on the spacecraft and subsequent testing of the XGRS flight system. Timing the arrival of the GRB event at spacecraft mutually separated by long baselines allows the triangulation and localization of the most powerful explosions in the Universe. An Inter Planetary Network (IPN) space network was established to localize these bursts[4]. The space network consisted of three sensors positioned in a tripod: Ulysses < 6 AU, NEAR < 2 AU, and those nearby: GGS-Wind, BeppoSAX and Rossi-XTE. IPN localizations of ~ 3 arc-minutes can be obtained (1 sec. timing over 1200 light-seconds separation), that was also improved by burst-profile fitting.

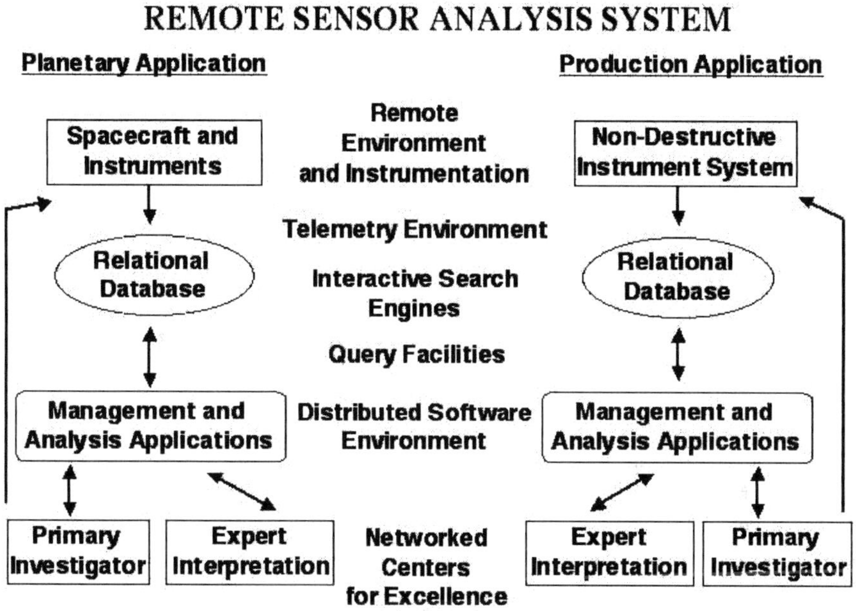

REMOTE SENSOR ANALYSIS SYSTEM

Planetary Application   **Production Application**

Planetary Application	Remote Environment and Instrumentation	Production Application
Spacecraft and Instruments		Non-Destructive Instrument System
Relational Database	Telemetry Environment / Interactive Search Engines	Relational Database
	Query Facilities	
Management and Analysis Applications	Distributed Software Environment	Management and Analysis Applications
Primary Investigator / Expert Interpretation	Networked Centers for Excellence	Expert Interpretation / Primary Investigator

FIGURE 1. Outline of a Remote System Analysis System

Precise source definitions enable other telescopes to pinpoint radio and optical associations (that decay over days to weeks) and to obtain spectral red shifts, thus determining the age of these sources. Between 2 September 1997 and 8 February 2001, the NEAR XGRS was part of the 3rd IPN for gamma-ray burst localization. XGRS was not designed to be a GRB detector, and as indicated above, the spacecraft was reprogrammed in flight to detect and downlink the GRB data. In the fifteen month period from October 1999 to February 2001, fifty-five bursts were localized both rapidly (~24 h) and accurately enough (~10 arc minute error boxes) to allow sensitive radio and optical searches for the burst counterparts. Not all of these bursts had positions that were actually suitable for follow-up observations, due mostly to proximity to the Sun. Of the bursts that were followed up, ten counterparts were identified, and of the ten, seven had their redshifts measured. The result was a 50% increase in the database of GRB counterparts, and a new record for a GRB redshift of 4.5 indicating an age of about ten to twelve billion years. The results are consistent with the belief that GRBSs are cosmological in origin. Thus the reprogramming allowed for a significant increase in the scientific yield of this spectrometer.

The major scientific goal of the NEAR XGRS was the determination of the elemental composition of the surface of the asteroid 433 Eros. The elemental composition is inferred from an analysis of the solar x-ray and cosmic-ray characteristic emission induced in the surface of the asteroid. Thus in the second case we consider, a major problem with the x-ray spectrometer which was found during system testing in the cruise phase of the mission. Overload pulses in the gamma-ray detector produced noise pulses in the low energy spectrometer channels of the x-ray spectrometer. If not eliminated, these noise pulses would have made it impossible to detect the magnesium, aluminum and silicon surface emission lines and a major objective of the experiment would have been lost. A software fix was developed on the ground and implemented on board the spacecraft. Whenever an overload pulse was detected in the gamma-ray detector, an anti-coincidence signal was used to block the noise pulse from being recorded in the x-ray spectrum. The software fix was uploaded and the problem eliminated. Successful measurements of the magnesium, aluminum and silicon elemental concentrations were achieved [5.]

Instrument calibration also can be accomplished remotely. The NEAR gamma-ray spectrometer was calibrated by measuring the natural background during periods when the asteroid was not in view of the asteroid. There are well-defined lines in the natural background so that gain and zero spectrometer changes were determined as a function of time, and this information was then used to correct the measured emission spectra before detailed analysis. Spectra were collected over five years of operation. Data obtained during these periods had spectral changes due to gain and zero shifts. Spectra had to be summed and compared over different time periods. The determination of the gain and zero shifts over the operation time of the mission allowed us to develop a system for performing automated system gain and zero corrections on accumulated spectra.

The natural background for the x-ray spectrometer did not contain discrete line features that could be used for calibration. A long-lived radioactive source was used. The radioactive source was rotated in position when the asteroid was not in the field of view of the detector and calibration data obtained. The source was, of course, shielded from the detector when not in the calibration mode

These data were also of importance in determining the health of the detector. Once the data was analyzed and problems determined then remedial actions were taken. For example, radiation effects can been seen in the degradation of spectral lines. The degradation can sometimes be annealed out and detector response somewhat recovered. An annealing capability is now included on the Mars Odyssey '01 gamma-ray spectrometer. After a major solar flare during that mission, the spectral line shapes were observed to be significantly degraded. The detector was then annealed and the spectral shapes partially recovered.

Telemetry Networks

Once data has been formatted and conditioned for transmission, the information must be transmitted from the remote site to a central data acquisition and management facility. We now describe the system outlined in Figure 1 as it operated during the NEAR mission. This type of distributed network system will be of great importance in application to remote and unattended sensor operations. The NEAR formatted data was transmitted from the spacecraft to the Earth and received at the NASA Deep Space Network operated by the Jet Propulsion Laboratory (JPL). The information was then transferred to the mission operation center at the John Hopkins Applied Physics Laboratory (JHUAPL). At JHUAPL the data was unpacked, sorted (engineering and science) and archived. The archived data was then accessed by the responsible instrument group and stored in their particular archive.

In the case of the NEAR XGRS experiment, the control and analysis operation was based at the Goddard Space Flight Center (GSFC), the data was archived at the University of Arizona and data analysis and visualization systems resident at GSFC. The XGRS science team included participants from National Laboratories and Universities in the United States and Germany. The science team members accessed the XGRS data using a query system developed by the team. The team developed extensive meteorite standards libraries for use in data interpretation. These systems were used in the cruise and orbital operations for calibration of the flight instruments, operation planning, and data analysis and interpretation. Considering gamma-ray burst detection, the network can indicate an interesting demonstration of the use of this distributed network. In this case, the NEAR network and the IPN network were coordinated. A GRB was detected in the NEAR system. A message was then sent to the IPN network and coordinated with events detected by other spacecraft near the Earth and Sun. If a coincidence occurred between the NEAR XGRS, detectors at the Sun and near Earth, a source position was triangulated and messages were sent to a large number of observatories throughout the world. Previous experience had shown that these spectral emissions decay rapidly within days of the observation of the GRB. Thus with this network, rather quick responses to GRBs were achieved, which allowed observation of the GRBs in the visible and sometimes radio spectral regions. Using this distributed network, many successful localizations of GRBs were accomplished.

Now we consider the applications of this distributed network approach to terrestrial applications. At present, there appears to be sufficient commercial network capability. These networks can be organized for transmission of data from remote or unattended sites to user sites for operational use. We will consider three cases: commercial applications where there is sufficient technical support within the company or such support is not cost effective; use in teleforensics in state or local operations; and application in national emergency home security systems.

Remote technologies using radiation detectors have been used in many industrial applications such as mapping the elemental composition down bore holes in mineral and oil exploration; measurement of component thickness in on line quality control processing; and measurement of fluid flow through various types of enclosed environments. Of course there are a myriad of applications of this type of technology in nuclear medicine. In the oil industry nuclear chemical profiling has been extensively used and when this technique was first used, the individual companies supported this capability within their corporate structure. This approach was found not to be cost effective because of the specialized staff that was required relative to the number of times the measurements were carried out. Furthermore, a number of the smaller companies were unable to support the required infra structure to maintain this capability. A number of companies were organized to perform these exploration functions on a lease basis. These companies then served the needs of a number of oil companies. They maintained the equipment, supported research and development programs, and thus sustained their capabilities. Furthermore, they were able to upgrade their capabilities to include modern technologies.[6] In a similar mode the advanced technologies developed for unattended and remote operations can be made available for application to government and industrial organizations without maintaining the required infrastructure in their groups. In fact, in the remote analysis case, a number of operations could be simultaneously carried out at different sites. The functions, for example, of calibration, maintaining the integrity of the operation system, research and upgrading of the technology would be carried out by the central facility. The data and operational information would be directly available to users and if necessary, security of the information maintained. A few application areas that might benefit from such systems are:

- Use of neutron/gamma techniques for studying the integrity of structures such as historic monuments, bridges and highways
- In situ studies of contamination of real estate for environmental impact certification
- Measurement of fat to meat ratios in the meat packing industry
- Monitoring water contamination down stream from industrial sites

We now consider a program in progress that addresses the development of a system for accumulation and use of information obtained in a field environment using remote technologies and applied in a forensic science scenario.[7] This program is part of a National Aeronautics and Space Administration (NASA)/National Institute of Justice (NIJ) program in Teleforensics. A Working Group of NASA scientists and law enforcement professionals has been established to develop and implement a "teleforensic" feasibility demonstration program. Teleforensics attempts to bring non-destructive analysis techniques to the crime scene.[6] The purpose of these field investigations is to: assist in the collection of forensic materials at the crime site; help maintain the chain of evidence; allow for experts at facilities to assist in the investigation of the crime site; maintain images and information obtained in the original investigation of the crime site; and have access to information from data bases to assist in the investigation. Following the outline shown in Figure 1, images and non-destructive systems such as x-ray fluorescent analysis (xrf) would be used. The xrf measurement might be used, for example, to determine the presence of gun shot, residue, blood, semen, cosmetics and bone. Characteristic signatures for these materials could either be stored on computers at the site or sent back to a crime laboratory to identify these signatures for their particular type. These evaluations could then be made available to the field investigators to assist in the collection of the evidence. In situ materials deemed significant can then be collected and returned to a crime laboratory for detailed study and results would be obtained which would better satisfy evidentiary criteria. Screening of materials in the field would allow for returning evidence that is most likely to be of importance in the investigation of the crime. This pre-screening would reduce the amount of non-relevant material returned to the already overworked laboratories, and produce significant cost savings for crime scene investigations. Information from the crime laboratory can also be sent to the field to assist in the investigation and proper collection of the materials, thereby reducing the bulk of unnecessary material collected. The non-destructive field analysis can indicate the most promising materials for further laboratory analysis. The information obtained at the remote site can be returned to an archive and query systems made available for later study. Parts of the networks needed for this project are already available. Some forensic materials such as the DNA and fingerprints are already accessible in networked systems, but the total system as outlined in Figure 1 needs to be designed and constructed. Parts of similar systems are being developed, for example, by the Department of Defense in their telemedicine program.A national network which is in place only in response to occasional emergency situations is not only not cost effective but could become outdated and subject to a greater failure rate due to the possible infrequent use of the response system. If one considers the development of a national network, the involvement of local and national resources will be required. Finally, there appear to be sufficient network capabilities currently available. The problem lies in organizing these networks so they may be accessed across agencies and nationally linked. The teleforensic program described above and other systems developed, for example, in telemedicine, medical and forensic data banks and highway safety programs by the National Institutes of Health, Department of Justice, Federal Bureau of Investigation, and Department of Transportation can become components in a National Security Network (NSN). These systems address directly the needs for homeland security. These various networks would be used regularly, assuring that they would be verified in operation and upgraded continuously. Their integrity could thus be maintained over long periods of time. Some uniform security and encryption systems could be included in the various networks so that in case of an emergency response, the networks could be safely connected. The NSN would require a system that would monitor the integrity and safety of the total network, and would have switching capabilities so that alternate routes could be connected if parts of the network site drop out. This might be similar to the switching network system developed by NASA and used to monitor and control space flight programs during operation.

Finally, autonomous networks that can operate without human intervention will be important in the development of remote analysis systems. When large numbers of radiation, chemical and biological detectors systems are deployed, autonomous systems will be needed to monitor the health of the network, accumulate and assimilate data to determine the background and allow for the detection of abnormal activity. A system that is being developed in the NASA space program is described in reference 8.

SUMMARY

The use of unattended and remote detection systems for use in telemedicine, teleforensics and applications to operations in homeland security and nuclear non proliferation programs will require the development of public information network systems. With the availability of such networks the deployment of relatively inexpensive sensor systems can be achieved. These systems would connect field operations with central operation and information systems. The detector systems at remote sites could then be calibrated, health of the operation system determined, remotely repaired, and data collected and transmitted to the central stations. The information can then be interpreted compared with data banks and operational information returned to the field to help in operations. Many such systems are in operation in the military, space, and medical programs, but more comprehensive systems are required for secure public information operation. These systems must be affordable so that they can be made available on national and local levels.

REFERENCES

1 Trombka, J. I., Squyres, S. W., Brückner, J., Boynton, W. V. Reedy, R. C, McCoy, T. J.,. Gorenstein, P, Evans, L. G., Arnold, J. R., Starr, R. D., Nittler, L. R., Murphy, M. E., Mikheeva, I., McNutt, R. L., McClanahan, T. P., McCartney, E., Goldsten, J. O., Gold, R. E. , Floyd, S. R., Clark, P. E., Burbine, T. H, Bhangoo, J. S., Bailey, S., *Science* 289, 2101-5, (2000)

2. Goldsten, J. O., McNutt, Jr., R. L., Gold, R. E., Gary, S. A., Fiore, E., Schneider, S. E., Hayes, J. R., Trombka, J. I., Floyd, S. R., Boynton, W. V., Bailey, S., Brückner, J., Squyres, S. W., Evans, L. G., Clark, P. E., and Starr, R., *Space Sci. Rev.* 82, 169(1997).

3. McClanahan, T.P., Trombka, J.I., Floyd, S.R, Boynton, W.V. , Mikheeva, I., Bailey, H., Liewicki, C., Bhangoo, J., Starr, R., Clark, P.E., Evans, L.G, Squyres, S., McNutt, R., and Brückner, J., *Nucl. Inst .and Meth. In Phys. Res. A* 422 582-585., (1999).

4. Hurley, K., Cline, T., Mazets, E., Aptekar, R., Golenetskii, S., Frederiks, D., Frail, D., Kulkarni, S., Trombka, J. , McClanahan, T., Starr, R., and Goldsten, J., *Astrophys. J. (Letters)* 534 23, (2000).

5. Nittler, L. R.; Starr, R. D.; Lim, L.; McCoy, T. J.; Burbine, T. H.; Reedy, R. C.; Trombka, J. I.; Gorenstein, P.; Squyres, S. W.; Boynton, W. V.; McClanahan, T. P.; Bhangoo, J. S.; Clark, P. E.; Murphy, M. E.; and Killen, R., Meteoritics.Planet. Sci. 36, 1673-1695. (2001).

6. Schweitzer, J.S., and Joel L. Groves, J. L., *"Subsurface Remote Sensing"* (in this publication).

7. Trombka, J., Schweitzer, J., Selavka, C., Dale, M., Gahn, N., Floyd, S., Marie, J., Hobson, M., Zeosky, J., Martin, K., McClannahan, T., Solomon, P., and Gottschang, E., *"Space Age Teleforensics: Non-Destructive Analysis of Forensic Evidence at Crime Scenes"*, (accepted for publication in the *International Journal of Forensic Sciences*).

8. McClanahan, T., Trombka, J., Floyd, S., Starr, R., and Evans, L., *"System Template for Remote Sensing Experiments"*, (in this publication).

Distributed Sensor Network With Collective Computation For Situational Awareness

Jared S. Dreicer[*], Anders M. Jorgensen[+], and Eric E. Dors[#]

[*] Space Data Systems Group, NIS-3, Los Alamos National Laboratory
[+] Space Instrumentation and System Engineering Group, NIS-4, Los Alamos National Laboratory
[#] Space and Atmospheric Sciences Group, NIS-1, Los Alamos National Laboratory

Abstract. Initiated under Laboratory Directed R&D funding we have engaged in empirical studies, theory development, and initial hardware development for a ground-based Distributed Sensor Network with Collective Computation (DSN-CC). A DSN-CC is a network that uses node-to-node communication and on-board processing to achieve gains in response time, power usage, communication bandwidth, detection resolution, and robustness. DSN-CCs are applicable to both military and civilian problems where massive amounts of data gathered over a large area must be processed to yield timely conclusions. We have built prototype hardware DSN-CC nodes. Each node has self-contained power and is 6"x10"x2". Each node contains a battery pack with power feed from a solar panel that forms the lid, a central processing board, a GPS card, and radio card. Further system properties will be discussed, as will scenarios in which the system might be used to counter Nuclear/Biological/Chemical (NBC) threats of unconventional warfare. Mid-year in FY02 this DSN-CC research project received funding from the Office of Nonproliferation Research and Engineering (NA-22), NNSA to support nuclear proliferation technology development.

INTRODUCTION

Sensors play an important role in many of the nation's security programs and may be critical in the future for counter-terrorism and homeland defense missions. Applications include monitoring the movement of personnel or materials and environmental changes (air, water, or soil) at critical facilities, cities, international borders, or geographic regions. Indications and warnings of events can assist greatly in operational planning and response, and can enhance protection against nuclear, biological, or chemical (NBC) threats.

Ubiquitous *in-situ* computing and sensing devices are inevitable. They may soon exist in large numbers embedded everywhere in civilian systems and are of interest in current military, non-proliferation and other national security programs. Research and development is progressing in the area of sensor hardware technology and miniaturization so that that thousands or even millions may be deployed affordably. This will enable a revolutionary change in remote sensing capability. The traditional approaches to ground-based surveillance are largely inadequate for homeland protection from NBC threats since the initial detection of potential threats must be automated. Further, the automated surveillance systems to be developed must be

CP632, *Unattended Radiation Sensor Systems for Remote Applications*, edited by J. I. Trombka et al.
© 2002 American Institute of Physics 0-7354-0087-3/02/$19.00

unattended, cheap, compact, reliable, fault tolerant, and consume minimal power. Such systems will need to be fielded across a range of foreign and domestic locations.

This paper discusses the Distributed Sensor Network with Collective Computation (DSN-CC) research project that is developing an *in-situ* sensing and processing capability. Such a capability can support a layered defense and provide real-time alarms. This new DSN-CC technology uses communication among sensors to enable adaptive information collection and signal-processing capabilities. We believe DSN-CC may play an important role in the automated detection and tracking of proliferation or terrorism threats.

DISTRIBUTED SENSOR NETWORK WITH COLLECTIVE COMPUTATION

A DSN-CC is a network of smart sensor nodes that communicate with neighboring nodes in order to cooperatively solve a sensing problem. Devices capable of sensing, computing, communicating and possibly actuating will be widely available in the near future. Hardware research is ongoing to develop powerful and inexpensive devices that can serve as nodes in a DSN-CC, but the fundamental understanding of the DSN-CC has not been developed [1-4]. Science and technology challenges include organizing networks, inter-communicating, and using distributed computation to achieve the goals of collective behavior, autonomy, and adaptability. A system composed of hundreds to thousands of such devices may be used for surveillance and monitoring applications. As extremely large networks of such sensors begin to communicate, we enter a new realm of collective behavior with unexplored dynamics that can be harnessed. Such networks will be similar to biological systems in their robustness, reliability and adaptability.

In a DSN-CC individual devices will interact strongly on a local scale, sharing data and computation, and adapting to a changing environment on a local and global scale. Collective behavior can reduce an otherwise massive data-collection effort by allowing conclusion, rather than data, to be disseminated. In a DSN-CC raw data is distributed only locally, and conclusions are arrived at by negotiation among sensors. This eliminates the need for a central processing station, which is significant. Better and more efficient signature isolation can be achieved when neighbor sensors share data ("compare notes"). Single point of failure and expensive long-range communication are eliminated.

Not only does DSN-CC represent significant gains over traditional sensor arrays, recent results open up the opportunity for new applications that were heretofore impossible because of massive communications and central processing requirements. Straightforward gains are available in real-time event detection and classification, robustness to failure exhibited by coherent k-out-of-n systems, efficiency due to decreased resource requirements, decreased false alarms, and reduced possibilities of deception. However, the greatest gains will only be realized as we learn to build new systems that were previously impossible to achieve with conventional sensor arrays.

An example application for a DSN-CC is local event detection in a geographic area illustrated in Figure 1. Sensor nodes near the event exchange measurements and negotiate a concise conclusion, which is rapidly propagated across the network. This is but one simple example of a DSN-CC, and tremendous gains are realized over classical sensor arrays.

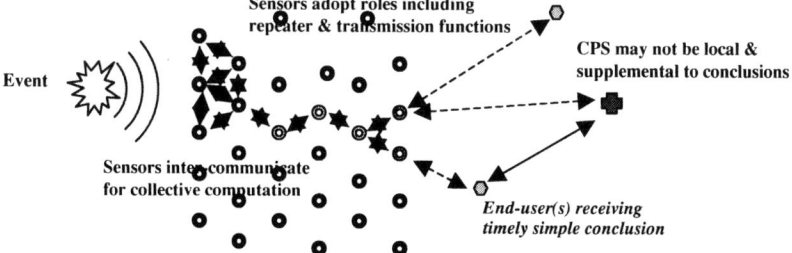

Figure 1. DSN-CC nodes detect a local event and communicate with near neighbor nodes.

Figure 2 shows three DSN-CC prototype nodes, each measures 6"x10"x2". The right side of Figure 2 shows the internals, from left to right, the battery pack with power feed from the solar panel that forms the lid, the central processing unit (CPU) top with global positioning system (GPS) card below, and radio (RF transmission) card. The sensor (not shown) is connected to the CPU board. These nodes have been used with acoustic sensors to determine the location and time of a sound impulse by exchanging data about the time of arrival of the impulse at each node.

Figure 2. Three prototype DSN-CC nodes and the internal hardware components.

DSN-CC Theory, Simulation, and Results

Classical sensor array technology utilizes a central processing station, designed using the star topology. This is a networking scheme in which data from a number of independent sensors are transmitted directly to a central processor (the hub of the star), where they are processed, combined, and delivered. The individual sensors do not directly communicate with each other. The conclusions are then redistributed to users. There are several problems with this approach. Transmitting the raw data long distances requires large amounts of power and bandwidth, or large amounts of time. Central processing incurs delays and leaves the system vulnerable to single point failures. In contrast, DSN-CCs operate with short-range transmissions between neighbor nodes only and *in-situ* data processing. This approach eliminates single point failures, provides redundancy, saves bandwidth, and delivers conclusions rapidly to users. With large numbers of sensors the star-topology paradigm breaks down.

DSN-CC Theory

There are significant scaling problems associated with the classical star-topology [3]. The classical star-topology is successful for arrays with small dimensions and few sensors, but for networks with a large number of sensors it is likely to fail for several reasons. First, for limited bandwidth and growing numbers of sensors, N, the time it takes for all sensors to transmit grows linearly with N. This decreases the sampling rate, a problem for time-critical applications. Secondly, transmitter power scales as the square of the transmission distance, so long-range communications are very expensive. In a classical sensor array where all information is transmitted to a hub, the energy one sensor needs to transmit increases as N (for an array with constant separation between sensors the transmission distance increases as $N^{1/2}$, thus the energy one sensor needs to expend on a transmission scales as N). The energy needed to extract data from all N sensors grows as N^2. A sensor can only transmit with a duty cycle of 1/N, so the time-averaged energy consumption of a sensor is independent of array size.

By contrast, we have found that significant savings are possible with an approach in which neighbor communication takes place. Suppose information still propagates to a hub, which is some sensor in the network. Information is passed from sensor to sensor and merged with a sensor's own information before it is passed on. In that case the time to extract will be a small constant times $N^{1/2}$ (scales as N in the classical case). Each sensor transmits once with energy independent of the size of the network, so total energy scales as N (scales as N^2 in the classical) [3]. Clearly, in simple applications where DSN-CC is used to merely improve the speed and efficiency of information transmission, there are significant quantitative gains over classical sensor arrays. For a DSN-CC peak power is independent of the size of the network. For the classical case the sensor must be designed for peak power operation, which is dependent on the size of the array.

DSN-CC Simulation

Simulations of randomly generated sensor networks with different numbers of sensors in them were performed. The simulation model was based on a sensor that includes communications inputs from neighboring sensors, local measurements, output, and a state machine. This architecture was used to demonstrate cooperative data analysis and detection. The simulation model contained sources and sensors. The sources emitted an omni-directional beacon, and the sensors could measure only the direction to the source if it was within detection range. The sensors communicated only with neighbors that were within transmission range. In order to compute the source locations, the sensors need to exchange measurements. The goal was to have the source locations computed and to have that information distributed across the network. Details are reported in [3].

DSN-CC Results

Simulations were run for networks of 100 sensors to 20000 sensors. One node in the network was monitored as a proxy for when detection information had been transmitted across the network. The simulations were performed to determine the energy and time for detection information to be sent across the network to the exfiltration[1] node, and how energy and time scale as a function of network size (number of sensors). Results were compared to the classical array approach of having each sensor in turn transmit information to a central hub (exfiltration node) where information is processed.

The left graph in Figure 3 shows the exfiltration energy as a function of the network size. Total exfiltration energy is the transmission energy used from the time the sources first appear until the time when the exfiltration node knows the position of all sources. The total exfiltration energy in the classical case is shown in the top curve, which scales with the square of the number of sensors (N^2). Exfiltration energy for the DSN-CC case is shown in the bottom curve, which scales as the number of sensors to the 3/2 power. This means that the total exfiltration energy per sensor scales as number of sensors to the 1/2 power. So the total amount of energy expended per sensor grows with network size, N. However, this energy is expended over a time interval that also grows with the square root of the number of sensors. The plot reflects total exfiltration energy divided by total exfiltration time and the total number of sensors to obtain the average exfiltration energy per time step per sensor. For the DSN-CC case the energy expended per sensor per time step is independent of the size of the network, N.

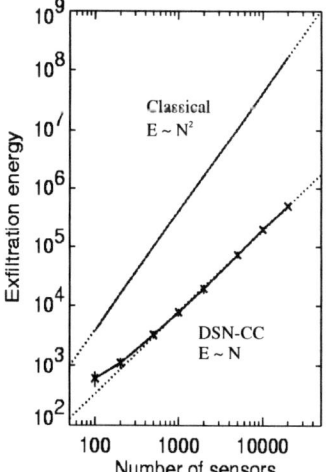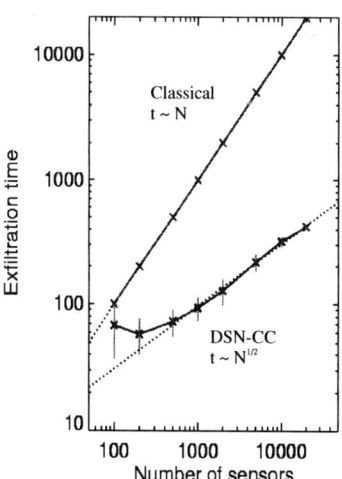

Figure 3. Simulation results for total exfiltration energy (left) and time (right) as a function of number of sensors in the classical case top and the DSN-CC case bottom. Dashed lines are the theoretically expected best exfiltration energy and time for the classical and DSN-CC cases.

---

[1] Exfiltration node is a proxy for the central hub in the classical case to which detection information has been transmitted from across the network to facilitate comparison of time and energy.

The right graph in Figure 3 shows the exfiltration time as a function of the network size. Exfiltration time is the time it takes from when the sources first appear until the exfiltration node knows the position of all the sources. Exfiltration time in the classical case, in which every sensor has to report directly to a central hub, is shown in the top curve. The exfiltration time scales linearly with the number of sensors (N). Exfiltration time for the DSN-CC case is shown in the bottom curve. The points are the median exfiltration time and the vertical bars are the RMS distribution of exfiltration times around that value. For large network sizes the exfiltration scales as $N^{1/2}$. For small network sizes the exfiltration time appears to increase and the spread around the median increases. For small network sizes the trend of a larger exfiltration time than that suggested by the scaling is being investigated.

RADIATION DETECTION SITUATIONAL AWARENESS

DSN-CC capabilities provide the technical underpinnings to address national security missions. Applications include situational awareness, monitoring, or defense of critical facilities or geographic regions (homeland defense and proliferation detection). DSN-CCs also support a military configured in mobile, fast-striking forces to execute missions ranging from peacekeeping to conflict (special operations and counter terrorism). Sensor suites depend on the application, but may include anything from miniaturized *in-situ* sensors to remote sensors on robots, UAVs, aircraft, or satellites. Many signatures can be detected only by placing sensors in close proximity to the target. These include vibrations in the earth, short-range radio and acoustic signals, production facility effluents, spores from biological weapons, and radiation from nuclear material. These signatures can be vital to identifying and understanding mobile targets and hard targets such as underground production facilities. However a number of basic research and development issues must be resolved to successfully field *in-situ* DSN-CCs for specific applications. In particular, radiation detection situational awareness will require continued development of DSN-CC capabilities and nuclear detector technology. Both are research areas that LANL is actively investigating.

DSN-CCs are pertinent to radiation detection situational awareness for the following applications: the protection of critical facilities, cities, or regions; the search and diagnoses of nuclear material in wide area radiation surveys; the response to nuclear accidents or terrorist acts requiring monitoring for dispersal of radioactivity; safeguards for the protection of nuclear material in storage; and finally, remote sensing addressing space exploration and planetary science questions.

The physical signatures of interest for nuclear material are known. The difficulty of the problem varies depending on the specific radiation detection application. Each application has specific problem requirements and boundary conditions that significantly impact the complexity of the problem. For example, the requirements for detection may vary with respect to the intensity and mobility of the signatures, the topography, population density, and proximity of physical structures in the detection area, and the number of dispersal or transit pathways.

Radiation Detection Challenges and DSN-CC Concept

There are two primary radiation detection challenges that must be addressed. Nuclear sensors have short ranges since the signals are small and the backgrounds are large. The solution to these challenges is to either reject the large background signal by imaging or to bring the detector to the source. Bringing the sensor closer to the source via DSN-CC is the concept proposed. This is achieved by utilizing a hybrid DSN-CC that employs both statically placed and mobile detectors over a large area (application dependent).

The DSN-CC provides a generic approach to different applications of this problem. Depending on the application, the physical configuration of static deployment of the DSN-CC can provide dense 0-dimensional (point), 1-dimensional (line, e.g. along a roadway), 2-dimensional (field, e.g. geographic region) and even 3-dimensional (e.g. at different altitudes within the atmosphere) measurements, which can be monitored over time. Thus, the DSN-CC concept for radiation detection situational awareness is the integration of mobility (ground or air based detector systems). An illustration of the hybrid DSN-CC system is presented in Figure 4.

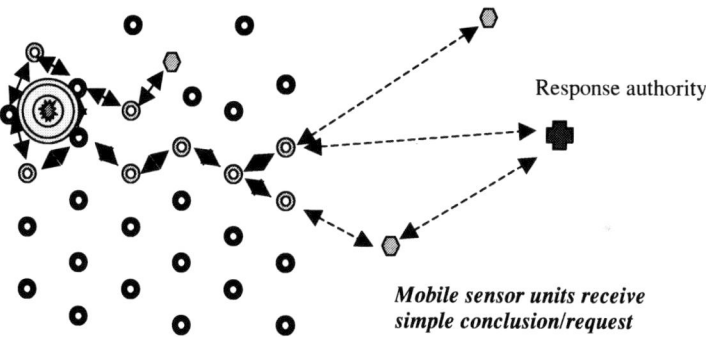

Response authority

Mobile sensor units receive simple conclusion/request

Figure 4. Concept for radiation detection situational awareness utilizes a hybrid DSN-CC approach composed of static nodes that cue mobile nodes (waiting or moving randomly).

The mobile detectors may be stationary units (waiting to be requested) or randomly moving units (that can be called to a location). Since the background signal varies over a geographic area, mobile detector backgrounds will vary, which causes a problem when trying to distinguish the natural variations from that caused by a real target. The DSN-CC can resolve the unknown background for a mobile detector by transmitting the "calibrated" or local background from the nearest static detector in real-time. An alternative approach is to characterize the geophysical and manmade backgrounds and their variations in a geographic region and correlate the data to the detector position as it moves.

The hybrid DSN-CC concept relies on deploying enough static and randomly moving radiation detectors so that one is in the right place at the time a target passes. Each detector must be large enough to have a reasonable neutron or gamma detection efficiency. But the system may be effective because of the large number of detectors and not necessarily because of the detector performance. The network density can be increased by orders of magnitude to realize a large sensitivity gain, if the cost of the detectors is inexpensive enough.

DSN-CC with neutron or gamma ray detectors will be crucial to future radiation detection scenarios. Hundreds of thousands of sensors may have to work unattended over large areas in a harsh environment over many years. Ruggedness, simplicity, low power consumption, and low manufacturing and maintenance costs are critical. A brief description of two detector development efforts that address a number of these requirements and that may provide appropriate detection technology in the future is presented.

The first project is a gamma detector technology research and development effort investigating a cadmium zinc tellurium (CdZnTe) device supported by NA-22, NNSA. CdZnTe is a new compound semiconductor that is being developed for gamma ray spectroscopy. Since the resistivity of the material is high, it is ideally suited for gamma ray spectroscopy at room temperature. LANL is leading the development of this new technology in several areas including: research on device physics and electronic properties; and development of specialized detector probes, electronics, and spectrum analysis software [5]. Figure 5 provides a picture of CdZnTe gamma device.

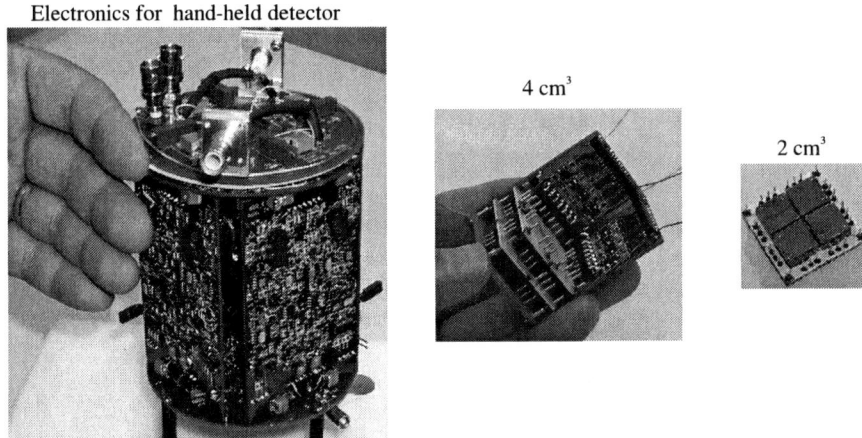

Figure 5. Gamma ray detector based on CdZnTe compound semiconductor material.

The second project is a neutron detector technology research and development effort that is investigating a ^{6}Li ionization chamber. The concept is an inexpensive and robust neutron detector that can be easily mass-produced. It is based upon a pulse mode ionization chamber lined with ^{6}Li as an active substance. Preliminary calculations show a relative neutron intrinsic efficiency of about 5% per cell. The gas purity is not critical for the detector's operation, which dramatically reduces the cost. The appropriate fill gas (Ar, N, Xe, air) can be selected for the specific application. Figure 6 is a cross section of the design and a picture of a preliminary ^{6}Li device. Results from an early feasibility experiment demonstrated a charge collection efficiency over 90% indicating a beneficial signal to noise ratio [6].

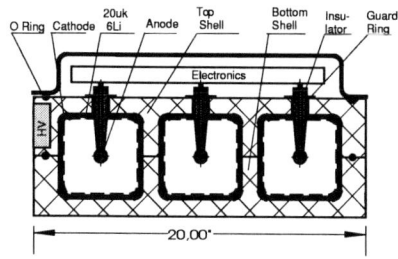

Slab cross-section.

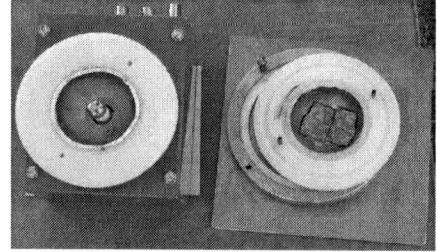

Preliminary working feasibility experiment using a 2" diameter spherical detector 5"X5".

Figure 6. Neutron detector concept based on ^{6}Li pulse mode ionization chamber.

CONCLUSION

DSN-CC's may provide tremendous gains over classical sensor arrays. DSN-CC's operate with short-range transmissions only and *in-situ* data processing. This approach eliminates single point failures, provides redundancy, saves communication bandwidth, and rapidly delivers conclusions to users. For typical situation assessment problems, response time and energy consumption scales as $N^{1/2}$, and N for a DSN-CC of N nodes, a significant improvement over scaling of N, and N^2 for a classical array of N nodes. DSN-CC is a **revolutionary expansion** in sensing capability, **not** a limited incremental increase over existing capabilities. The DSN-CC radiation detection situational awareness concept employs static and mobile detectors. Such a system increases coverage, sensitivity gain, and provides a deterrent value. However, a better understanding of the technical decisions and trade off between cost, simplicity, detection efficiency, and network density is required. The required detector efficiency must be determined within the constraints of the application and system design using the lowest-cost, -power, and smallest size.

REFERENCES

1. Dreicer, J.S., "Theory Development for Sensor Array Anomaly Analysis," Los Alamos National Laboratory report LA-UR-98-4331 (1998).
2. Kahn, J., Katz, R. and Pister, K., "Next Century Challenges: Mobile Networking for Smart Dust," 5th Annual Joint ACM/IEEE International Conference on Mobile Computing and Networking (MOBICOM'99), 271-278 (1999).
3. Jorgenson, A. M. and Dreicer, J. S., "Simulation of a Distributed Sensor Network with Collective Computation," Los Alamos National Laboratory report LA-UR-01-2077 (2001).
4. Dertouzos M., "The Future of Computing," *Scientific American* **281**(2), 52-55 (1999).
5. Prettyman, T., Private communications 2000-2002, Los Alamos National Laboratory, Contact: tprettyman@lanl.gov
6. Ianakiev, K., Private communications 2001-2002, Los Alamos National Laboratory, Contact: ianakiev@lanl.gov

NEWNet: Web-Based Unattended Monitoring in the Public Domain

Michael W. McNaughton, Allen Treadaway, Kevin Anderson,
and M. William Johnson

Los Alamos National Laboratory, Los Alamos, New Mexico 87545

Abstract. The Neighborhood Environmental Watch Network, NEWNet, is a system of environmental radiation monitors near Los Alamos National Laboratory and other sites that provides data at 15-minute intervals on the Internet. For details see http://newnet.lanl.gov. The system originated at the Nevada Test Site and was transferred to Los Alamos in the mid-1990s. In this paper we describe recent improvements in sensitivity and enhancements of capabilities. For durations of days or months, the radiation measurements are limited by variations in the natural background. The daily fluctuation of radon decay products causes the dose rate to change by several nanogray per hour (nGy/h), and a major rainstorm or snowstorm can cause a change of as much as 100 nGy/h. Nevertheless, NEWNet was able to monitor activated air carried by the wind to a station 600 m from the source. At present, we are testing extensions of the NEWNet system to include environmental alpha detectors. A status report on this work will be provided, and possible future applications of the enhanced technology will be discussed.

INTRODUCTION

The Neighborhood Environmental Watch Network, NEWNet, is a network of radiation and meteorological sensors developed and deployed by Los Alamos National Laboratory to provide communities with a means for continuous monitoring of the radiation environment of the Laboratory and other selected locations. NEWNet has been functional at sites in New Mexico, Alaska, and (previously but not currently) elsewhere since 1993, and therefore provides a relatively long-running example of a system deployed to perform unattended radiation sensing. The interested reader is referred to the NEWNet Web site, http://newnet.lanl.gov, for more information than can be provided in this paper.

A very important characteristic of NEWNet is that it is a working, real system rather than a hypothetical one requiring "additional research"—indeed, it is a working system that has been adopted enthusiastically by the communities it services. While development and installation of NEWNet has been performed by Los Alamos National Laboratory, individual sensor stations in the field are maintained and serviced by members of the communities within which the sensors are located. Los Alamos provides training for the "system managers" who perform this maintenance. Much of the descriptive text presented here is based on material used in the training courses given to the system managers.

CP632, *Unattended Radiation Sensor Systems for Remote Applications*, edited by J. I. Trombka et al.
© 2002 American Institute of Physics 0-7354-0087-3/02/$19.00

In this paper we describe the NEWNet system, starting with its historical roots and operational concept and proceeding to a description of the technology. Examples of the presentation of NEWNet information follow, with some interpretation to show how information obtained and archived by NEWNet might relate to broader problems. Finally, some enhancements are identified that might enable a next-generation system with greater capabilities and direct usefulness in national-security missions and elsewhere.

NEWNET HISTORY AND CONCEPT

The NEWNet system had its roots in the realization, following the Three-Mile Island incident of 1979, that nuclear facilities lacked fully satisfactory means of informing nearby communities when events involving releases of radioactive materials occurred. The problem was twofold: warnings needed to be improved in the event of a severe release that threatened public well-being, while realistic information had to be provided if less threatening events occurred so that exaggerated rumors and resulting hysteria could be avoided. To meet this need would require sensors that met the dual requirement of being easy to use and maintain, yet affording adequate information to allow consumers of the information to make informed decisions.

A satisfactory instrumental basis for such a system was identified in the form of the Remote Area Monitoring System at the Nevada Test Site (NTS) [1]. This program was aimed at providing comparable information (and reassurance) to the so-called "downwinders" who might have been exposed in the event of a radiation release resulting from breach of containment in an NTS nuclear test. NEWNet monitoring stations were emplaced around the NTS perimeter as a follow-on to this program. By 1993, however, it appeared that NTS nuclear testing was unlikely in the near future; emphasis therefore shifted to other locations, notably those near nuclear facilities at Los Alamos (the "Plutonium Facility" at Technical Area [TA] 55, the Los Alamos Critical Experiments Facility [LACEF] at TA-18, and so on). Around this time it also became feasible to use the Internet as a means of data dissemination, solving the problem of getting the information to the community. These considerations gave NEWNet its basic concept, which remains in use, although the NTS monitoring sites have been decommissioned.

The NEWNet concept, as currently implemented, remains consistent with these roots. Information is collected by the nuclear and meteorological sensors in the field stations, and transmitted to a central server, whereupon it becomes available to the public at the abovementioned Web site. Although data are logged within each field station on a 1-second basis, the information presented on the Web site is composed of 15-minute averages. The 15-minute time blocks are intended as a compromise between providing sufficiently timely information if a radioactivity release occurs, and avoiding information overload that would stress the communications system and/or create difficulties in interpreting the data.

TECHNICAL DESCRIPTION

A typical NEWNet sensor station is shown in Figure 1. A NEWNET station consists of a data collection platform (DCP) in combination with a radio transmitter for data transmission and a power source, usually a solar panel and storage battery, although a few stations draw directly from 120 V AC power. The typical DCP has instruments to measure wind direction, wind speed, ambient temperature, barometric pressure, humidity, and gamma radiation. A few stations have rain gauges. The radiation detector used in the typical configuration is a very simple, commercially available ionization chamber (usually Reuter-Stokes Model 100) coupled to a data logger (Reuter-Stokes Model 1013 in most DCPs, although some DCPs contain other, functionally equivalent modules). Recently, an environmental continuous air monitor (ECAM) has been demonstrated in the DCP that extends the radiation-detection capabilities to include monitoring for alpha emitters, and some information from the ECAM is presented below. The data presented at the newnet.lanl.gov site do not yet incorporate information from the ECAM; however, the redundant newnet2.lanl.gov site includes ECAM information. The meteorological sensors are of comparably simple, commercially available designs.

FIGURE 1. NEWNet station located at Cochiti Pueblo, New Mexico.

Data are collected by the data logger at 1-second intervals, averaged in 15-minute blocks, and transmitted to a processing station in Los Alamos. For most of the lifetime of NEWNet to date, communications have been accomplished by satellite, exploiting the Geostationary Operational Environmental Satellites (GOES) network. Presently NEWNet receives data via the Domestic Satellite (DOMSAT), a GOES satellite in a geostationary orbit at a longitude of 105 degrees east. In addition, the capability to transmit information from the DCPs via telephone link has been demonstrated, and offers some advantages in transmission speed, etc., where such links are feasible. Two redundant DOMSAT receiver stations are located at Los Alamos. At present these systems use the SCO UNIX Version 5.05 operating system and are configured with Franklin Telecom ICP188 boards to connect the receivers.

246

Presentation of the compiled information to the public is by means of Windows NT servers that in turn draw upon an AlphaServer 1000 computer from Digital Equipment Corporation on which the data are archived. Total archival storage capabilities are presently in the 25-gigabyte range. The Oracle relational database package is used for maintenance of the data reported by the sensors, and for storing information on the sensor locations and descriptions. This package allows simple reconfiguration so that sensors can be added or their locations changed. The NEWNet Web site resides on two identical Windows NT servers that communicate with this machine. Server NEWNET hosts the Web site and server NEWNET2 is a redundant copy for fault tolerance in the event of NEWNET hardware/software failure.

DATA AND OPERATING EXPERIENCE

Consistent with the primary NEWNet mission, most of the sensor stations are presently located in northern New Mexico, with a few outliers in Alaska. Locations of the New Mexico stations are shown in Figure 2. The concentration of sites on Los Alamos National Laboratory grounds reflects the main nuclear-material handling facilities at the Laboratory. The frequent movement and usage of nuclear materials in these areas, along with frequent use of radiography sources and other radiation sources of an industrial nature, afford opportunities to examine the capability of the NEWNet system to detect even small changes in the radiation environment of the sensors.

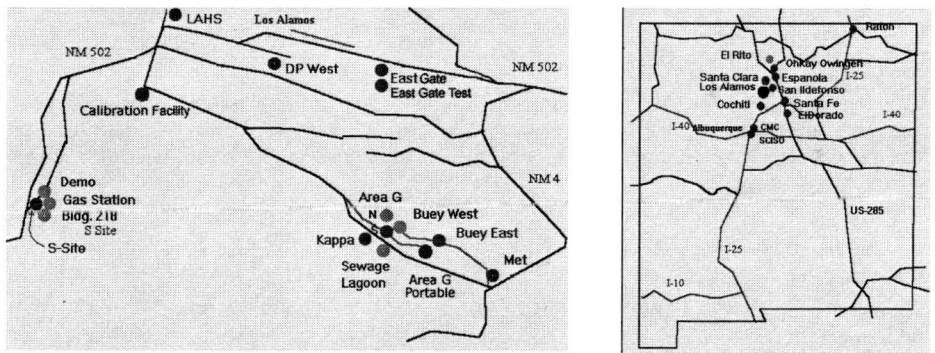

FIGURE 2. NEWNet station locations at Los Alamos National Laboratory (left) and throughout New Mexico (right) as presented on the Web site. Active stations are shown on the Web site as red dots; green dots denote stations that have been retired.

Figure 3, extracted from the Web site (and therefore somewhat difficult to read), shows an example of the information presented to a visitor to the Web who asks for long-term information from a particular sensor station. The meteorological data show the expected diurnal variations, superimposed on long-term variations as frontal systems pass through. The radiation field is much more nearly constant, but minor cyclic variations are still observed. These variations result from radon that remains near the ground at night and is dispersed by convective air currents during the day. Radon is a significant contributor to the ambient background, and its effects can be seen clearly in times of rainfall; rain "washes out" radon decay products from the atmosphere and this results in a transient increase in the ambient radiation field. The variations in temperature and pressure shown in Figure 3 were not associated with significant precipitation, and accordingly variations in the radiation field due to radon were minor during this period of time.

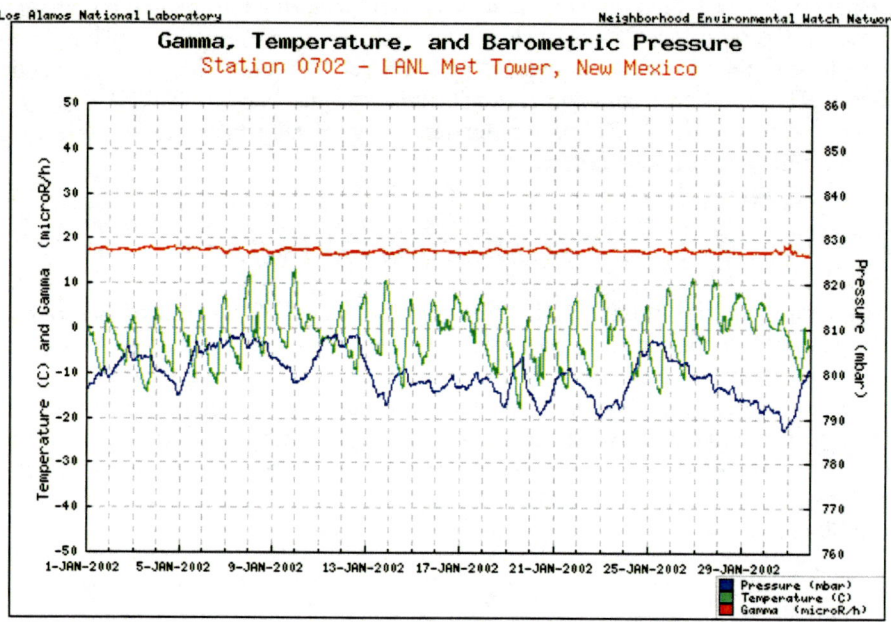

FIGURE 3. Portion of a Web page obtained from the newnet.lanl.gov site, showing a 31-day history of environmental information at the "LANL meteorological tower" station. Gamma dose rate, ambient temperature, and barometric pressure are shown on the site in the red (upper), green (middle), and blue (lower) traces, respectively.

A different situation is seen in Figure 4, showing data taken over a 3-day period at the "East Gate" site near the Los Alamos Neutron Science Center (LANSCE) Facility, a research site where large radiation sources are occasionally handled. The radiation field showed a sudden 10–15% increase on the evening of the second day that lasted approximately two hours, after which the level returned to normal. On investigation, it was found that during this period a 1.5-kCi, industrial ^{137}Cs source was in use at LANSCE, a few hundred meters away. (The intervening space is open air above a canyon, so that a clear line of sight between source and sensor existed.) The fact that the increase in radiation field was sharp-sided rather than gradual is an indication to the experienced observer that the source was anthropogenic rather than rainout of radon and its decay products.

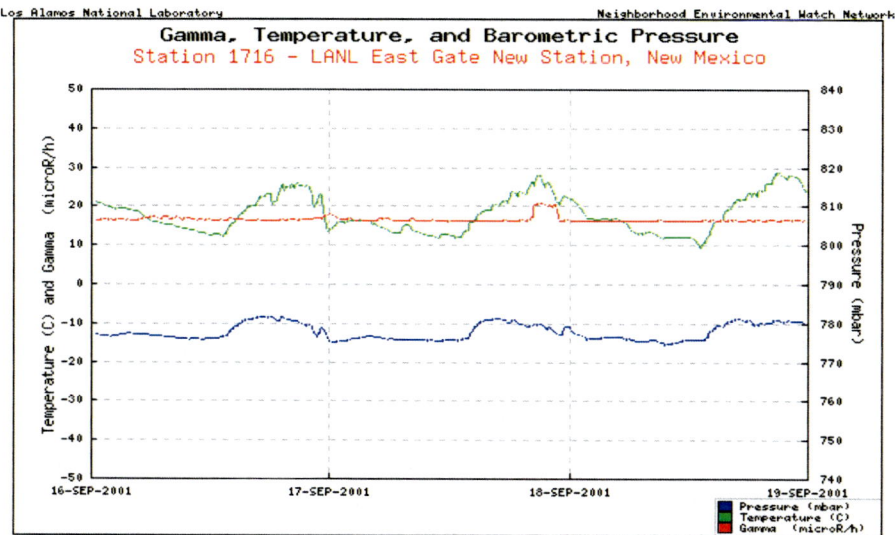

FIGURE 4. As in Figure 3, but showing a three-day archive for the "East Gate" site. The sudden, transient rise in dose rate (red) to the right of center is due to the ^{137}Cs source described in the text.

Finally, Figure 5 shows a capability that has been added to NEWNet recently: means to acquire, transmit, and analyze spectrometric information, specifically α-particle energy spectra obtained using the ECAM. The ECAM system has a limited spectrometric capability (FWHM ~0.3 MeV, but with a low-energy tail) sufficient to distinguish between 7.7-MeV α particles emitted by ^{214}Po, the most readily identified alpha emitter in the naturally occurring 4n+2 uranium decay chain, and the broad collection of lines in the energy spectrum near 5.5 MeV, produced by anthropogenic isotopes such as ^{239,240}Pu and ^{241}Am [2]. A simple analysis routine based on regions of interest (ROIs) extracts numbers of net counts for ROIs spanning these two parts of the energy spectrum, plus a third near 6.1 MeV that is useful for correcting for tailing and presents the net count numbers to the user in the same fashion as the dose information from the ionization chambers. The data shown here make the point that the α activity detected at this station is due entirely to the 4n+2 decay chain because the net counts in the 5.5-MeV ROI are statistically indistinguishable from zero. (The scatter in the results is due to subtraction of an underlying background from the tails of the 6.1-MeV and 7.7-MeV peaks.) In the event of a release of anthropogenic alpha emitters of interest from a public-health standpoint, the count rates in the 5.5-MeV ROI would be expected to be significantly different from zero.

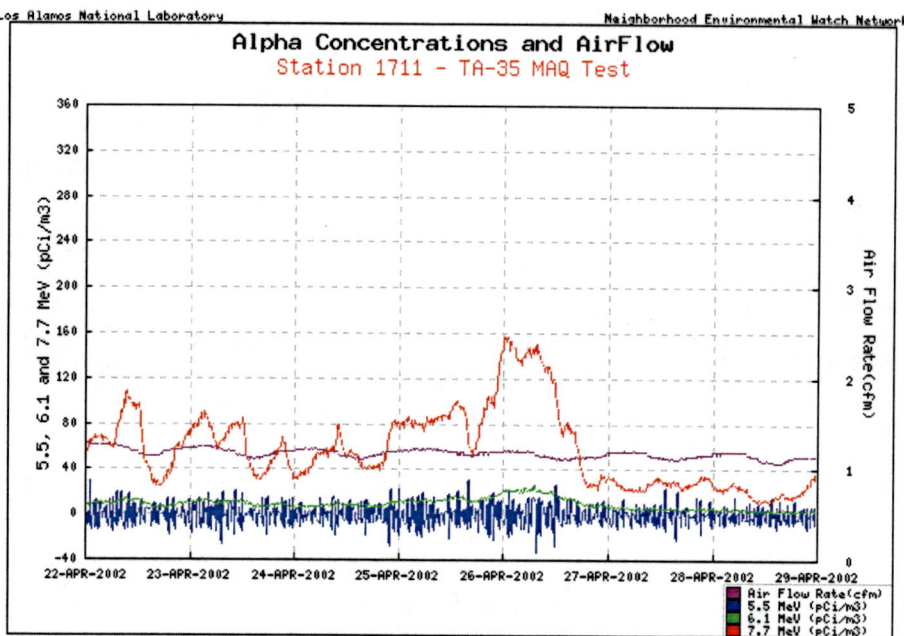

FIGURE 5. Portion of a newnet2.lanl.gov Web page showing a 7-day archive of ECAM data. The 7.7-MeV, 6.1-MeV, and 5.5-MeV ROIs are used to generate the activity levels plotted in red, green, and blue, respectively. Air flow through the ECAM is also tabulated and plotted.

CONCLUSION

The public-awareness mission of NEWNet has shaped the system configuration as described above, but we believe that the basic NEWNet architecture could be adapted to a broader range of problems of national interest. We conclude with a brief discussion of some possible enhancements that are consistent with the architecture and how these enhancements might fit into that broader range of problems. Shorter integration times would be required for applications involving detection of smuggled nuclear materials or weapons. The architecture is not limited, in principle, in the integration times it can accommodate, and this requirement should therefore be easy to meet. Detecting smuggled nuclear materials would also be expedited by a more capable sensor suite, which can easily be accommodated within the existing architecture. The relatively inefficient ionization chamber actually used in the DCP is all that is required to obtain the desired health-physics information, and more efficient detectors are readily available and compatible with the remainder of the system. The ability to transmit spectral information from more capable detectors also exists, but in many applications it would be desirable to perform some amount of spectrum preprocessing at the DCP [3] rather than after transmission of the spectra to the home site. Isotope identifiers currently being researched might suffice for this purpose, but would have to be incorporated into the architecture, a task that has not yet been done. Finally, the only real-time analysis capability currently in NEWNet is the α-particle discrimination routine exhibited above. Real-time methods for correlating signals among multiple detectors are a subject of current research, including some ideas presented at this conference, and would represent a useful enhancement to a security-oriented NEWNet successor.

ACKNOWLEDGMENTS

The authors wish to acknowledge the pioneering work done on NEWNet by Larry Sanders, Orval Hart, and James Ogle. Clayton Watson supplied valuable information on the NEWNet software and hardware.

REFERENCES

1. Sanders, Larry D., and Hart, Orval F., "Department of Energy Nevada Test Site Remote Area Monitoring System," Los Alamos National Laboratory report LA-UR-93-2201, June 1993.
2. Lederer, C. Michael, and Shirley, Virginia S. (eds.), *Table of Isotopes, Seventh Edition,* John Wiley & Sons, Inc., New York, 1978.
3. Scott, H. L., "Radiation Identification System," in *Arms Control and NonProliferation Technologies,* edited by David Spears, Department of Energy publication DOE/NN/ACNT-96AB, 1996, p. 28.

COMMERCIALIZING
NEW TECHNOLOGIES

A Small Business Model for Technology Transfer

D.K. Steinman

Steinman Consulting, Missouri City, Texas 77459

Abstract. Government funding of technology transfer and technological innovation through the Small Business Innovative Research program has had many successes during the nearly two decades of its operations. A means is suggested here to enhance the conditions for successful commercialization of both existing government technology transfer programs and for developing new technology that address technology needs of the various agencies and departments. Small businesses are critical sources of new technology and have been for many years. A business model that will foster technology innovation and help small businesses be successful developing technology and making the technology commercially viable will enhance the value of the program to the government and small business community.

It is recommended that additional processes to promote small, technology-based businesses be attempted. One such process is to establish a program that works in conjunction with SBIR/STTR programs, but focuses on solving high priority problems for which an end-user can be identified. End-users must have a commitment to the development from the beginning of the project.

INTRODUCTION

Currently several programs exist for transferring government developed technologies and for fostering technology innovation in industry. Such programs as the SBIR, the STTR are aimed directly at fostering technology innovation in small and start-up businesses. CRADA and Work for Others programs are aimed at transferring government developed technology to industry. Many other programs developed and supported by individual government laboratories have also attempted to encourage use of government technology in industry and support for small businesses in the local communities around major government laboratories. Such programs as Thermo Technology Ventures at the INEEL and Savannah River sought to commercialize lab technology by getting the individual inventors to "spin out" from the lab to start local ventures.

Most of these programs have some level of success. There are few statistical data, however, that demonstrate that the success rate is affected by the programs over and

CP632, *Unattended Radiation Sensor Systems for Remote Applications*, edited by J. I. Trombka et al.
© 2002 American Institute of Physics 0-7354-0087-3/02/$19.00

above what would be expected by normal human drive that led to the formation of many "spin out" business before the advent of these programs. That the current programs have some severe limitations for small business is clear. Firstly, CRADA programs and Work for Others often require a significant investment by the industrial partners, and hence are mainly of benefit to companies with the resources to participate. These tend to be large businesses that need a particular technology as part of a larger in-house project. The SBIR program often leads to the existence of "SBIR Houses" whose principal business is winning SBIR grants. Very often these companies have outstanding technical expertise but little commitment to commercialization of the technology. Often the funding sources for the new technology is not sufficient to carry it to the next step in commercialization.

From the standpoint of commercialization of existing technologies, agencies often have open house events and open publications that showcase available technologies, and people from industry frequently attend such events or read the open literature in search of opportunities. Although the mechanisms are often clumsy because of intellectual property concerns by industry and procurement regulations on the government side, very often the good technologies from the government labs are brought into industrial usage through investment by large companies.

Commercializing new technology by small businesses is more problematic. For example, on the Department of Defense SBIR web site[1] the following statistics are shown: From 1983 through 2001

- More than $5.4 billion was spent on SBIR projects

- 19,413 Phase I awards were made

- 7,562 Phase II awards were made

- Under the heading of "Successes" there are 21 entries

The "Successes" were, in fact, major successes with companies gaining revenues in the tens of millions from commercial deployment of their original SBIR grants. There were probably many other companies that enjoyed lesser levels of success. However, in the private sector this level of business investment would not be considered successful. Indeed, venture capital and other start-up funding sources often entertain highly risky new technology and business ideas, and their rate of success is in the range of 40%[2] or more. Although this statistic is for business start-ups as opposed to new product introductions, start-up companies generally bring something new to market, and success rates cannot be much less than the business success rate, or the business would not survive.

---

[1] http://www.acq.osd.mil/sadbu/sbir/success/success.htm

[2] Headd, Brian, "Business Success: Factors Leading to Surviving and Closing Successfully," U.S. Small Business Administration, paper on the internet at : http://www.ces.census.gov/paper.php?paper=101614

ANOTHER APPROACH

The principle reason for higher success in private ventures is that funding organizations recognize that the true cost of commercialization does not stop at the development of the technology. That is just the second milestone; the true cost is creating a path to market. (The first milestone is recognition of a market need or opportunity.) Industrial corporations both in the U.S. and worldwide spend on the average of $7\%^3$ of their revenues on research and development. Often a significant proportion of this amount is directed towards engineering and technical support of existing product lines, hence only a small part of this amount is used for advancing technology. The major new product expenses are in distribution, sales, and manufacturing. Often selling expense is 30% or more revenues for manufacturing companies. Creating the channel to market so that buyers are aware of company products and companies are aware of the needs of the buyers is the most expensive component of product development.

In programs such as CRADA and Work for Others, participating companies have already established their links to markets, and they know where the particular technology target of a particular project will fit. They also have a firm estimate of the cost and value of the project. For programs such as SBIR, the commercial considerations are explicitly ignored except for a requirement that the grantees show some level of commercial interest in the technology as part of the proposal. Given the existence of many "SBIR Houses" this exhibition of commercial interest is often merely *pro-forma*.

The purpose of the present paper is to suggest a process whereby commercial interest is the mechanism that sets the goals of each project and directs the execution of the project. The suggested method is to link technologists with businesses and government agencies having specific technology needs and to have the participation of the end-users of the technology in development projects from its outset.

It is assumed that the goals of the technology transfer and small business innovation programs are:

- Fostering innovation and growth by small technology-based companies,

- Development of technologies that address industry and agency needs,

- In today's environment to develop suppliers and support for advanced security products and services.

Significant funding is being directed into technology-based solutions for homeland security problems, and involvement of the small business community in those solutions is very desirable. Many new ideas will come from that quarter, and means must be developed to ensure that these ideas are developed into workable products. However, once the product has been created, it must be supported, and the source of supply must be

---

[3] This represents the target rate for Research and Development Expense and the method of allocating tasks to that category by several major instrument companies in the U.S. The numbers are management guidelines and may vary depending on business conditions.

robust so that future supplies and product support will be available in times of national need.

The source of supply is both the heart of the problem and the seed of the solution. Industry today relies on outsourcing to assure supplies of products and services. There are almost no vertically integrated companies, and therefore suppliers have truly become the "ball bearings" of the American economy. When technology needs arise (either in industry or government agencies) one mechanism to address them is to involve a small technology business as the supplier of the new technology. The business community encourages such assistance. Indeed, many large companies will help small business get started or enter a new business area (in which the small company has some competence) when a competitive advantage will accrue to the larger company. While helping the large company with new technology, the large company gives the small company a huge advantage of reducing many market entry expenses because the sponsoring end-user becomes the initial market for the new technology.

The most important factor in the commercial success of any product is the existence of a market for the product or service. Technology developed within the government and technology developed for the government has one known market: the government. The government often has specialized technology needs for which there are no exact commercial or industrial counterparts. The Nuclear Weapons Complex is a case in point. Clearly, the needs of the Weapons Complex are unique, and security considerations forbid direct solicitation of industrial contributions. In this program, many parts, devices, components, and subsystems were designed and fabricated in-house. NASA, in contrast, has consistently acquired equipment from outside vendors; however, the configuration of the equipment is unique for space flight.

Making small businesses successful in commercializing new technology is challenging. Accomplishing this goal is the main focus of this new approach. The basic premises of the process that will channel funding to solve industrial and agency technology-based problems, that will both, bring small companies into the solution process, and give the participating small business an excellent chance for commercial success are:

1. Solve real problems,

2. Provide end-user direction and guidance of the project to ensure the solution is targeted and stays on target, and

3. Provide a path to market to assist in starting production and short-term commercial viability.

An important question to be addressed is the following:

Does the government wish to promote growth of small companies by giving them opportunities to leverage government grants and contracts into viable commercial enterprises, or is the purpose of the SBIR/STTR grants to keep small companies in business until they eventually find one project that will be a commercial success?

The present discussion assumes a strong commitment to the former premise, and hence, there is a willingness to improve the system to enhance the likelihood of success. Given the record of success cited above, many people would consider it timely to try other processes. However, one aspect of the current program that is highly beneficial is that many high-risk projects are undertaken through the SBIR process that may not otherwise be investigated. Hence, the recommendation here is not total abandonment of the current process. It is to try additional methods to determine whether the success rate can be improved with some reasonable modifications of the process.

THE PROCESS

The process begins by identifying real and important problems in government agencies and industry that require new technology in order to:

- Reduce costs in products, processes, or services;

- Open opportunities for new markets that enhance the competitiveness of U.S. industry;

- Enhance homeland security;

- Enhance national security;

- Enhance the quality and quantity of health care

- Address other important national needs.

In a word, many of the same needs that show up in standard solicitations. If the problems are selected from actual government solicitations or if industrial companies seek active participation in the process, the problems is, by definition, real because:

- The end-user is willing to commit funds to develop solutions.

In addition, directing these procurements to small companies will show the commitment of the government to foster small business development. The process uses agency funding and end-user funding from the outset to ensure that the project's technical goals are channeled correctly and that the project is managed with the commercialization goal in sight. The end-user should be involved at the outset by participating in the selection process. Furthermore, the end-user must participate in the entire process including making a commitment to technical and business reviews and providing cash support to the project.

The boundaries of the process can be laid out once the premise is established. First, an organization must locate companies and agencies that have needs for new technologies together with the financial resources to fund projects that address the needs. These needs, outlined above, should have the following characteristic:

1. The technology to be developed is not core to the end-user (company or agency), but it is required in order to enable the venture or program being pursued. For example, in oilfield services, whenever a new type of nuclear detector is required,

oilfield service companies work with detector manufacturers instead of developing in-house capabilities.

2. The solution to the problem must not require a scientific break-through. The solution should be based on well-known science and engineering and should be achievable by employing good technical practices. Engineering new materials, processes, and techniques is within the scope of these projects as with current SBIR granting criteria.

Subsequently, there must be a:

3. Small businesses with strong technical expertise in the target technology (product, service, or process).

The end-user must have the following characteristics

1. A budget to address its need and a willingness to spend the budget in this manner,

2. A willingness to be involved in the development process as exhibited by a commitment to assign personnel to monitor the project in detail and to review and provide direction to the characteristics of the solution.

3. When the project demonstrates engineering viability, a willingness to commit to purchase enough of the product to establish the small business as a viable supplier for a period of two or more years.

Other features of the process are that the project can begin funding at any point in the development cycle. The project can begin before proof of principle, during establishment of engineering feasibility, or at the stage of adapting an existing technology to a novel application. Because there is always risk, the need for continuing funding will exists. Moreover, whatever the state of development of the technology, the end-user must contribute financially. Should the proof of principle or engineering feasibility not be satisfactory for the intended application, the agency could be authorized to refund a portion of the end-user's contribution. As long as there is progress, the fraction of the financial support supplied by the end-user should grow as each new phase of the project begins.

It will also be necessary to establish a suitable arrangement for sharing intellectual property developed during the project. The end-user (if a private company) will wish to protect its investment, and the funding government agency has an obligation to serve the public. This can be addressed by retaining a field-of-use protection for the end-user while allowing the developer to hold rights in other well defined areas of application of the technology. Other strategies can also be employed.

Another point on the boundary of the process is to provide funding that is tied to the reality of producing a product. Currently, SBIR grants are fixed in size and duration. It is clear that the funding is often inadequate to propel the technology into commercial use because no one ever asks for less than the amounts fixed in the statutes. In order to enhance the probability of success, a flexible funding concept must be established. Although the funds into the project would be higher because of end-user participation, there is little certainty about the total required development resource requirement until the technology is quite well down the development path. Additionally, gaps in the development retard progress out of proportion to their duration. A two-month funding gap can result in critical loss of personnel to the project, and the project is likely to take on a different character with new people, no matter how talented.

ESTABLISHING THE PROCESS

Establishing this process should itself be given a reasonable chance for success. Each agency currently funding SBIR programs could engage a contractor familiar with the relevant industries of the technology to be developed and the applications of the end-users. It must be recognized that development of a new technology from concept through commercial availability is seldom less than three years and often longer, and the process must be given sufficient time to shepherd a few projects through the process. The test of the process should run for a minimum of five years, and there should be goals for the success rate.

Clear measures of success and achievement must be established for the process. For example, one definition of success could be that:

> "a project that returns investment capital to the end-user and returns capital to the government in the form of taxes equal to or greater than the investment in the project."

Properly counting investment cost should include the cost of hiring and retaining the contractor that locates the end-user, the target technology, and the small business.

CONCLUSION

A process has been described that could be incorporated as an extra measure for the SBIR program to enhance the success rate for driving new technology to commercial viability. The process employs the concept that an end-user becomes involved from the beginning of the endeavor and helps guide and support a small business to produce value in the public interest.

The benefits of this approach are:

- Real problems are solved,

- Flexible funding is provided so that projects with promise do not conclude prematurely, and

- A market entry mechanism is provided so the small company has time to adjust to commercial business.

As a final note: for security projects, there is really only one end-user. The government should be prepared to fund the risky part of the development in much the way SBIR funding is currently provided. In addition, the end-user agency, i.e., the actual group within the government that will deploy the technology must contribute both funding and direction. Furthermore, the end-user must be prepared to build a stockpile of the technology that will help establish the small business as a viable supplier. It is very likely that the need for the technology will persist for many years, potentially for two decades. If the technology is successful, it will be in the government's interest to maintain support for it over these time periods.

Tapping into a Billion Dollar Resource, SBIR/STTR

Paul Mexcur and James Kalshoven

NASA, Goddard Space Flight Center, Greenbelt, MD 20771

Abstract. This presentation provides an overview of the Small Business Innovation Research (SBIR) and the Small Business Technology Transfer (STTR) Programs as implemented by the National Aeronautics and Space Administration (NASA). These programs, as mandated by Congress, provide an opportunity for small, high technology companies and research institutions to participate in Government sponsored research and development (R&D) efforts in key technology areas. This presentation describes the background and operation of these two programs and discusses what factors a business should consider in making the decision to participate.

SBIR and STTR Program Overview

The SBIR program was established by Congress in 1982 to: (1) stimulate U.S. technological innovation; (2) use small businesses to meet federal research and development needs; (3) increase private-sector commercialization of innovations derived from federal R&D; and (4) foster and encourage participation in technological innovation by women-owned and disadvantaged small business concerns.

Legislation enacted in December 2000 extended and strengthened the SBIR program and increased its emphasis on pursuing applications of SBIR project results. The SBIR program has been re-authorized through September 30, 2008.

The SBIR research topics and subtopics focus on the highest priority technology development needs of the Agency as defined by the 5 NASA Strategic Enterprises. This year, each Strategic Enterprise was allowed 24 subtopics to support their technology development needs.

The STTR program awards contracts to small business concerns for cooperative research and development with a non-profit research institution (RI), such as a University. The goal of the Congress in establishing the STTR program is to facilitate the transfer of technology developed by a RI through the entrepreneurialship of a small business to meet a government need and have a commercial application for the technology outside the government.

Legislation enacted in October 2001 extends the STTR program through September 30, 2009.

The STTR research topics are focused on Agency core competencies and are aligned with NASA's Centers of Excellence as defined in the NASA Strategic Plan. Because of STTR funding limitations, each Center of Excellence participates in this annual solicitation on a two-year cycle.

CP632, *Unattended Radiation Sensor Systems for Remote Applications,* edited by J. I. Trombka et al.
2002 American Institute of Physics 0-7354-0087-3

The structure of the SBIR program reflects the Congressional understanding that the processes of innovation and bringing new products to the market have a high degree of technical and financial risk. The programs, therefore, have three phases:

Phase I is the opportunity to establish the feasibility and technical merit of a proposed innovation. Selected competitively, NASA's SBIR Phase I contracts last for 6 months with a maximum funding of $70,000.

Phase II is the major R&D effort in SBIR. Phase II continues the most promising of the Phase I projects based on scientific/technical merit, expected value to NASA, company capability, and commercial potential. Phase II places greater emphasis on evidence of commercial development than Phase I, particularly for NASA uses. Phase II contracts are usually for a period of 24 months with a maximum funding of $600,000. NASA usually selects approximately 40% of the Phase Is to go on to a Phase II.

Phase III is the infusion of the Phase II results into regular NASA programs and/ or into the commercial market. Phase III projects are funded with money from a source other than the SBIR or STTR program. It is understood that further development of the product may be needed. NASA is able to accelerate its Phase III procurement process by recognizing that the federal competition in contracting requirements have been met by the Phase I and II competitions. In other words, the Phase III funding is granted based on the merits of the Phase II results without further need for competitive bids. Private-sector investment, in various forms, is also a vehicle for the Phase III process.

Significant Changes Included in the 2002 SBIR Solicitation

The following overview summarizes some of the more significant changes from last year's Solicitation:

The **Aerospace Enterprise** (AT) added substantially new or rewritten research areas including nuclear and intelligent propulsion technologies, air-traffic management, and nanotechnology. AT selected subtopics from six of NASA's field centers (ARC, DFRC, GRC, LaRC, MSFC, and SSC), the same group as last year's solicitation. Of its 24 subtopics, 7 are new this year. Of the 17 returning, 3 have new subtopic managers. Overall, 10 of this year's AT subtopics are either new or have been substantially revised.

The **Biological and Physical Research Enterprise** (BPR) added a new research topic area, Biomolecular Systems, Devices and Technologies, which included seven new subtopics. Other research areas receiving substantial rewrites include human adaptation to space and food science and processing for space. BPR has technology development subtopics from six of NASA's field centers (ARC, GRC, JSC, KSC, LaRC, and MSFC). Unlike 2001, JPL did not lead a subtopic this year, although it participates in several. Of its 24 subtopics, 10 are new this year, partially reflecting an increase in the total number of subtopics represented by BPR from 18 in 2001 to 24 this year. Of the 14 returning subtopics, 5 have new managers. Overall, 11 of 24 subtopics for BPR are either new or have been substantially revised.

The **Human Exploration and Development of Space Enterprise** (HEDS) added several new or substantially rewritten subtopic research areas including solar power generation and power management; vibroacoustic predication and simulation technologies; distributed architecture and operations for the International Space Station; and intelligent command, control, and monitoring systems for Spaceports. HEDS has technology development subtopics from six centers (GRC, JSC, KSC, LaRC, MSFC, and SSC). ARC led a subtopic last year, while LRC did not. Of its 24 subtopics this year, 7 are new, but 2 of these are transfers from last year's AT Enterprise subtopic list. Of the 19 returning subtopics, 6 have new managers. Overall, 13 of HEDS 24 subtopics are either new or substantially rewritten from last year.

The **Earth Science Enterprise** (ES) had substantial enhancements to its research areas involving technology developments for practical use of ES measurements for quality of life and education. Also added or rewritten this year were technologies sought for understanding the carbon cycle, improved on-board intelligent sensors and active optical systems. ES has technology development subtopics from five of NASA's field centers (ARC, GRC, GSFC, LaRC, and SSC) and JPL. Unlike 2001, MSFC is not leading a subtopic this year, although it participates in several. Of its 24 subtopics, 2 are new this year. Of the 22 returning subtopics, 6 have new managers. Overall, 9 of ES subtopics are either new or substantially revised from last year.

While not adding any new subtopics for 2002, the **Space Science Enterprise** (SS) substantially rewrote several subtopic research areas including deep space propulsion, distributed observing platforms, high contrast astrophysical imaging, and biological contamination in flight hardware and return-sample handling. SS has technology development subtopics from four NASA centers (ARC, GRC, GSFC, and MSFC) and JPL, the same as in 2001. Of the 24 returning subtopics this year, 4 have been substantially rewritten from last year.

Based on the current funding level of $115.1M, we anticipate awarding approximately 330 phase I and 150 phase II contracts. Approximately 750 small business concerns compete for these contracts each year with about a third of the awards going to firms who are totally new to NASA's SBIR or STTR program. A fully searchable electronic version of the Solicitation will be publicly released on June 6, 2002 at NASA's SBIR homepage: http://sbir.nasa.gov

Small Business Grants at the National Cancer Institute and National Institutes of Health

Houston Baker

Biomedical Imaging Program, Division of Cancer Treatment and Diagnosis, National Cancer Institute, National Institutes of Health, Bethesda MD 20892-7412, USA

Abstract. Ten Federal Agencies set aside 2.5% of their external research budget for US small businesses—mainly for technology research and development, including radiation sensor system developments. Five agencies also set aside another 0.15% for the Small Business Technology Transfer Program, which is intended to facilitate technology transfers from research laboratories to public use through small businesses. The second largest of these agencies is the Department of Health and Human Services, and almost all of its extramural research funds flow through the 28 Institutes and Centers of the National Institutes of Health. For information, instructions, and application forms, visit the NIH website's Omnibus Solicitation for SBIR and STTR applications. The National Cancer Institute is the largest NIH research unit and SBIR/STTR participant. NCI also issues SBIR and STTR Program Announcements of its own that feature details modified to better support its initiatives and objectives in cancer prevention, detection, diagnosis, treatment, and monitoring.

SESSION ON COMMERCIALIZING NEW TECHNOLOGIES: ROLE OF SBIR FUNDING

Overview and Task Assignment

The agenda for the commercialization session at the Workshop on Unattended Radiation Sensor Systems for Remote Applications (Carnegie Institution, Washington DC, April 15-17, 2002) called for perspectives on the two approaches—contracts and grants—used by federal agencies to implement the US Small Business Administration's Programs for Small Business Innovation Research (SBIR) and Small Business Technology Transfer (STTR).

Paul Mexcur, National Aeronautics and Space Administration, presented the contract approach, exemplified by NASA's implementation to procure small business services to meet some of their technology development needs. NASA has the third largest SBIR budget among Federal agencies.

This presentation covers the grant approach described in the publicly available Omnibus Solicitation used by the Public Health Service of the Department of Health and Human Services, which has the second largest Federal SBIR budget. DHHS participants include the Food and Drug Administration, Centers for Disease Control and Prevention, and 28 Institutes and Centers of the National Institutes of Health (NIH provides almost all of DHHS SBIR funding). This presentation also covers specific

CP632, *Unattended Radiation Sensor Systems for Remote Applications*, edited by J. I. Trombka et al.
2002 American Institute of Physics 0-7354-0087-3

program variations used by NIH's largest institute, the National Cancer Institute. The NCI also has additional initiatives for technology developments that are inspired by the SBIR and STTR models. These draw upon non-SBIR funds, making them available to all eligible research organizations, small businesses included. Contracts play a minor, but useful role in NIH SBIR funding.

I was asked to address six questions:
1. Importance of the SBIR Programs;
2. How they work;
3. Where they succeed;
4. Where they fail;
5. Ideas to improve the SBIR program; and
6. Ideas to improve the translation of prototypes into market-ready devices, systems and/or methods.

Importance of Federal, NIH, and NCI SBIR Funding to Small Businesses

Overall Federal Funding—Importance to Small Businesses

Ten Federal Agencies with external research budgets over $100 million per year are required by law to set aside 2.5% to support the Small Business Administration's Small Business Innovation Research Program (www.sba.gov/sbir/indexsbir-sttr.html). The five agencies with research budgets exceeding $1 billion are required to commit an additional 0.15% (0.3% starting FY2004) to the SBA's Small Business Technology Transfer Program. The "billionaires" with both SBIR and STTR Programs include the Department of Defense, Department of Health and Human Services (includes NIH, FDA, and CDC), National Aeronautics and Space Administration, Department of Energy, and National Science Foundation. The five "100-millionaire" agencies with SBIR Programs are the Department of Agriculture, Department of Commerce, Environmental Protection Agency, Department of Transportation, and Department of Education.

In Fiscal-Year 2001, their cumulative SBIR budgets totaled about $1.2 billion, and STTR budgets amounted to approximately $70 million. The significance of this money specifically set aside to support technology developments through small business Research and Development is hard to overestimate.

Since its 1982 inception, the SBIR Program has become a key source of enabling money for entrepreneurs starting businesses based on technology-driven products and services. By now the number of companies alive and productive after their fetal and pediatric phases because of SBIR money must number in many thousands. Business exhibitors at major medical meetings present significant numbers of products that started directly or secondarily with SBIR-funded grants and contracts. SBIR/STTR Programs are an important source of funding opportunity for small business R&D programs. Participation is limited to independent US companies with 500 or fewer employees.

267

Overall NIH Funding—Importance to Small Businesses

The overall budgets of the NIH and its confederation of 28 Institutes and Centers are as follows:

- FY2001 actual $20.5B
- FY2002 estimate $23.5B
- FY2003 proposed $27.3B

The proposed fiscal 2003 budget would support 9,854 new Research Program Grants selected through competition, and 35,920 grants overall—SBIR grants not included (http://www.nih.gov/news/budgetfy2003/2003NIHpresbudget.htm).

Estimates of the FY2003 SBIR share are as follows:

- FY2001 actual   ~$0.460B
- FY2002 estimate   ~$0.529B
- FY2003 proposed   ~$0.608B

Clearly NIH is an important source of funding opportunity for small business R&D programs.

Overall NCI Funding—Importance to Small Businesses

NCI, the largest of the 28 Institutes and Centers, has overall budgets as follows:

- FY2001 actual $3.8B
- FY2002 estimate $4.2B
- FY2003 proposed $5.7B

Estimates of the FY2003 NCI SBIR and STTR shares are as follows:

	SBIR	STTR
FY2001 actual	~$ 78M	~$4.7M
FY2002 estimate	~$ 87M	~$5.2M
FY2003 proposed	~$118M	~$7.1M

Clearly NCI is an important source of funding opportunity for small business R&D programs. It approaches the size of the NASA budget reported in the prior talk by Paul Mexcur.

How the SBIR Program Works

The essence of the SBIR Program is to fund by phases, thereby providing a tradeoff between cost and risk. The higher risk of Phase I feasibility work is offset by limiting budgets to a shorter funding period and lower dollar allowance. Transition to Phase II funding requires demonstration of successful Phase I performance and a Phase II plan that passes review. According to the 1982 legislation and continuing today, SBIR Phase III funding cannot be paid from SBIR funds. Details are provided on the SBA website. Phase I and II budget guidelines on times and dollar amounts vary from agency to agency, and within agency, *cf.,* NCI special SBIR program announcements differ in funding periods and guideline dollar amounts from the Department of Health and Human Services Omnibus SBIR/STTR solicitation.

A solo Phase I application followed upon completion by submission of a solo Phase II application represents the basic sequence for SBIR/STTR applications. Alternatively, applicants may choose to submit paired Phase I and II applications to be reviewed as one package, with both receiving a single, overall priority score. This is known as the "Fast Track" approach, and if the Phase I work succeeds, the project receives prompt Phase II funding and avoids the second cycle of submission and review for a separate Phase II application. The NIH implementation is described in http://grants.nih.gov/grants/funding/sbirsttr1/index.pdf, the Public Health Service Omnibus Solicitation, PHS 2002-2.

How They Work—NIH Implementation

The SBIR/STTR plans contemplate three phases of work. The version provided in the NIH Omnibus Solicitation is as follows:
- Phase I: Test feasibility and demonstrate company capacity to perform
  - SBIR: 6 months, ~$100k total costs (more time or money if justified—meaning that information supporting the budget request must be appropriate and convincing to both reviewers and program administrators),
  - STTR: 1 year, ~$100k total costs (more if justified).
- Phase II: Continuation R&D
  - SBIR: 2 years, ~$750,000 total costs (more if justified).
  - STTR: 2 years, ~$500,000 total costs (or more).
  - Both: a product development plan is required.
- Phase III: Commercialization of Results
  - No SBIR or STTR funding allowed,
  - Funding must come from sources other than small business set-aside funds.

How They Work—NCI Variants

The National Cancer Institute uses variations on the NIH implementation to meet specific program objectives. One example is PAR-01-102, an SBIR/STTR program announcement for Development of Novel Imaging Technologies for in vivo Imaging (see http://grants.nih.gov/grants/guide/pa-files/PAR-01-102.html). The differences are as follows:

- Phase I SBIR or STTR: 1-2 years (vs. 6 months), ~$100k total costs (more time or money if justified.
- Phase II SBIR or STTR: 1-3 years (vs. 1-2 years), no dollar limitations under this PA for Phase II budgets, but good justifications are required, and the amounts requested are subject to peer review recommendations, availability of funds, and Program priority. (STTR funds are scarcer, so larger budgets are more vulnerable to cuts.)

- A four-year limit for either Fast-Track Applications (separate Phase I and II applications bundled together, reviewed together, and given one score), or a Phase I and renewal Phase II application. Compare to the DHHS Omnibus Solicitation guidelines of 2.5 years total for Phase I and II SBIRs, and 3 years for STTRs.

Where the SBIR Program Succeeds

From NIH and NCI perspectives, this is a successful program. The basic parameters of this legislative series have remained essentially the same, likely because success provides only modest motivation to ask for major legislative changes. The Small Business Innovation Development Act of 1982 (P.L. 97-219), which set up the original parameters of the SBIR program, was reauthorized for 3 more years when its original 15-year sunset provision came due (P.L. 102-564, Small Business Research and Development Enhancement Act), and renewed again in 2000 for 8 years (P.L. 106-554, Small Business Reauthorization Act of 2000).

A search for SBIR success stories on the NIH web site yielded only eleven examples (http://grants.nih.gov/grnats/funding/sbir_successes/sbir_successes.htm), yet a brief effort of personal recall yielded an equal number of company successes, of which four are relevant to the objectives of this Workshop:

1. Advanced Detectors, Inc., and Photon Imaging, Inc., project titles: Silicon drift detectors for x-rays; Detector arrays for scintigraphy mammography; Solid state Anger-type camera; Detector for vulnerable plaque in coronary arteries. They also had some projects that did not pan out as products, but produced useful information that contributed to the success of projects that succeeded: Mercuric iodide intra-operative gamma camera; $CsI(Tl)/HgI_2$ detector arrays for scintigraphy; Fast silicon drift photodetectors for PET.

2. Creatv MicroTech, Inc: Fabrication of 2D x-ray anti-scatter grids; 2D focused x-ray anti-scatter grids; Field-emitter arrays with lenses for flat panel displays; Electron beam collimation for field emission displays.

3. Radiation Monitoring Devices, Inc: PbI_2-based imaging tube for digital mammography; Germanium detector for lung burden screening; Avalanche photodiode gamma camera; Digital mammography with CMOS readout; Neutron sensor for macromolecular crystallography; Intra-operative digital imaging probe; High resolution imaging detector for mammography; among many other titles.

4. X-ray Instrumentation Associates: High-speed processing for x-ray detector arrays; Portable spectrometer for mammography system calibration; 2D electron multiplier dynode material studies; Digital CZT gamma ray imaging detector; Linear detector for x-ray absorptiometry; Energy resolved digital HP-Ge imaging detector.

Several other (anecdotal) estimates of success are brought to the reader's attention, although they suffer from being neither quantitative nor objective:

1. During the 1996 audit of the SBIR program by the Government Accounting Office (GAO) to prepare an SBIR evaluation report for Congress prior to their reauthorizing P.L. 102-564 in 1997, an auditor visited NIH and their Medical Imaging

Technologies review group, Special Study Section 7. When invited to say something, he said that the NIH SBIR program was working better than the other 9 Federal Agencies because it had the highest ratio of product to market per SBIR dollars spent, and among NIH-funded products to market, medical imaging technology applications stood out. He said that his visit to this review group made it easy to see why: (1) careful discussion given each application, (2) evident reviewer diligence in doing their homework, (3) clear focus on trying to understand the problems posed and solutions proposed in each application, (4) assessing the importance of the problem addressed, and (5) providing a balanced, comprehensive, objective summary statement of each application's strengths and weaknesses—useful to NIH administrators and applicant investigators alike.

2. This talk's author needed less than a day to tour all of the Grantee exhibits of SBIR products at the Radiological Society of North America meeting in 1993 but could not complete them in 2.5 days in 2000, due to the large increase in their numbers.

Where the SBIR Program Fails

Problems, and Recommended Solutions or Mitigations

The SBIR Program has some problem areas, but no overt areas of frank failure. Experience has led NIH scientific review administrators of review committees and program officers responsible for funding decisions to identify areas of weakness and develop either solutions or mitigations for a number of them. Some of the problems and more successful responses are as follows:

Problem: The time from application submission to funding takes too long.
This can take up to a year. If an application fails to receive a fundable score, preparing an amended version typically takes 8 to 12 months more. This is especially problematic if the field is moving fast. Some companies say they don't apply because they cannot afford to waste personnel time on applications that won't yield a grant for one or more years—or never.
Solutions or Mitigations:
No obvious solution is apparent for shortening the deliberate process used by NIH and NCI review. Speeding it up might reduce the quality of review, and could reduce systematic protections that assure fair, high quality reviews.

Some companies submit a series of ideas over time to load their pipeline with grants. When one is out for review, another is available to fund their R&D work.

The foregoing are not solutions, but mitigations. Anyone with a new idea is invited to share it with us.

Problem: A successful Phase I feasibility study must precede a Phase II application.

The wait for review and processing of an application submitted after completion of Phase I work leads to a one-year hiatus in funding at best, and longer if a weak score leads to an amended application. This causes loss of research momentum, lets hard-earned know-how grow stale, and throws timing to market off schedule. Keeping the group together becomes problematic.

Solution or Mitigation: Whenever appropriate, submit paired Phase I and II applications under the Fast-Track option (see above) that enables prompt transition to Phase II funding after successful completion of Phase I work.

Problem: Reviewers perceive Fast-Tracks as riskier and give fundable scores at a lower rate.

Recent NCI experience in the imaging technology development area is that Fast-Track applications get funded at a 17% rate, whereas funding rates are 23% for solo Phase I applications and 42% for solo Phase IIs. The perception that Fast-Tracks are risky shows in their weaker priority scores. (The number of Fast-Tracks for which reviewers choose to score only the Phase I application could not be ascertained because the identifying code gets changed in the database to a Phase I code. However, their funding success rate probably is similar to the solo Phase I rate.)

Solution or Mitigation: Require that applications state objective and preferably quantitative performance milestones to be met in Phase I to receive Phase II funding. The NCI Program Announcement PAR-01-102 requires milestones in Fast-Track submissions to provide targets by which to judge Phase I work. Work that falls short won't get Phase II funding. This transition hurdle appears to ease reviewer concerns because priority scores have jumped to 56% in the short time this requirement has been in place (don't over-interpret—low numbers!). We shall see if this promising result holds up as we accrue larger numbers of Fast-Track applications. In addition, the following approaches are employed by NCI to improve review of PAR-01-102 applications:

(a) Reviewers are advised to consider an application according to engineering standards and its potential societal value, not limiting themselves to hypothesis driven research standards.

(b) Reviewers are asked to specifically judge proposed milestones and suggest revisions if they are inadequate.

(c) The Program Director negotiates revised milestones to incorporate reviewer suggestions.

Problem: Technology development applications on average do poorly in review groups that lack reviewers with technology development experience.

Solutions or Mitigations:

The NCI study section that reviews PAR-01-102 applications recruits to have substantial technology development representation. Ben Franklin's report on Mssr. Montgolfier's 18th century hot air balloon experiment provides a classic illustration of the main difference between technology users and developers:

Anonymous Reviewer: "What good is it?"
Franklin: "What good is a newborn baby?"

Take-home lesson for technology development reviewers: *Expect a prototype to show it can get off the ground. Requiring a Zeppelin is premature.*

Problem: Many engineers, physicists and scientists don't work for an SBIR-eligible small business.

Solution: PAR-01-101, Development of Novel Imaging Technologies for in vivo Imaging (R21/R33). July 16, 2002 is the last published receipt date for applications, but there are plans to renew it for another two years. This program announcement is inspired by the Fast-Track SBIR/STTR approach, with differences (see http://grants.nih.gov/grants/guide/pa-files/PAR-01-102.html):

It is open to large and small businesses, academic and non-profit institutions, foreign and domestic.

It uses a single application limited to a 25-page research plan that must address both the R21 and R33 phases (like Phase I and II of the SBIR). The R21 addresses the exploratory—feasibility phase, and the R33 addresses the developmental—prototype phase.

Unlike the SBIR prohibition against starting with a Phase II application, applicants who have already established feasibility may apply directly for an R33 developmental grant.

Budgets for the R21 phase may request 1 or 2 years, usually not to exceed $100,000 in direct costs. There is no guideline for R33 Phase II budget amounts, but budgets in excess of $500,000 per year require a request for permission for submit, addressed to the Program Director listed in the Program Announcement. Solo R33 applications may request up to 3 years of support. A combined R21-R33 application is limited to 4 years.

Problem: Inconsistent scoring within and between study section review meetings weakens comparisons of merit among applications and adds noise to funding decisions.

In the NIH system, each reviewer's score is recorded in tenths from 1 to 5 (1.0, 1.1, . . .4.9, 5.0), strongest to weakest, then averaged and multiplied by 100 to generate the score reported in the Summary Statement. Many study sections use schemes of adjectives to guide reviewers, *e.g.,* some say outstanding = 1.0 to 1.5 but others say 1.0 to 2.0; and so on for excellent, very good, and the yield has more variability than is desirable.

Solution: Apply the NIH scoring system in a manner consistent with engineering practice, as follows: On the premise that a ruler or thermometer with uniform gradations will yield linear, intuitively obvious metrics that allow better comparisons, so will uniform application of the NIH scale to scoring each application. In other words, score in the range of 1 to 2 if the application is judged to be among the first quarter of applications under consideration, and 2 to 3 for the second quarter, and so on.

Some NIH study sections have used this approach to achieve stable results that remain consistent from meeting to meeting.

Other Problems and Few Answers:

Unusual projects with a small potential market, even if they address an important need, tend to fare poorly, *e.g.,* x-ray detector array with fast front-end electronics for protein EXAFS crystallography on a synchrotron beam line. (Is market saturation likely to exceed 10?) The Societal Value concept can help reviewers find appropriate scores for this sort.

Exotic projects that strain reviewer expertise, *e.g.,* Alan Cormack's unfunded 1962 project to develop and experimentally demonstrate the feasibility of his line integral equations to solve for holes in a wooden block and density differences in an aluminum sphere. These experiments were key to modern CT imaging systems (Nobel Prize, 1979). No clear answer here: some reviewers thought he was crazy. However, it may help if the application's cover letter states the kinds of expertise needed for the review, and offers to suggest candidate reviewers.

Vulnerable projects due to poorly justified budgets that produce a recommendation for cuts that doom completion of the work. Answer: Poor grantsmanship; needs coaching.

Low Participation from many US states: Fewer than half account for most of the grants. Answers: Outreach to inform people of grant opportunities helps, but has not made much of a dent. NIH SBIR conferences help most who attend, but these conferences attract a fraction of potential applicants.

Low Participation by non-medical researchers, many of whom don't know how to begin a medically oriented development. There are many developments in other fields that appear suitable to adapt and migrate to medical imaging, and vice versa. Answers: Get acquainted with medical physicists, biomedical engineers, and other medical scientists who can provide insight into what outstanding problems need solution. This applies to both hardware and software solutions for detection, image reconstruction and pre- and post-processing to extract information and meaning from them. Try submitting an application and pay attention to useful criticisms in the resulting summary statement: there is a wealth of information in negative results identified by good reviewers.

Translating Laboratory Prototypes into Marketable Prototypes

Progressing to Phase III Funding: Not Covered by SBIR Funds

Funding for SBIR/STTR Phase III work is actually another problem that might have been appropriately addressed above. Many promising Phase II prototypes fail to progress to the commercialization phase.

Solution: Neither NIH nor NCI have a good solution to this problem. Many reasons may contribute to failure to progress: No SBIR funding is allowed, the team of inventors may lack the interest, skills, funding, R&D expertise, clinical study methods, and business know-how to conduct the translation of a laboratory prototype to clinical feasibility, FDA approval, and successful marketing. However, some investigators attract the skills and investor capital or other funding to do the engineering refinements and other preparations necessary to field a marketable product. Some of that money can come from non-SBIR grant applications.

Concept for Translational Research and Development—RFA CA-03-002, Network for Translational Research: Optical Imaging

Barriers to translating working prototypes to clinical feasibility have come into focus as an important problem and are the subject of a concept now under development at the NCI. From time to time the NCI develops and fields implementations of concepts to target funding at under-represented communities of researchers or to accelerate progress in lagging areas, *e.g.*, the well-received experiment of PAR-01-101 to provide Fast-Track-SBIR-like funding to device and methods developers who are ineligible for Small Business set-aside funds.

We have a new concept ready to try on the problem of translational research and development. Our intent is to develop a way to improve the translation of prototypes from demonstration of laboratory feasibility to clinical feasibility (stopping short of the large, expensive, double-blinded clinical trials that most effectively demonstrate clinical utility that becomes the basis for a shift in the standard of medical care). It is the Request for Applications (RFA), CA-03-002, Network for Translational Research: Optical Imaging, which is in the final stages of editing prior to publication (http://www3.cancer.gov/bip/concepts.htm).

The purpose of this RFA is to provide assistance and support to the extramural research community to establish Specialized Research Resource Centers, each of which will participate as a member of a network of inter-disciplinary, inter-institutional research teams for the purpose of supporting translational research in optical imaging and/or spectroscopy in vivo. The network will operate under the guidance of a Steering Committee (SC).

The goal of this RFA is to organize a consortium with flexibility in scope, funding, and incentives to encourage inter- and intra-team collaborations on translational cancer research. The objective is to accelerate the pace of translational research by developing a consensus process to improve methods for system integration, optimization and validation of next-generation in vivo optical imaging and/or spectroscopy methods and technologies, including contrast agents. It encourages investigators to include in vivo molecular imaging methods that give information about molecular events in cells or the extra-cellular milieu. The research scope includes feasibility studies for the detection of pre-cancerous lesions, cancer detection and diagnosis, and measurement or prediction of response to therapy.

Each team would include at least two organizations—academic institutions, national laboratories, large and/or small businesses, and NIH-intramural programs. The intent is to for the investigators to assemble broad, multi-disciplinary teams. They will be encouraged to include a breadth of expertise sufficient to address the translation problem, including for example, molecular biologists, chemists, physicists, optical and computer engineers, imaging scientists, mathematicians and physicians. Involvement of scientists who may not have a history of cancer research, but who have the potential to provide critical experience for the success of this network is encouraged. Partnerships with industry are encouraged to gain the benefit of their experience in translating medical device prototypes to market-ready systems.

How well our implementation succeeds will be seen. Wish us well as we field and try to iterate this concept into a useful funding mechanism. If it works, we can use it as a model for developing another RFA—Network for Translational Research: (picture your modality here).

Closing Advice to Engineers

If you can't make it better, try making it worse because there is a lot of information in negative results. And if you can't make it worse, you are twiddling the wrong knob!
Francis J. Clauser, Dean of Applied Science and Engineering, retired, California Institute of Technology

ACADEMIC PERSPECTIVES

Current and Historic Trends in Physics and Related Fields

Roman Czujko

American Institute of Physics, Statistical Research Center
College Park, MD 20740

Abstract. This talk provides a statistical overview of the supply side in physics and related fields. Data on current and historic trends are presented in selected fields at both the bachelor's and PhD levels. Several of the major factors driving enrollment patterns in higher education are discussed. This paper concludes with an examination of issues related to both supply and demand. The AIP Statistical Research Center has been collecting data on enrollments and degrees in physics for 40 years and they are the source for the physics data presented. The sources for the data on related fields are the National Science Foundation in the case of PhD data and the U.S. Department of Education for bachelor's level data. Finally, this paper includes data on the size of the bachelor's degree classes of 1999 and 2000. These data were not available when the talk was presented.

BACHELOR'S DEGREES

In 2001, about 1.2 million bachelor's degrees were awarded across all fields in the U.S. (see Figure 1)[3]. After nearly a decade of declining undergraduate majors, the number of physics bachelor's awarded has increased in each of the last two years to just over 4,100 for the class of 2001 [2]. However, physics is a comparatively small field. Of every 1,000 bachelor's degrees awarded in the U.S., only 3.4 are in physics.

Physics lost market share during the late 1980's and most of the 1990's. By way of example, in 1985, 5.4 out of every 1,000 bachelor's degrees awarded in the U.S. were in physics. One of the reasons why fewer students were majoring in physics is that the academic environment for undergraduates has become very competitive. Even though the number of people earning bachelor's degrees increased during the last 15 years, their options increased even faster. In other words, today's students have far more majors to choose from than did students 10 or 15 years ago.

Thus, it is no longer possible for departments to sit back and expect the best students to come to them. In order for a department to survive, let alone thrive, it must become proactive in recruiting potential majors and in developing both activities and programs that keep those majors through to the bachelor's degree.

CP632, *Unattended Radiation Sensor Systems for Remote Applications,* edited by J. I. Trombka et al.
© 2002 American Institute of Physics 0-7354-0087-3/02/$19.00

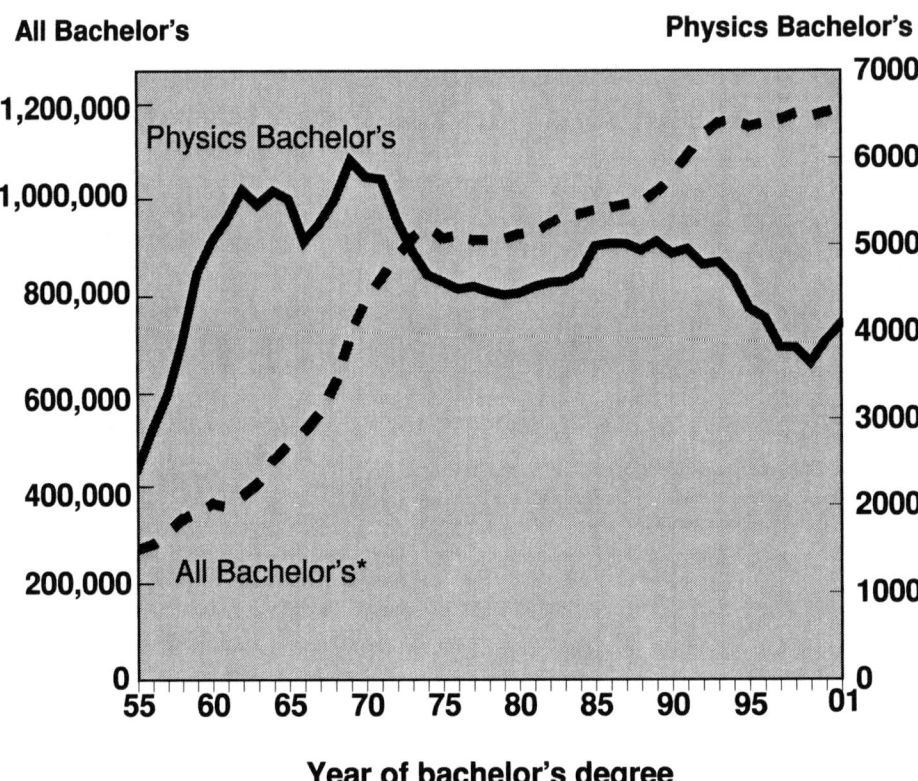

FIGURE 1. Physics bachelor's and total bachelor's produced in the U.S., 1955 to 2001 [2] [3].

The trends depicted in Figure 1 are not unique to physics. The decade of the 1990's was very volatile in terms of undergraduate education for many fields. Figure 2 illustrates the bachelor's degree production in selected fields from 1985 through 2000. Each of these fields portrays a different pattern. Many calculus-based fields, such as physics, engineering, and mathematics, lost majors during the decade of the 1990's. In fact, the engineering bachelor's class of 1998 was the smallest in 17 years [1].

Chemistry reversed the decline in their bachelor's degree production before 1995. In part, this was due to an increased emphasis on biochemistry within the undergraduate chemistry curriculum and, in part, it was because chemistry discovered opportunities in the pharmaceutical industry.

The number of students earning bachelor's degrees in computer science also went through a large cycle. The bachelor's class of 1996 was the smallest in 14 years and the dot com boom of the late 1990's had a dramatic effect on enrollments in computer science departments. In fact, the number of bachelor's degrees awarded in computer science has increased by nearly 8,000 in just the last two academic years.

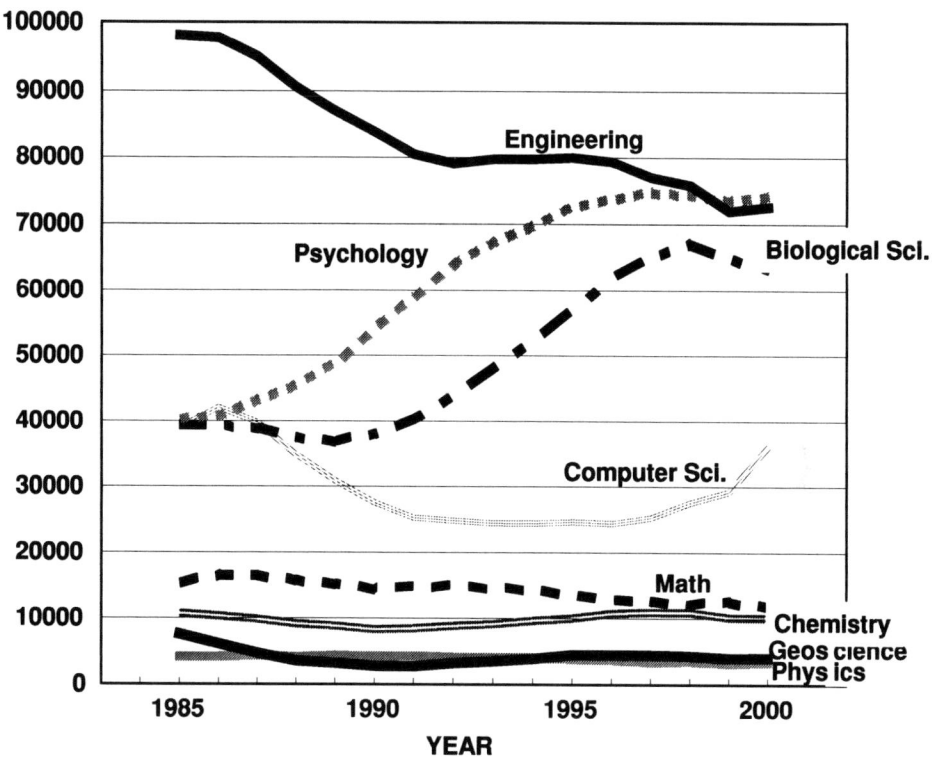

FIGURE 2. Total number of bachelor's degrees granted by discipline, 1985 to 2000 [1].

At the bachelor's level, the two fastest growing fields during the 1990's were biology and psychology [1]. Among the things that these fields have in common are the obvious connection to improving human life and women students. By way of example, 75% of psychology majors are women [3].

Thus, the decisions that women undergraduates make are, to a significant degree, driving higher education. In 1998, about 55% of all bachelor's degrees awarded in the U.S. were earned by women. The U.S. Department of Education projects that this trend will continue to grow and that women will represent 58% of the bachelor's class of 2010 [3].

Women have gradually increased their representation among physics and engineering bachelor's, but at levels far below the overall trends. The physics bachelor's class of 1999 was 21% women, the first time that it had ever passed 20%. The physics class of 2001 was 22% women [2].

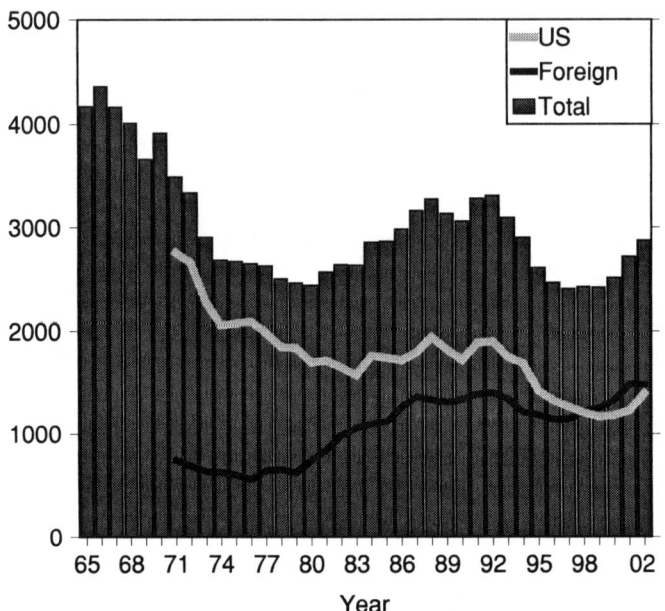

FIGURE 3. First-year US and foreign graduate physics students, 1965 to 2002.[1][2]

FIRST-YEAR GRADUATE STUDENT ENROLLMENTS

This discussion will focus exclusively on physics as it is among the few fields that regularly collect data on first-year graduate students. Figure 3 illustrates the cyclical pattern of graduate student enrollments over the last 38 years. The height of the bars illustrates the number of students admitted into physics graduate programs each year.

The two lines superimposed across the bars reflect the number of U.S. citizens and the number of foreign citizens entering graduate study in physics. The number of foreign citizens in graduate physics programs increased dramatically during the 1980's and 1990's. Over the last few years, enrollments increased among both US citizens and foreign citizens.

PHYSICS PHD DEGREES

Figure 4 depicts the number of physics PhDs awarded in the U.S. each year from 1900 through 2001. You don't have to be a physicist to recognize that this is a system under stress. Similarly, you don't have to be a social scientist to recognize that there is more going on in this graph than simply a reflection of the growing and waning interest in physics.

---

[1] A change in wording on the 2001 questionnaire resulted in more accurate data on first-year graduate students. This change was responsible for 3% of the reported 8% increase in total first-year students between 2000 and 2001.

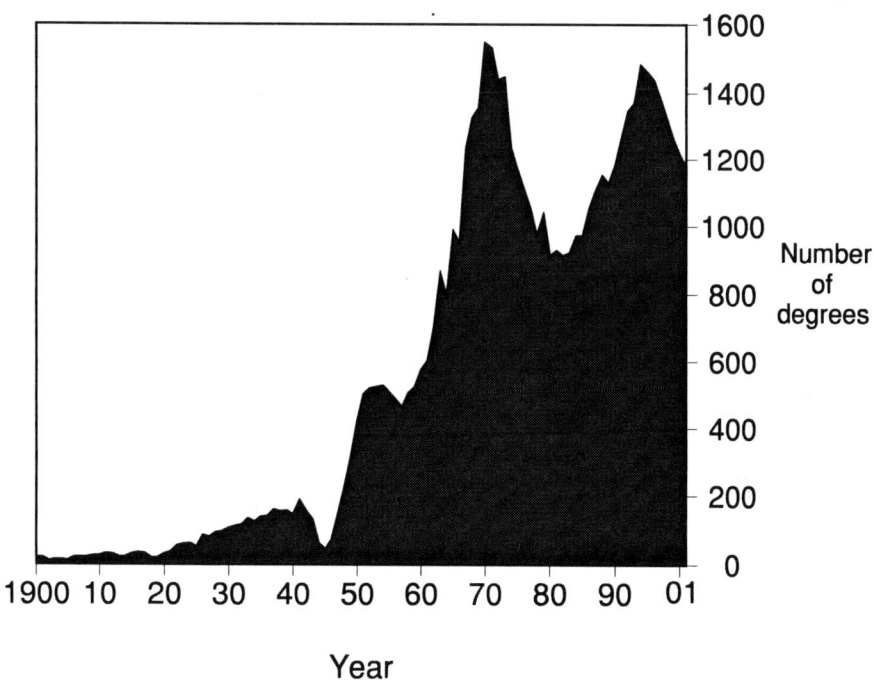

FIGURE 4. Number of Physics PhDs conferred in the US, 1900 to 2001 [2].

Economic and political events in both national and international arenas affect the number of physics PhDs awarded in the U.S. Physics PhD production plummeted during World War II and after the international recessions of 1970 and 1990. Conversely, physics PhD production increased after Sputnik was launched and again after President Kennedy announced the race to the moon.

After the People's Republic of China opened its doors to the West, many Chinese students came to the U.S. to attend graduate programs in physics and related fields. These Chinese students contributed significantly to the increase in physics PhD production during the 1980's. Similarly, after the Soviet Union dissolved in 1990-91, many of their students came to the U.S. to attend graduate programs. If not for the students from the Former Soviet Union, the decline in physics PhD production during the second half of the 1990's would have been more severe.

The number of physics PhDs awarded has been declining since 1994 at a rate of nearly 4% per year. Based on first-year student enrollment trends during the 1990's, we expect that physics PhD production will continue to decline until the PhD class of 2003. The number of U.S. citizens earning physics PhDs, however, is expected to decline until the class of 2005, at which point fewer U.S. citizens will have earned physics PhDs than in any year since 1965.

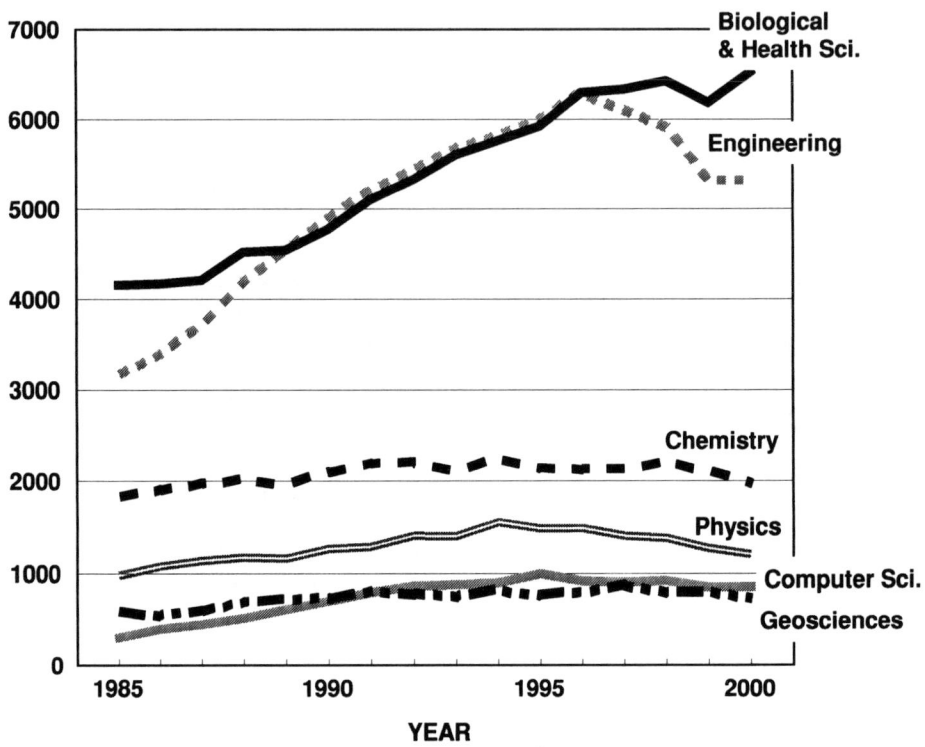

FIGURE 5. Total number of PhDs granted by discipline, 1985 to 2000 [1].

PhD Degrees in Selected Fields

Graduate education in the U.S. is arguably the best in the world and over 40,000 PhDs are awarded each year in the U.S. [3]. Figure 5 illustrates the annual production of PhDs in selected fields from 1985 through 2000. In each case, the number of PhDs awarded in 2000 was greater than in 1985. The growth in the life sciences and in engineering is especially noteworthy.

While the pattern of change during those 15 years varies by field, the annual PhD production in most of the fields displayed in this graph has either stabilized or declined during the last 4 or 5 years. The total number of new PhD's awarded in engineering has dropped by nearly 1,000 during the last part of the 1990's.

Since many of the issues discussed at this conference have a national defense component, citizenship is an important consideration. Figure 6 depicts the number of U.S. citizens earning PhDs in the same set of selected fields from 1985 though 2000. More U.S. citizens earn PhDs in each field in 2000 than in 1985 and, once again, the pattern of change during the intervening 15 years varies by field.

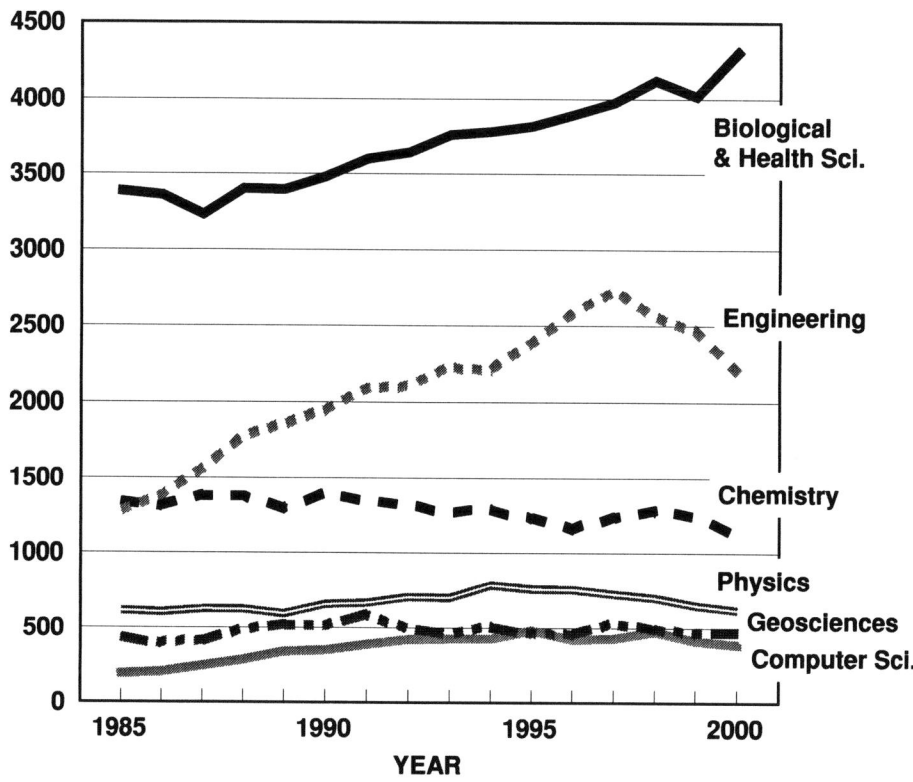

FIGURE 6. Number of PhDs awarded by discipline, US citizens only, 1985 to 2000 [1].

As indicated in Figure 5, engineering and life sciences overlap in total PhD production during the late 1980's and early 1990's. However, when citizenship is taken in account, thousands fewer U.S. citizens earn PhDs in engineering each year than do U.S. citizens in the life sciences. This discrepancy is because nearly two-thirds of life science PhDs are U.S. citizens whereas fewer than half (about 45%) of engineering PhDs are U.S. citizens [3].

Figures 7 and 8 depict the number of PhDs earned by U.S. citizens in subfields that are directly related to the issues discussed at this conference. Obviously, these graphs reflect small numbers. In order to factor out some of the random noise in the system, we have presented the data as two-year averages. Once again, in each case a greater number of U.S. citizens earned PhDs in each of these subfields in 1999 than did a decade earlier. However, in most cases, the peak in PhD production was during the mid 1990's with a drop off in the most recent years for the number of U.S. citizens earning degrees [1].

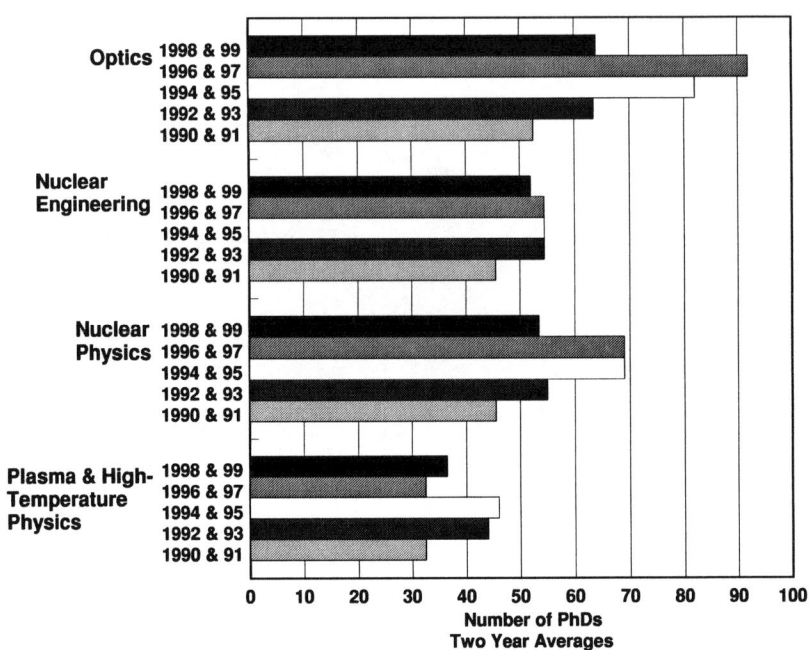

FIGURE 7. Number of PhDs granted by specific subfield, U.S. citizens only, 1990 to 1999 [1].

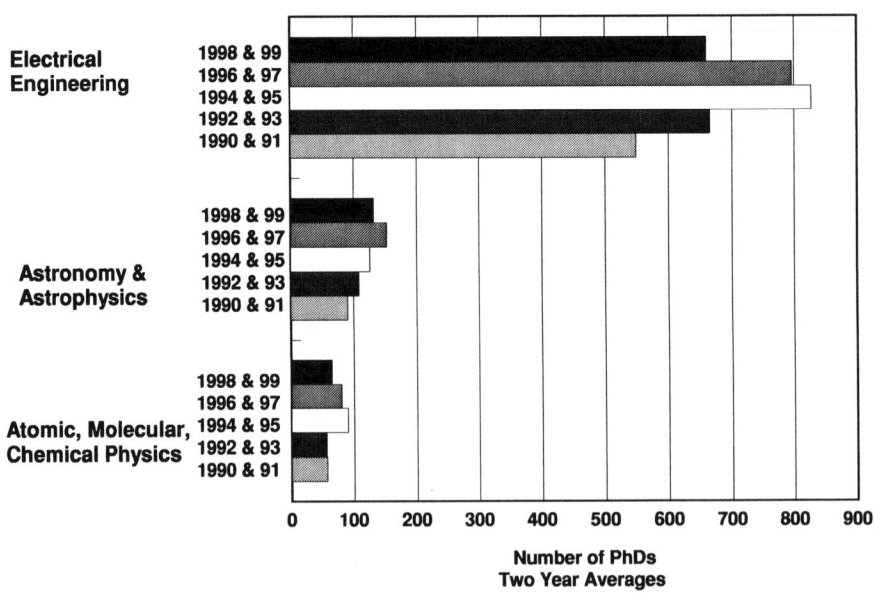

FIGURE 8. Number of PhDs granted by specific subfield, U.S. citizens only, 1990 to 1999 [1].

During this conference, there has been considerable concern expressed about the need for greater numbers of people trained in areas of direct relevance to our national security at the same time that we are seeing declining numbers of U.S. citizens pursuing degrees in these fields. The prospect of dramatically increasing the number of degrees awarded in selected fields can seem daunting. However, these are comparatively small fields in light of the total number of PhDs awarded each year. By way of example, to double the number of degrees awarded in nuclear engineering would mean changing the intended fields of study of less than 2% of the U.S. citizens who are likely to earn PhDs.

CONCLUDING REMARKS

The data presented in this paper have focused on the supply side, that is, on degree production in relevant fields over time. Equally important are issues related to the demand side and, in particular, the issue of whether we are facing a shortage. To increase the number of people in any field, in the author's opinion, involves three basic themes: encouragement, opportunity and resources.

To increase the number of people who may choose a particular field of study, we must encourage students, that is, heighten their awareness of the importance, relevance and excitement of the field. However, we must also provide them with the opportunities both to pursue advanced education in the field and to find meaningful, productive and challenging employment in the field. Such opportunities require resources and, in this case, the word resources has a dollar sign in front of and behind it.

Increasing research support of faculty has the inevitable effect of increasing the number of graduate students who are trained in a particular discipline. While this is necessary for increasing the supply, it is not sufficient. It is equally important to earmark the resources that will assure employment opportunities for these students after they graduate. To put money into supply without thought to the demand leads to the dissatisfaction and disillusionment of the students that you encouraged and trained. The latter can result in negative repercussions and a decrease in supply in the long run. In other words, demand is not simply a data issue, it is a policy issue.

REFERENCES

1. NSF (National Science Foundation) WebCASPAR Database System, <http://caspar.nsf.gov/> March 2002.
2. Nicholson, S., and Mulvey, P. J., *Roster of Physics Departments with Enrollment and Degree Data, 2000*. College Park, MD: American Institute of Physics, 2001.
3. Snyder, T. D., and Hoffman, C. M. *Digest of Education Statistics, 2000*, Washington, DC: U.S. Department of Education. National Center for Education Statistics, 2001.

FORENSIC APPLICATIONS

A Neutron Sensor for Detection of Nuclear Materials in Transport

David M. Gilliam, Alan K. Thompson, Jeffrey S. Nico

National Institute of Standards and Technology, Gaithersburg, MD 20899

Abstract. Spent fuel from nuclear reactors, even weapons-grade plutonium, and certain weapons-initiator materials emit large numbers of neutrons from spontaneous fission or (alpha, n) reactions. These are materials which terrorist groups could fairly easily obtain for construction of radiation dispersion weapons or fission weapons. Because the U.S. borders are fairly easily crossed in many wilderness areas and because these materials could be obtained from sites within the U.S., it is not sufficient to inspect baggage and cargo at points of entry. Very sensitive neutron detectors coupled with surveillance cameras could be positioned at highway choke points on the outskirts of major cities or near other sensitive locations to detect neutrons from contraband nuclear components and provide warnings to security personnel to try to prevent such an attack. The major advantage of neutron detectors over beta and gamma detectors is that the background rates and incidences of false positives would be much lower. There is only a very low level of neutron background due to cosmic rays, much lower than the natural beta/gamma background. As a spin-off from a neutrino research collaboration with a Russian team, NIST physicists are developing a very-high-sensitivity neutron spectrometer which could be configured for unattended, remote sensing duty. This spectrometer is based on photon emission from recoil protons in a liquid scintillator followed by thermal neutron capture. It uses a segmented geometry to improve energy resolution. Some examples of applications of a spectrometer of this type for highway surveillance are given.

INTRODUCTION

It is conceivable that a terrorist could attack a city in the U.S. with a nuclear or radiological weapon without ever having to pass nuclear material through a point-of-entry surveillance check. The nuclear material might be stolen from some inadequately secured installation within U.S. borders or the material could be smuggled in at a poorly protected border area or coastal area. In either case, it might be possible to foil the attack by employing sensitive radiation detectors at highway choke points on the approaches to U.S. cities. Beta and gamma ray detectors and spectrometers are already being employed in applications of this kind, but this paper describes possible applications of a new high-sensitivity neutron spectrometer to this interdiction effort.

There are several advantages to be gained by including neutron detectors and spectrometers. Fast neutrons are emitted at high rates (1) by most of the easily

CP632, *Unattended Radiation Sensor Systems for Remote Applications*, edited by J. I. Trombka et al.
2002 American Institute of Physics 0-7354-0087-3

obtained fissile materials, (2) by certain isotopic materials, which could potentially be employed as initiators in a fission weapon, and (3) especially by spent reactor fuel, which could be employed in a radiological dispersion weapon or "dirty bomb." These fast neutrons are a penetrating form of radiation which requires a kind of shielding that is entirely different from that required for gamma rays. With a combination of gamma ray and neutron spectrometers to evade, the shielding requirements would force the enemy to use both heavy and voluminous shielding. This combination of shielding could only be carried by a large, heavy vehicle, reducing the class of vehicles that might have to be pulled aside for direct inspection and increasing the cost and complexity of the enemy's task. Furthermore, the neutron background from cosmic rays, uranium minerals, and fallout is very low compared to beta and gamma ray background, so that false alarms due to background would be very low for a neutron detector. There is no neutron emission from medical isotope sources, which are frequently shipped with minimal shielding and which remain in high concentration in patients bodies for many days after a nuclear medicine procedure. Finally, if the neutrons are well moderated and absorbed by shielding materials, characteristic gamma ray lines are usually given off when the neutrons are absorbed, adding a new easily recognized signal to be seen by gamma ray spectrometry unless additional heavy metal shielding surrounds the moderator/absorber.

DESCRIPTION OF THE NEUTRON SPECTROMETER

The neutron spectrometer under discussion was initially developed by a Russian team for use in the "SAGE" Russian-American neutrino experiment [1]. The spectrometer was able to measure neutron fluence rates as low as 10^{-7} neutron cm^{-2} s^{-1} in the presence of a much higher gamma ray background [2]. The initial design consisted of a cylinder of about 15 liters of NE213 liquid organic scintillator with 19 re-entrant tubes containing 4-atmosphere ^{3}He proportional counters. Three photomultiplier tubes viewed the scintillator. In the latest design, the liquid scintillator volume is separated optically into 16 distinct sections, with dissolved ^{6}Li replacing the ^{3}He detectors. The purpose of the segmented design is to improve the energy resolution of the scintillator [3]. The sections are small enough that there is usually only one recoil proton from a single neutron within a given section. This separation allows the non-linear, but single-valued light emission curve to be corrected to directly give a linear energy response and thus improved resolution. No unfolding is required to derive the neutron spectrum. The energy resolution is of the order of 10% at 14 MeV. The efficiency of the initial spectrometer model (the number of neutron counts registered per neutron entering the spectrometer) has been measured as about 9% for Pu-Be neutrons. Somewhat higher efficiency may be obtainable with the dissolved ^{6}Li design. Almost all of the neutrons entering the spectrometer will ultimately be captured by either a ^{6}Li or a H nucleus or will leak out before being captured.

THE NEUTRON BACKGROUND FROM COSMIC RAYS

At sea level the total cosmic ray fluence rate is typically of the order of $0.024 \text{ cm}^{-2} \text{ s}^{-1}$, of which about 75% is muons [4]. The muons interact with the detector and its surroundings to produce a fast neutron fluence rate of typically $0.02 \text{ cm}^{-2} \text{ s}^{-1}$ [5]. This neutron background, the muon background, and the gamma background can be reduced in several ways. Most of the muon background and the directly related neutron background can be eliminated by operating with a plastic scintillator muon detector underneath and around the neutron spectrometer in anticoincidence. Much of the gamma background is removed by Compton anticoincidence using the same system. Finally, because the duration of cosmic ray showers is very short ($\sim 200 \text{ } \mu s$) compared to the counting time for a source on a truck passing the detector at highway speeds (~ 230 ms), one can eliminate most multiple counts from a given shower by rejecting all neutron counts that occur within 200 μs of another neutron count. This is expected to reduce the background by about a factor of 2 without appreciable addition of dead time.

APPLICATION EXAMPLES

We shall consider the case of a vehicle moving past a pair of fixed spectrometers at 97 km/h (60 mph), carrying a neutron source emitting 64000 neutrons/s. The pair of detectors monitor a single lane of traffic and can be close together or separated by as much as 25 meters. Separating the detectors by 25 meters would introduce a delay time of almost 1 second between the detection intervals and would completely prevent any possibility of having both counting intervals affected by the same cosmic ray shower. The minimum source to detector spacing is taken as 3 m (~ 10 ft). The detector surface area is of the order of 1 m^2 with an efficiency such that the probable number of source neutrons detected in a single pass is 6.17 within a 230 ms time window. The total cosmic ray neutron background is reduced by an assumed additional factor of 4 by limiting the detection window to the energy interval 1 MeV – 5 MeV [6]. The average background count is then 0.31 within a 230 ms time window.

Now we can see how different choices of threshold settings will affect the chance of missed detection or false alarm probability, assuming Poisson distribution of the background and signal counts.

If the detectors are operated independently and we require that both detectors have one or more counts, then the probability of detection is 0.9967 and the probability of a false alarm is 0.0711. This false alarm rate would be unacceptably high except during a time of extremely high estimated threat. Alternatively if we require that both detectors register two or more counts, then the probability of detection drops to 0.9772 while the false alarm probability falls to 0.0015. These would be a better compromise during a time of moderate estimated threat.

On the other hand, if we just add the counts of the two detectors together, and treat the pair as a single larger detector, then for a threshold of 5 counts, we would get a detection probability of 0.9940 and a false alarm probability of 0.000465, apparently a better choice than either of the above if the common-shower counts are sufficiently suppressed by the 200 μs spacing criterion without physically separating the detectors.

CONCLUSIONS

Based on the assumptions noted above, it appears to be possible to achieve reasonable levels of detection probability without incurring so high a false alarm rate that one would paralyze traffic during periods of moderate threat. The addition of neutron detection could add a significant improvement in early detection of some of the more probable modes of terror attacks.

Appendix: Another Application

As a remote sensor, we have only discussed passive neutron detection applications, but the spectrometer could also be used in certain important applications with active neutron or gamma ray interrogation. The most important of these is the detection of highly enriched uranium (HEU). If hidden HEU is exposed to pulsed neutrons or energetic gamma ray, delayed fission neutrons from the HEU can be detected unambiguously for times up to a minute after the interrogating pulse. The more sensitive the neutron detector employed, the smaller the interrogating pulse required to reveal the contraband.

REFERENCES

1. Abdurashitov, D. N., Gavrin, V. N., Efimov, G. D., Kalikhov, A. V., Shikhin, A. A., and Yants, V. E., *Instruments and Experimental Techniques* **40**, pp. 741-752 (1997).
2. Abdurashitov, D. N., Gavrin, V. N., Kalikhov, A. V., Klimenko, A. A., Osetrov, S. B., Shikhin, A. A., Smolnikov, A. A., Vasiliev, S. I., Yants, V. E., and Zaborskaya, O. S., *Physics of Atomic Nuclei* **63**, 2000, pp. 1276-1281.
3. Abdurashitov, D. N., Gavrin, V. N., Kalikhov, A. V., Matushko, V. L., Shikhin, A. A., Yants, V. E., Zaborskaya, O. S., Adams, J. M., Nico, J. S., and Thompson, A. K., *Nucl. Instr. and Meth. in Phys. Res.* **A476**, 2002, pp. 318-321.
4. Kelly, R. L., "Elementary Particles," in *Physics Vade Mecum*, edited by H. L. Anderson, New York, American Institute of Physics, 1981, p. 158.
5. Lindstrom, R. M., Lindstrom, D. J., Slaback, L. A., and Langland, J. K., *Nucl. Instr. and Meth. in Phys. Res.* **A299**, 1990, pp. 425-429.
6. Allkofer, O.C., and Grieder, P. K. F., "Cosmic Rays on Earth," in *Physics Data* **24-1**, 1984, p. 80.

APPENDICES

Gorgiana Alonzo
Lawrence Livermore National Laboratory.
P. O. Box 808, L-389
7000 East Avenue
Livermore, CA 94550
Phone: 925-424-6100, Fax: 925-423-4563
Email: alonzo1@llnl.gov

Justin Anderson
SAIC, 1-7-3
1710 SAIC Drive
McLean, VA 22102
Phone: 703-676-6743, Fax: 703-676-2722
Email: andersonjus@saic.com

Mark Antkowiak
NETL-Energentics
2414 Cranberry Square
Morgantown, WV 26508
Phone: 304-594-1450, Fax: 304-594-1485
Email: mark.antkowiak@en.netl.doe.gov

Houston Baker
National Cancer Institute
6130 Executive Boulevard
EPN Room 6060
Bethesda, MD 20892-7412
Phone: 301-594-9117, Fax: 301-480-3507
Email: Bakerhou@mail.nih.gov

John Becker
Lawrence Livermore National Laboratory.
P. O. Box 808, L-414
7000 East Avenue
Livermore, CA 94550
Phone: 510-865-2997, Fax: 925-422-5940
Email: jabecker@llnl.gov

Adam Bernstein
Lawrence Livermore National Lab.
P. O. Box, L-290
7000 East Avenue
Livermore, CA 94551
Phone: 925-422-5918, Fax: 925-424-5512
Email: bernstein3@llnl.gov

CDR W. Skeen Blair
180 Fieldstream Lane
Idaho Falls, ID 83404
Phone: 301-669-2561, Fax: 301-669-3120
Email: wblair@nmic.navy.mil

James Bogard
Oak Ridge National Laboratory
P.O. Box 2008
Oak Ridge, TN 37831-6480
Phone: 865-574-5851, Fax: 865-574-1778
Email: bogardjs@ornl.gov

James Bonomo
RAND
1200 South Hayes Street
Arlington, VA 22202
Phone: 703-413-1100, Fax: 703-413-8111
Email: James_Bonomo@rand.org

Barry Botnen
EERC
P.O. Box 9018
Grand Forks, ND 58201
Phone: 701-777-5073, Fax: 701-777-5181
Email: bbotnen@undeerc.org

Michael Briggs
3722 N. Pershing Drive
Arlington, VA 22203
Phone: 703-248-9635, Fax:
Email: briggsmb@hotmail.com

Jay Brown
INEEL
P.O. Box 1625
Idaho Falls, ID 83415-3710
Phone: 208-526-0980, Fax: 208-526-6802
Email: JTB@INEL.gov

Debbie Bush
The MITRE Corporation
M/S CHAN
1820 Dolley Madison Blvd.
McLean, VA 22102-3481
Phone: 703-808-5480, Fax: 703-808-1497
Email: dbush@mitre.org

Joseph Callerame
American Science & Engineering
829 Middelsex Tpke.
Billerica, MA 1821
Phone: 978-262-8698, Fax: 978-262-8801
Email: jcallerame@AS-E.com

Eugene Cheney
USAF
TMA
1030 S Hwy A1A
Patrick AFB, FL 72925
Phone: 321-494-1776, Fax: 321-494-8521
Email: genec@aftac.gov

Muren Chu
Fermionics
4555 Runway Street
Simi Valley, CA 93063
805.582.0155
fax: 805.582.1623
EMail M.chue@fermionics.com

Donald Creighton
Pacific Northwest National Laboratory
P.O. Box 999, K7-22
Richland, WA 99352-0999
Phone: 509-375-2333, Fax: 509-375-3641
Email: don.creighton@pnl.gov

Roman Czujko
American Institute of Physics
One Physics Ellipse
College Park, MD 20740
Phone: 301-209-3080, Fax: 301-209-0843
Email: rczujko@aip.org

Mark Cunningham
Lawrence Livermore National Laboratory
P. O. Box 909, L-270
7000 East Avenue
Livermore, CA 94551
Phone: 925-423-2269, Fax: 925-424-5512
Email: cunningham1@llnl.gov

Sheng Dai
Oak Ridge National Lab
Chemical Sciences
Bldg. 4500 N, MS 6201
Oak Ridge, TN 37831
Phone: 865-576-7307, Fax: 865-576-5235
Email: dais@ornl.gov

Rebecca Detwiler
Bechtel Nevada
4600 N Hollywood, B12211
Remote Sensing Laboratory
Las Vegas, NV 89129-6403
Phone: 702-295-8613, Fax: 702-794-1038
Email: detwilrs@nv.doe.gov

Arden Dougan
Lawrence Livermore National Laboratory
P.O. Box 808, L-175
7000 East Avenue
Livermore, CA 94550
Phone: 925-422-5549, Fax: 925-422-6434
Email: dougan1@llnl.gov

Jared Dreicer
Los Alamos National Laboratory
Mail Stop E541
Los Alamos, NM 87544
Phone: 505-667-0005, Fax: 505-665-4197
Email: jdreicer@lanl.gov

Steven Ebstein
Lexitek, Inc.
14 Mica Lane
Wellesley, MA 02481-1708
Phone: 781-431-9604, Fax: 781-431-9605
Email: ebstein@lexitek.com

Gerald Entine
Radiation Monitoring Devices, Inc.
44 Hunt Street
Watertown, MA 2472
Phone: 617-926-1167, Fax: 617-926-9980
Email: GEntine@RMDInc.Com

Larry Evans
NASA-Goddard Space Flight Center
Code 690.2
Greenbelt, MD 20771
Phone: 301-286-5759, Fax: 301-286-1681
Email: larry.evans@gsfc.nasa.gov

Dennis Fargo
McNeil Technologies Inc.
6564 Loisdale Court
Springfield, VA 22150
Phone: 703-921-1671, Fax: 703-921-1610
Email: dfargo@mcneiltech.com

Samuel Floyd
NASA-Goddard Space Flight Center
Code 691
Greenbelt, MD 20771
Phone: 301-286-6881, Fax: 301-286-0212
Email: ulsrf@lepvax.gsfc.nasa.gov

Larry Franks
Bechtel Nevada
5520 Ekwill Street, Suite B
Special Technologies Laboratory
Santa Barbara, CA 93111
Phone: 805-681-2426, Fax: 805-681-2241
Email: Franksla@nv.doe.gov

Gerard Garino
Bechtel Nevada
P.O. Box 380
Suitland, MD 20752
Phone: 301-817-3381, Fax: 301-817-3411
Email: garinog@nv.doe.gov
David Gilliam
NIST
100 Bureau Drive, #8461
Gaithersburg, MD 20899-8461
Phone: 301-975-6206, Fax: 301-926-1604
Email: david.gilliam@NIST.gov

Anshel Gleyzer
PhotoPeak, Inc.
10180 Queens Way, Units 3E4
Chagrin Falls, OH 44023
Phone: 440-543-1197, Fax: 440-543-1632
Email: photoa@alltel.net

Bruce Glick
eV PRODUCTS
373 Saxonburg Blvd.
Saxonburg, PA 16056
Phone: 724-352-5236, Fax: 724-352-4435
Email: bglick@ii-vi.com

C.R. Goetzman
U. S. Dept. of Energy
WSRC Bldg 735-A
Aiken, SC 29808
Phone: 803-725-2215, Fax: 803-725-4478
Email: rudy.goetzman@srs.gov

Lee Grodzins
AS&E
Phone: 978-262-8655, Fax: 978-262-8801
Email: lee@grodzins.com

Mark Grohman
Sandia National Laboratories
Mail Stop 1361
P.O. Box 5800
Albuquerque, NM 87185-1361
Phone: 505-844-0350, Fax: 505-284-5437
Email: magrohm@sandia.gov

Kaye Hart
Embassy of Australia
1601 Massachusetts Ave. NW
Washington, DC 20036
Phone: 202-797-3042, Fax: 202-797-3089
Email: ansto-washington@ansto.gov.au

Zhong He
The University of Michigan
3038 PML 2100
Nuclear Engr. & Radiological Sci.
2355 Bonisteel Blvd.
Ann Arbor, MI 48109-2104
Phone: 734-764-7130, Fax: 734-763-4540
Email: hezhong@umich.edu

Peter Heimberg
Bechtel Nevada
RSL Andrews
Andrews AFB, MD 20762
Phone: 301-817-3463, Fax: 301-817-3401
Email: heimbepc@nv.doe.gov

Sam Hitch
Advanced Measurement Technology, Inc
801 South Illinois Ave.
Oak Ridge, TN 37931
Phone: 865-483-2191, Fax: 865-481-2438
Email: sam.hitch@ortec-online.com

Gary Hoggard
Global Technologies Inc.
2265 East 25th Street
Idaho Falls, ID 83404
Phone: 208-523-6763, Fax: 208-523-6843
Email: ghoggard@gtiusa.net

Ethan Hull
Lawrence Berkeley National Laboratory
1 Cyclotron Road, Mail Stop 62-313
Berkeley, CA 94720
Phone: 510-495-2312, Fax: 510-486-7557
Email: elhull@lbl.gov

Joanna Ingraham
DynaCorp/DTRA
2550 Hungtington Ave. S300
Alexandria, VA 22303
Phone: 703-253-1496, Fax:
Email: Joanna.Ingraham@DynCorp.com

Kenneth Inn
NIST
Ionizing Radiation, #8462
100 Bureau Drive
Gaithersburg, MD 20899-8462
Phone: 301-975-5541, Fax: 301-869-7682
Email: kenneth.inn@nist.gov

Jan Iwancyzk
Photon Imaging Inc./Radiant Detector
Technologies
19355 Business Center Drive
Northridge, CA 91324
Phone: 818-709-2468, Fax: 818-709-2464
Email: iwanczyk@compuserve.com

Ralph James
Associate Director
Energy, Environment and National Security
Brookhaven National Laboratory
Upton, NY 11973-5000
Phone: 631-344-8633, Fax: 631-344-5584
Email: rjames@bnl.gov

William Johnson
Los Alamos National Laboratory
P. O. Box 1663, MS D460
Los Alamos, NM 87545
Phone: 505-665-4465, Fax: 505-665-3657
Email: mnjohnson@lanl.gov

Christopher Joines
Bechtel Nevada
P.O. Box 98521
Las Vegas, NV 89193-8521
Phone: 702-295-8926, Fax: 702-295-8794
Email: joinescj@nv.doe.gov

John Jones
National Nuclear Security Administration
232 Energy Way
Las Vegas, NV 89030
Phone: 702-295-0532, Fax: 702-295-1810
Email: jonesjb@nv.doe.gov

Judith Kammeraad
Lawrence Livermore National Laboratory
Analytical & Nuclear Chemistry, L231
P.O. Box 808
Livermore, CA 94550
Phone: 925-423-6757, Fax: 925-422-3160
Email: kammeraad1@llnl.gov

Lawrence Karch
ITT Industries
2560 Hungtington Ave.
Alexandria, VA 22303
Phone: 703-253-1484, Fax: 703-960-2070
Email: larry.karch@dyncorp.com

Martin Keillor
USAF
TMA
1030 S. Hwy A1A
Patrick AFB, FL 32925
Phone: 321-494-5352, Fax: 321-494-8521
Email: krkmek@earthlink.net

Raymond Kimble
DOJ/NIJ
Investigative Forensic Science
810 Seventh St., NW
Washington, DC 20531
Phone: 202-305-4638, Fax: 202-307-9907
Email: kimbler@ajp.usdoj.gov

John King
Global Technologies Inc.
2265 East 25th Street
Idaho Falls, ID 83404
Phone: 208-523-6763, Fax: 208-523-6843
Email: jking@gtiusa.net

Andrew Klein
NIST/Adv. Technology Prog.
MS 4730
100 Bureau Drive
Gaithersburg, MD 20899-4730
Phone: 301-975-4292, Fax: 301-921-6319
Email: andrew.klein@nist.gov

Glenn Knoll
University of Michigan
NERS Dept. Cooley Bldg.
2355 Bonisteel Blvd.
Ann Arbor, MI 48109
Phone: 734-936-0121, Fax: 734-763-4540
Email: gknoll@umich.edu

Richard Kouzes
PNNL
National Security Directorate, #88-20
P.O. Box 999
Richland, WA 99352
Phone: 509-376-2320, Fax: 509-376-3868
Email: rkouzes@pnl.gov

Steven Kreek
NNSA/NA-22
Nonproliferation Research & Engineering
1000 Independence Ave, SW
GH-068
Washington, DC 20585
Phone: 202-586-5539, Fax: 202-586-2612
Email: steven.kreek@hq.doe.gov

Robert Kremens
Rochester Institute of Technology
Center For Imaging Science
54 Lomb Memorial Drive
Rochester, NY 14623
Phone: 585-475-7286, Fax: 585-475-5988
Email: kremens@cis.rit.edu

Richard Kroeger
Naval Research Laboratory
Mail Stop 7650
4555 Overlook Avenue, SW
Washington, DC 20375
Phone: 202-404-7878, Fax: 202-707-6473
Email: kroeger@nrl.navy.mil

James Kurfess
Naval Research Laboratory
Code 7650
4555 Overlook Avenue, S.W.
Washingon, DC 20375
Phone: 202-767-3182, Fax: 202-767-6473
Email: kurfess@gamma.nrl.navy.mil

Tanya Kuritz
Oak Ridge National Lab
P.O. Box 2008, MS 6194
Oak Ridge, TN 37831-6194
Phone: 865-241-6013, Fax: 865-574-1275
Email: kuritzt@ornl.gov

Simon Labov
Lawrence Livermore National Laboratory
P.O. Box 808, L-270
7000 East Avenue
Livermore, CA 94551
Phone: 925-423-3818, Fax: 925-424-5512
Email: slabov@llnl.gov

Carolyn Lehner
Nuclear Engineering & Radiol. Science
University of Michigan
2355 Bonisteel Blvd.
Ann Arbor, MI 48109-2104
Phone: 734-936-0123, Fax: 734-763-4540
Email: clehner@umich.edu

Zhichao Lin
NIST
C114, Building 245
Gaithersburg, MD 20899
Phone: 301-975-5645, Fax: 301-926-7416
Email: zh

Jon Losee
SPAWAR
2744
53560 Hull Street
San Diego, CA 92152-5001
Phone: 619-553-2820, Fax: 619-553-1915
Email: losee@spawar.navy.mil

James Lund
Sandia National Laboratories
P.O. Box 969, Mail Stop 9671
Livermore, CA 94551-0969
Phone: 925-294-3871, Fax: 925-294-1489
Email: jlund@sandia.gov

Kelvin Lynn
College of Engineering and Architecture
Washington State University
P.O. Box 642711
Pullman, WA 99164-2711
Phone: 509-335-1131, Fax: 509-335-4145
Email: kal@wsu.edu

Derek Mahin
Defense Threat Reduction Agency
6801 Telegraph Road
Alexandria, VA 22310
Phone: 703-325-7389, Fax: 703-325-6226
Email: derek.mahin@dtra.mil

Anthony Marti
CMTCO
1030 S Hwy. A1A
Patrick AFB, FL 32925
Phone: 321-494-6087, Fax: 321-494-2680
Email: anthony.marti@aftac.patrick.af.mil

John Mattson
Bechtel Nevada
P.O. Box 380
Suitland, MD 20752
Phone: 301-817-3458, Fax: 301-817-3411
Email: mattsoje@nv.doe.gov

Timothy McClanahan
NASA Goddard Space Flight Center
Code 691
Greenbelt, MD 20071
Phone: 301-286-6748, Fax: 301-286-0212
Email: xrtpm@lepxgrs.gsfc.nasa.gov

Robert McLaren
ITT Industries, AES
2560 Huntington Avenue
Alexandria, VA 22303
Phone: 703-253-1481, Fax: 703-960-2070
Email: bob.mclaren@dyncorp.com

Paul Mexcur
NASA Goddard Space Flight Center
Code 712
Greenbelt, MD 20771
Phone: 301-286-8888, Fax: 301-286-0321
Email: wmexcur@pop700.gsfc.nasa.gov

Michael Momayezi
X-Ray Instrumentation Associates
8450 Central Ave.
Newark, CA 94560
Phone: 510-494-9020, Fax: 510-494-9040
Email: Momayezi@xia.com

Monica Montague
NASA Goddard Space Flight Center
Code 713
Greenbelt, MD 20771
Phone: 301-286-7957, Fax: 301-286-0301
Email: mmontagu@pop200.gsfc.nasa.gov

Beth Moore
US Dept. of Energy, EM-52
1000 Independence Avenue S.W.
Forrestal Bldg, Room 3E066
Washington, DC 20585
Phone: 202-586-6334, Fax: 202-586-1492
Email: beth.moore@em.doe.gov

Eric Moore
Bechtel Nevada
P.O. Box 380
Andrews AFB, MD 20762
Phone: 301-817-3461, Fax: 301-817-3411
Email: mooreet@nv.doe.gov

Marko Moscovitch
Georgetown University
3970 Reservior Road NW
Washington, DC 20007
Phone: 202-687-8993, Fax: 202-687-2221
Email: moscovim@georgetown.edu

Elaine Mullen
The MITRE Corporation, MS W940
1820 Dolley Madison Blvd.
McLean, VA 22102
Phone: 703-883-7110, Fax: 703-883-4585
Email: emullen@mitre.org

Sarah Mullen
U.S. Dept of State
State/VC/TA, # 2833
2201 C Street NW
Washington, DC 20520
Phone: 202-647-0854, Fax: 202-647-1407
Email: mullensa@T.state.gov

Luc Murphy
Bechtel Nevada
RSL Andrews
RSL AO
Andrews AFB, MD 20762
Phone: 301-817-3347, Fax: 301-817-3411
Email: murphyly@nv.doe.gov

John Nemeth
Oak Ridge Associated Universities
P.O. Box 117
130 Badger Ave, MS-29
Oak Ridge, TN 37831-0117
Phone: 865-576-1898, Fax: 865-576-3816
Email: nemethj@orau.gov

Suzanne Nguyen
Dept. of Energy
EM-52, 3.00E-66
1000 Independence Ave, SW
Washington, DC 20585
Phone: 202-586-5528, Fax:
Email: suzanne.nguyen@em.doe.gov

Brian Nordmann
US Dept of State
State/VC/TA
2201 C St., NW
2833
Washington, DC 20520
Phone: 202-647-2408, Fax: 202-647-1407
Email: nordmanBr@T.state.gov

Michael O'Connell
National Nuclear Security Org.
Nonproliferation Res. & Engr.
1000 Independence Avenue
NA-22, GH-068
Washington, DC 20585
Phone: 202-586-1766, Fax: 202-586-2612
Email: Michael.O'Connell@HQ.doe.gov

Robert Okagawa
DoD
3090 Defense Pentagon, Rm E3121
Washington, DC 20301-3090
Phone: 703-695-9292, Fax: 703-697-9532
Email: robert.okagawa@osd.mil

Ann Parsons
NASA Goddard Space Flight Center
Code 661
Greenbelt, MD 20771
Phone: 301-286-1107, Fax: 301-286-1684
Email: parsons@lheamail.gsfc.nasa.gov

Bradley Patt
Photon Imaging Inc./Radiant Detector
Technologies
19355 Business Center Drive
Northridge, CA 91324
Phone: 818-709-2468, Fax: 818-709-2464
Email: bradpatt@compuserve.com

David Penn
Veridian Systems
1400 Key Boulevard
Arlington, VA 22209
Phone: 703-807-5681, Fax: 703-524-2420
Email: david.penn@veridian.com

Dave Peterson
Sandia National Laboratories
Department 1738, #1073
P.O. Box 5800
Albuquerque, NM 87185
Phone: 505-844-6009, Fax: 505-844-7001
Email: Petersdw@sandia.gov

Jeff Pfohl
6601 Tennyson Street NE
Albuquerque, NM 87111
Phone: 505-299-9516, Fax: 706-568-9124
Email: pfohl.nucalf.physics.fsu.edu

Gary Phillips
Georgetown University
Research Bldg., Room E202A
3970 Reservoir Road NW
Washington, DC 20007
Phone: 202-687-6337, Fax: 202-784-1519
Email: phillipg@georgetown.edu

Bernard Phlips
Naval Research Lab
SSD, Code 7650
4555 Overlook Avenue SW
Washington, DC 20375
Phone: 202-767-3572, Fax: 202-767-6473
Email: Phlips@gamma.nrl.navy.mil

Leticia Pibida
NIST
8462, C114
100 Bureau Drive
Gaithersburg, MD 20899
Phone: 301-975-5538, Fax: 301-926-7416
Email: leticia.pibida@nist.gov

George Powers
U.S. NRC
T-9F31
Research
Washington, DC 20555
Phone: 301-415-6212, Fax:
Email: gep@nrc.gov

Don Price
NAVIGANT
Suite 500
1801 K NW
Washington, DC 20006
Phone: 202-973-7212, Fax:
Email: Dprice@NvigantConsulting.com

Carolyn Pura
Sandia National Laboratories
Department 8120, #9104
P.O. Box 969
Livermore, CA 94550
Phone: 925-294-2811, Fax: 925-294-1377
Email: CAPURA@sandia.gov

Charlie Ramos
DME Corp.
12889 Ingenuity Drive
Orlando, FL 32826
Phone: 407-381-6062, Fax: 407-381-6063
Email: cramos@DMEcorp.net

Marc Rasmussen
SPAWARSYS Center
53605 Hull Street, #2744
San Diego, CA 92152
Phone: 619-553-1861, Fax: 619-553-1915
Email: rasmu@spawar.navy.mil

Nihar Ray
U.S. Dept of State
VC/NA
2201 C St. NW
Washington, DC 20520
Phone: 202-647-8695, Fax: 202-736-7634
Email: raynk@t.state.gov

Brian Rees
Los Alamos National Laboratory
P. O. Box 1663, MS J562
Los Alamos, NM 87545
Phone: 505-667-9980, Fax: 505-665-1147
Email: Brees@lanl.gov

Edgar Rhodes
Johns Hopkins Applied Physics Lab
11100 John Hopkins Road
Laurel, MD 20723-6099
Phone: 240-228-3829, Fax: 240-228-7636
Email: ed.rhodes@jhuapl.edu

Rex Richardson
SAIC
16701 West Bernardo Drive
San Diego, CA 92127
Phone: 858-826-9664, Fax: 858-826-9224
Email:

David Roelant
Florida International University
10555 W. Flagler Street
CEAS-2100
Miami, FL 33174
Phone: 305-348-6625, Fax:
Email: Roelant@HCET.FIU.EDU

Stanley Roeske
Bechtel Nevada
P.O. Box 98521
Las Vegas, NV 89193
Phone: 702-794-1663, Fax: 702-295-8794
Email: roeskedb@nv.doe.gov

Peter Roman
Anser Institute of Homeland Security
Suite 800
2900 South Quincy
Arlington, VA 22206
Phone: 703-416-1304, Fax:
Email: Peter.Roman@anser.org

Janis Romo
U.S. Dept of Energy NNSA
Technology Development, #505
Las Vegas, NV 89193
Phone: 702-295-0838, Fax: 702-295-1810
Email: Romo@nv.doe.gov

Benny Rose
Sandia National Laboratories
Science & Technology Partnerships
P.O. Box 5800
MS 0874
Albuquerque, NM 87185-0874
Phone: 505-844-5272, Fax: 505-844-7011
Email: bhrose@sandia.gov

Henry Rutkowski
Lawrence Berkeley National. Laboratory
1 Cyclotron Road
Mail Stop 7-1J
Berkeley, CA 94720
Phone: 510-486-4113, Fax: 510-495-2323
Email: hlrutkowski@lbl.gov

Suree Saengkerdsub
Oak Ridge National Lab
P.O. Box 2008, MS 6201
Oak Ridge, TN 37831-6201
Phone: 865-574-7232, Fax: 865-576-5235
Email: suree@novell.chem.utk.edu

Richard Salomon
Vantage Point Consultants
Suite 200
5 Revere Drive
Northbrook, IL 60062
Phone: 312-460-8200, Fax: 312-460-8300
Email: vpclexcare@aol.com

Colin Sanderson
DOE-EML
201 Varick Street,D139 5th Floor
New York, NY 10014-4811
Phone: 212-620-3642, Fax: 212-620-3600
Email: colin.sanderson@eml.doe.gov

Jeffrey Schweitzer
University of Connecticut
Department of Physics, Unit 3046
2152 Hillside Road
Storrs, CT 06269-3046
Phone: 860-486-6010, Fax: 860-486-3346
Email: schweitz@phys.uconn.edu

304

Hugh Scott
Westinghouse Savannah River Company
SRTC, Building 735-7A
Aiken, SC 29808
Phone: 803-725-5416, Fax: 803-725-4478
Email: hugh.scott@srs.gov

Carl Selavka
Massachusetts State Police Crime Laboratory
59 Horse Pond Road
Sudbury, MA 01776
Phone: 508-358-3101, Fax: 508-358-3111
Email: cselavka@pol.state.ma.us

Christine Shannon
Lawrence Livermore National Laboratory
P. O. Box 808, L-270
7000 East Avenue
Livermore, CA 94551
Phone: 925-423-6683, Fax: 925-424-5512
Email: shannon1@llnl.gov

Brad Shoup
3309 Westminister Ave.
Dallas, TX 75205
Phone: 214-373-9869, Fax:
Email: bcshoup@yahoo.com

Glenn Sjoden
USAF
1030 S Hwy A1A, TM
Patrick AFB, FL 32925
Phone: 321-494-4955, Fax: 321-494-8522
Email: glenn.sjoden@patrick.af.mil

Eric Smith
Pacific Northwest National Laboratory
902 Battelle Blvd.
Richland, WA 99352
Phone: 509-376-1599, Fax: 509-376-3868
Email: eric.smith@pnl.gov

Pam Solomon
NASA Goddard Space Flight Center
Code 691
Greenbelt, MD 20771
Phone: 301-286-8797, Fax: 301-286-1681
Email: xrphs@lepvx3.gsfc.nasa.gov

David Spears
DOE/NN-20
Office of Research and Eng.
1000 Independence Ave. SW, GH068
Washington, DC 20585
Phone: 202-586-1313, Fax: 202-586-2612
Email: david.spears@hq.doe.gov

Nancy Sprinkel
The MITRE Corporation
M/S W940
1830 Dolley Madison Blvd.
McLean, VA 22102-3481
Phone: 703-883-6739, Fax: 703-883-4585
Email: sprinkel@mitre.org

Richard Starr
NASA Goddard Space Flight Center
Code 690.2
Greenbelt, MD 20771
Phone: 301-286-5073, Fax: 301-286-1681
Email: richard.starr@gsfc.nasa.gov

Donald Steinman
Steinman Consulting
1810 Timber Creek Drive
Missouri City, TX 77459
Phone: 281-835-6364, Fax: 281-835-3009
Email: dksteinman@aol.com

Bart Stephens
2374 R. Vallejo Street
San Francisco, CA 94123
Phone: 415-315-5503, Fax: 706-568-9124
Email: bstephens@ivancorp.com

James Stewart
System Planning Corporation
1000 Wilson Blvd
Arlington, VA 22209-3901
Phone: 703-351-8319, Fax:
Email:

Wolfgang Stoeffl
Lawrence Livermore National Laboratory
P. O. Box 808, L-414
7000 East Avene
Livermore, CA 94551
Phone: 925-422-7312, Fax: 925-422-5940
Email: stoeffl1@llnl.gov

Joel Swanson
Lawrence Livermore National Laboratory
Analytical & Nuclear Chem., L-232
7000 East Avenue
Livermore, CA 94551
Phone: 925-423-8584, Fax: 925-424-2298
Email: swanson12@llnl.gov

Sevag Terterian
Fermionics Corporation
4555 Runway Street
Simi Valley, CA 93063
Phone: 805-582-0155, Fax: 805-582-1622
Email: sevt@att.net

Robert Thompson
Pacific Northwest National Laboratory
P8-01
P.O. Box 999
Richland, WA 99352
Phone: 609-376-9216, Fax: 609-372-0672
Email: rc.thompson@pnl.gov

Michael Tobin
Lawrence Livermore National Laboratory
NIF, L-472
P.O. Box 808
Livermore, CA 94551
Phone: 925-423-1168, Fax: 925-423-6212
Email: tobin2@llnl.gov

Jacob Trombka
NASA/Goddard Space Flight Center
Code 691
Greenbelt, MD 20771
Phone: 301-286-5941, Fax: 301-286-1648
Email: jacob.trombka@gsfc.nasa.gov

Michael Unterweger
NIST
C114 Radiation Physics Bldg.
100 Bureau Drive, Stop 8462
Gaithersburg, MD 20899-8462
Phone: 301-975-5536, Fax: 301-926-7416
Email: unterweg@nist.gov

Dan Upp
ORTEC
8015 Illinois Ave.
c/o Advanced Measurement Tech.
Oak Ridge, TN 37931
Phone: 865-483-2106, Fax: 865-481-2438
Email: dan.upp@ortec-online.com

Lodewijk vandenBerg
Constellation Technology Corporation
Suite 100
7887 Bryan Dairy Road
Largo, FL 33777-1498
Phone: 727-547-0600, Fax: 727-545-6150
Email: lvdberg@contech.com

Peter Vanier
Brookhaven National Laboratory
Building 197C
Upton, NY 11973
Phone: 631-344-3535, Fax: 631-344-7535
Email: vanier@bnl.gov

Kai Vetter
Lawrence Livermore National Laboratory
P. O. Box 808, L-231
7000 East Avenue
Livermore, CA 94550
Phone: 925-423-8663, Fax: 925-422-3160
Email: kvetter@llnl.gov

Richard Vondrak
NASA-Goddard Space Flight Center
Laboratory for Extraterrestrial Physics
Code 600
Greenbelt, MD 20771
Phone: 301-286-8112, Fax: 301-286-1683
Email: rvondrak.1@gsfc.nsa.gov

Tzu-Fang Wang
Lawrence Livermore National Laboratory
P.O. Box 808, L-231
7000 East Avenue
Livermore, CA 94551
Phone: 925-422-9666, Fax: 925-422-3160
Email: wang6@llnl.gov

David Waymire
Sandia National Laboratories
P.O. Box 5800, MS 1207
Albuquerque, NM 87185-1207
Phone: 505-844-1175, Fax: 505-844-6729
Email: drwaymi@sandia.gov

Stephen Weeks
BN/STL
5520 Ekwill St.
Santa Barbara, CA 93111
Phone: 805-681-2262, Fax: 805-681-2241
Email: weekssj@nv.doe.gov

David Wehe
University of Michigan
Nuclear Engr. & Radiological Sci.
3038 PML 2100
Ann Arbor, MI 48109-2100
Phone: 734-764-6215, Fax: 734-763-4540
Email: dkw@umich.edu

Robert Whitlock
Naval Research Lab
Chemistry
4555 Overlook Avenue, S.W.
Washington, DC 20375
Phone: 202-404-4321, Fax: 202-767-3321
Email: robert.whitlock@nrl.navy.mil

Mitchell Woodring
Radiation Monitoring Devices, Inc.
44 Hunt Street
Watertown, MA O2172
Phone: 617-926-1167, Fax: 617-926-9980
Email: mwoodring@rmdinc.com

Christopher Yeaw
US Dept of State
VC/TA
2201 C St. NW
Washington, DC 20520
Phone: 202-647-8126, Fax: 202-647-1407
Email: yeawct@T.state.gov

George Yoakum
BIO Technologies
2521 Meadowview Circle
Windermere, FL 34786
Phone: 407-523-4111, Fax: 407-292-7040
Email: yoak013@earthlink.net

Frederic Ze
Lawrence Livermore National Laboratory
P.O. Box 808, L-186
7000 East Avenue
Livermore, CA 94450
Phone: 925-422-5466, Fax: 925-423-2759
Email: ze1@llnl.gov

Ron Zeszut
ORTEC
8015 Illinois Ave.
c/o Advanced Measurement Tech
Oak Ridge, TN 37931
Phone: 216-328-1888, Fax: 216-328-1887
Email: ron.zeszut@ortec-online.com

Klaus Ziock
Lawrence Livermore National Laboratory
P.O. Box 808, L389
7000 East Avenue
Livermore, CA 94551
Phone: 925-423-4082, Fax: 925-423-5998
Email: ziock1@llnl.gov

AUTHOR INDEX

A

Anderson, K., 244

B

Baciak, J., 113
Baker, H., 266
Berry, J., 209
Biegalski, S., 17
Bolozdynya, A., 71
Burks, M. T., 118, 129
Bushaw, B. A., 173

C

Chiu, N., 182
Clark, P. E., 158
Cork, C. P., 118, 129
Craig, W., 118
Czujko, R., 279

D

Dai, S., 220
Dantzler, A. A., 135
Deng, W., 200
Detwiler, R., 79
DeVito, R., 71
Dors, E. E., 235
Dreicer, J. S., 235
Du, Y., 209

E

Eckels, D., 118
Eisen, Y., 142
Evans, L. G., 158

F

Fabris, L., 118, 129
Floyd, S. R., 135, 142, 158, 227
Forman, L., 37

G

Gallagher, A. J., 190
Gilliam, D. M., 291
Gleyzer, A., 216
Goldstein, W. H., 60
Goldsten, J., 101
Groves, J. L., 153

H

He, Z., 101, 113, 209
Holloway, J. P., 101
Hull, E. L., 118, 129

I

Im, H.-J., 220

J

Johnson, M. W., 244
Jorgensen, A. M., 235

K

Kalshoven, J., 263
Knoll, G. F., 209
Kremens, R. L., 190
Kuritz, T., 167

309

Other Related Titles from AIP Conference Proceedings

616 Experimental Cosmology at Millimetre Wavelengths: 2K1BC Workshop
Edited by Marco DePetris and Massimo Gervasi, May 2002, 0-7354-0062-8

608 Space Technology and Applications International Forum - STAIF 2002: Conference on Thermophysics in Microgravity; Conference on Innovative Transportation Systems for Exploration of the Solar System and Beyond; 19th Symposium on Space Nuclear Power and Propulsion; Conference on Commercial/Civil Next Generation Space Transportation
Edited by Mohamed S. El-Genk, February 2002, 0-7354-0052-0
CD-ROM: 0-7354-0053-9

559 Spectral Line Shapes, Volume 11, 15th ICSLS
Edited by Joachim Seidel, April 2001, 1-56396-991-2

552 Space Technology and Applications International Forum - 2001: Conference on Space Exploration Technology; Conference on Thermophysics in Microgravity; Conference on Innovative Transportation Systems for Exploration of the Solar System and Beyond; Conference on Commercial/Civil Next Generation Space Transportation; 18th Symposium on Space Nuclear Power and Propulsion; Space Radiation and Environment Effects Track
Edited by Mohamed S. El-Genk, February 2001, 1-56396-980-7
CD-ROM: 1-56396-981-5

499 Small Missions for Energetic Astrophysics: Ultraviolet to Gamma-Ray
Edited by Steven P. Brumby, December 1999, 1-56396-912-2

385 Robotic Exploration Close to the Sun: Scientific Basis
Edited by Shadia Rifai Habbal, February 1997, 1-56396-618-2

To learn more about these titles, or the AIP Conference Proceedings Series, please visit the webpage **http://proceedings.aip.org**